AMERICAN **PESTS**

AMERICAN **PESTS**

THE LOSING WAR ON INSECTS FROM COLONIAL TIMES TO DDT

James E. McWilliams

COLUMBIA UNIVERSITY PRESS   NEW YORK

Columbia University Press
Publishers Since 1893
New York Chichester, West Sussex
Copyright © 2008 James E. McWilliams
All rights reserved

A Caravan book. For more information, visit www.caravan.org.

Library of Congress Cataloging-in-Publication Data
McWilliams, James E.
  American pests : the losing war on insects from colonial times to DDT / James E. McWilliams.
    p. cm.
  Includes bibliographical references and index.
  ISBN 978-0-231-13942-7 (cloth : alk. paper)—
  ISBN 978-0-231-51136-0 (electronic)
1. Insect pests—Control—United States—History.
2. Pesticides—Environmental aspects— United States.
3. Pesticides—Political aspects— United States.
4. Ethnoentomology— United States. I. Title.
II. Series.

SB950.2.A1M39 2008
632'.70973—dc22
                                    2008001756

Columbia University Press books are printed on permanent and durable acid-free paper.

Printed in the United States of America

c 10 9 8 7 6 5 4 3 2 1

For Owen

CONTENTS

Acknowledgments   ix

Introduction. "the dunghill of men's passions":
THE INSECT PARADOX   1

1. "the insect tribes still maintain their ground":
   INSECTS AND EARLY AMERICANS   5

2. "there is no Royal Road to the destruction of bugs":
   THE RISE OF THE PROFESSIONALS   26

3. "Let us conquer space": BREAKING THE PLAINS AND
   FIGHTING THE INSECTS   56

4. "a great schemer": CHARLES V. RILEY AND THE
   BROKEN PROMISES OF EARLY INSECTICIDES   81

5. "let us spray": MOSQUITOES, WAR, AND CHEMICALS   111

6. "vot iss de effidence?": RESIDUES, REGULATIONS,
   AND THE POLITICS OF PROTECTING INSECTICIDES   144

7. "complaints are coming in": A YEAR IN THE LIFE
   OF AN INSECTICIDE NATION, 1938   168

8. "Let's put our heads together and start a new
   country up": SILENT SPRINGS AND LOUD PROTESTS   194

Epilogue. "Some very learned men are the greatest
fools in the world": IN PRAISE OF LOCALISM   221

Notes   225
Bibliography   257
Index   287

ACKNOWLEDGMENTS

My list of debts is too extensive to publish a parade of thanks. So let it be said that many wise individuals have helped me write this book, and let it be said that I thank them profusely for their unbounded intellectual and emotional generosity. Institutions have helped out as well, and I would like to thank especially the Program in Agrarian Studies at Yale University, which provided me with time, colleagues, and resources to finish this book. My home base in the Department of History at Texas State University continues to offer unstinting support in ways tangible and intangible, and I remain ever grateful for both forms of help. And then there is that very special institution, the family, to whom I owe more than I am able to express. In particular, though, I must single out and thank my son, Owen, for spending much of his toddling years doing little more than studying the ground for insect discoveries. It might seem untoward to dedicate a book on pests to one's dear son, but there is untold inspiration in a child's curiosity. I feel privileged to have been moved by its innocence.

# AMERICAN **PESTS**

INTRODUCTION

"the dunghill of men's passions":
THE INSECT PARADOX

The professional fight against insects in the United States began with a man who refused to ignore his passion. Thaddeus William Harris was born on November 12, 1795, in Dorchester, Massachusetts. His father, the Reverend Thaddeus Mason Harris, was widely known for his personal manifesto, *The Natural History of the Bible,* a book that became popular in educated circles. Thaddeus's mother, Mary Dix Harris, also approached nature with considerable reverence. The young Thaddeus, in fact, was able to grow up wearing silk shirts because his mother cultivated silkworms in their backyard. From an early age, the boy was immersed in the rhythms and realities of the natural world, a world he savored at every opportunity.

Thaddeus's parents recognized early on that their only child had inherited their love of nature. But, silk shirts notwithstanding, they also knew that they lacked the resources to allow him to become a "gentleman naturalist" in the vein of men of privilege. Being a naturalist was, among other things, an expression of leisure. As Cadwallader Colden had put it in the 1760s, the true student of science was a man able to escape "the dunghill of men's passions" and live in splendid isolation from the mundane concerns of daily life. Becoming part of this elite cadre of New Englanders—men who spent their days collecting, identifying, and drawing the creatures that would come to enthrall the young Harris—was out of the question for a boy whose family's blood did not run blue enough to support the expensive demands of such activities. Harris would have to earn a living. Dutifully, he chose to study medicine at Harvard. By 1824, he was practicing in Milton, Massachusetts; had married his business partner's daughter; and, so it seemed, had settled into a stable life as a country physician mired in "the dunghill of men's passions."[1]

But it was not long before Harris developed an interest in the insect world, and as his interest intensified, so did his restlessness. What especially tormented him as he went through the motions of checkups and minor operations was that, unlike European entomologists, he was living amid an entomological gold mine. "The conditions of a new country," wrote T. W. Higginson, "imply also a great wealth of material." Whereas "in older countries it is rare to discover a new species," in America ambitious entomologists were "like many scientific Robinson Crusoes, each with the insect wealth of a new island at his disposal." Probably more often than he should have (given his pressing medical responsibilities), Harris absconded into the woods to gather rare and often undiscovered insect specimens. He took full advantage of the surrounding biodiversity, amassing an impressive collection and tending it with remarkable fastidiousness. In spite of his official designation as a medical doctor, Harris was quietly, almost secretly, becoming a self-taught entomologist.[2]

There was something different about Harris's entomology. He was not satisfied simply to compile and identify his discoveries, as had the leisured naturalists of previous generations. He did not view naturalism as an entrée into an elite social world or a verification of his intellectual status. Instead, he desperately wanted to *apply* his knowledge. In 1828, Harris began to contribute articles to the *New England Farmer,* a rare thing for a scientist to do. The goal behind these articles was even more revealing: he aimed to foster connections between farmers and scientists. This was a radical departure from conventional practice, about which Harris was slightly defensive. "I am aware," he wrote to the old-school naturalist Nicholas Hentz in 1828, "that the 'New England Farmer' is not likely to be much circulated among men of science, and will therefore not be considered the best authority." Nevertheless, reaching the scientific establishment mattered less to Harris than reaching American farmers.[3] Thus the *New England Farmer* was the ideal publication for Harris to use to initiate his entomological career, a career that, in only two decades, would profoundly shape the future of insect control in the United States.

Harris's crowning achievement was *A Report of the Insects of Massachusetts Injurious to Vegetation,* published in 1841. At the core of this pioneering book was the assumption that knowledge of insect behavior was power. The farmer, he wrote, "ought to be acquainted with the transformations and the habits of [insect enemies], in all their states, so that he may know how and when most successfully to employ the means for preventing their ravages." It was in the spirit of this idea that Harris would dedicate his book to "those persons whose honorable employment was the cultivation of the soil."[4]

On December 10, 1841, only ten days after Thaddeus William Harris professed his duty to help the noble cultivators, Solon Robinson, a Connecticut farmer who had recently resettled in Indiana, published a telling article in the *American Agriculturalist*. Responding to east coast readers who repeatedly remarked "I do wish I could see a prairie," Robinson regaled his audience with a response that captured the pioneer's view of the frontier:

> In the first place then, my dear reader, I also wish you could see a prairie. You would feel as you never felt before. You would feel as I once did, when for the first time I stood upon the edge of the prairie upon which I now reside. It was about noon of a beautiful October day, when we emerged from the wood, and for miles around stretched forth one broad expanse of clear, open land. I stood alone, wrapped in that peculiar sensation that man only feels when beholding a broad rolling prairie for the first time—an indescribable delightful feeling.

In case anyone might have mistaken Robinson's rapture as appreciation for something as innocent as natural beauty, he added this clarification: "Oh, what a rich mine of wealth lay outstretched before me!"[5]

It was a telling remark. At the very moment when Harris was preparing to teach east coast farmers the fine points of insect control, new waves of eager settlers were crashing upon the sod of the Old Northwest. Agricultural publications helped promoters such as Robinson entice people to leave the east coast, claiming how easy it was for a man "with his own hands" to build a profitable venture inland. Robinson's enthusiasm proceeded with remarkable confidence, a confidence not unlike that expressed by his seventeenth-century British American ancestors as they built the nation's earliest farms and plantations. Like thousands of American pioneers before and after him, Robinson migrated west to pursue many goals, but they all fell under the spell of Senator John Calhoun's bold insistence: "Let us conquer space."[6]

"Almost the whole of that mine of wealth," Robinson continued for his loyal readers, "holds its forbidden and unsought for treasures." A virgin-soil mentality mixed with Manifest Destiny—the idea that Americans were chosen by providence to move west—was an ideological brew that intoxicated Robinson's dispatches: "No plough or spade has broken the sod of ages; no magician has appeared with the husbandman's magic wand and said to the coarse and useless grass that has grown for centuries, 'Presto, be gone,' give place to the lovely Ceres with her golden sheaves." The land of which he spoke had been under extensive use for centuries by Indians engaged in the fur trade, but even by 1841 few plows and spades had broken ground, nor had many farmers made such firm declarations. But that, as Robinson's article suggested, was about to change.[7]

Watching the careers of Thaddeus William Harris and Solon Robinson evolve is a little like watching two ships pass in the night. Harris, after years of studying American agriculture as it developed in the early republic, began to introduce farmers to entomological tools that would help them effectively manage the pests that were attacking their crops. At the same time, Robinson, after years of planning his move into the American West, joined with thousands of other expansionists to transform agricultural practices in ways that Harris could never have predicted or recognized. In other words, just when Harris was pioneering relevant solutions, Robinson was redefining the problems in more expansive and commercialized terms.

Herein lies a paradox—a very American paradox—central to this book. The men and women who settled and developed the United States sought to control the environment. At the same time, however, they strove to achieve economic goals through the kind of agricultural expansion that undermined that control. It is only within the framework of this paradox that the following story does more than reveal the past. Indeed, it is only within this framework that it speaks to the present, where our futile quest to remove ourselves from the inexorable realities of the insect world confirms our ongoing quest to strike a compromise between economic ambition and environmental responsibility.

CHAPTER 1

"the insect tribes still maintain their ground":
INSECTS AND EARLY AMERICANS

The insect paradox began well before the time of Thaddeus William Harris and Solon Robinson. Unfamiliar insects greeted the first European settlers of North America with unprecedented authority. Oliver Goldsmith, in *An History of the Earth, and Animated Nature,* summed up the popular attitude toward these creatures: "Even in a country like ours, where all the noxious animals have been reduced by repeated assiduity, the insect tribes still maintain their ground, and are but the unwelcome intruders upon the fruits of human industry."[1] Not only were indigenous insects making life difficult for planters, but even those that they inadvertently had brought with them from Europe intensified their noxious activity to levels never before experienced. The reasons for this escalation were largely lost on the colonists, who generally lacked the ecological knowledge to make the connection between their environmental behavior and the insect-related consequences. Despite their relative scientific ignorance, however, farmers did not shy away from fighting back. Through ad hoc experimentation, routine cooperation, and an admirable sense of humility, they devised homegrown, localized solutions to the insects problems they had helped create. Planters responded to the paradox by striking a balance with the insect world.

"fatal to the hopes of the husbandman"

It must have been a daunting experience. Insect pests routinely, immediately, and irreverently invaded the bodies, clothing, food, and houses of the colonists. None of these attacks, however, measured up to the dire panic the insects evoked when ravaging crops. It is important to recall that free white British Americans built their colonies on the promise of cash crops. Entire geographical regions oriented their economies around one or two dominant crops—be it wheat, tobacco, corn, or rice—or around an elaborate system of mixed agriculture. Vibrant channels of Atlantic trade integrated the colonial farms and plantations into the most thriving economic network in the world—the transatlantic economy—which was known for reaping handsome dividends. Plantation owners, fueled by these opportunities, bent landscapes to their will, staking claim over vast acreage and marking it with furrowed fields, sturdy fences, stables, smokehouses, and dwellings for servants, slaves, and their immediate families. The entire project was based on what farmers could take from the soil.

Development in the New World was rampant. Systems of roads, ocean channels, and ferries soon connected the interior to the coast, the backcountry to navigable rivers, and northern entrepôts to southern ports. Social relations, patterns of interaction, hierarchies of status and wealth, notions of cultural superiority—all in one way or another came down to one thing: agricultural competence. It was due to the capitalistic pursuit of cash crops that planters bought slaves, hired servants, apprenticed children and teenagers, and promoted the virtues of patriarchy. And thus, when the insects struck, farmers felt the foundation of life as they knew it shift. Samuel Deane, a New England farmer, spoke for his entire profession when he bemoaned the fact that insects were "often fatal to the hopes of the husbandman."[2] This was not rhetorical exaggeration. Should the husbandman falter, so, too, would the larger colonial mission that he served.

Grappling with crops that had been thinned by insect predators became a chronic reality of commercial life. Tobacco planters were the most besieged of these farmers, as tobacco proved to be especially susceptible to a wide array of pests, including flies, green worms, cutworms, and hornworms. "The tobacco flea beetle," wrote Joseph Clarke Robert, "attacked the young shoots and sometimes did such damage as to cut short a projected full planting." William Tatham, whose eighteenth-century tobacco guide assisted nearly every large planter in the Chesapeake, repeatedly pleaded that farmers—like soldiers—become hypervigilant of the "very dangerous enemy" lurking in their midst. He mentioned that there were "several species of the worm, or rather grub genus, which prove injurious to the culture of tobacco." However, the pest "which is most destructive, and consequently creates the most employment, is the horn worm, or large green tobacco worm." The hornworm—"the worm that never dies"—is, as the naturalist Mark Catesby wrote, "about four inches long, besides the head and tail; it consists of ten joints . . . of a yellow color; on the head, which is black, grows four pair of horns." Complementing the hornworm in its thorough infestations of tobacco were flies. "After all his trouble and care," Tatham continued, "the planter's hopes are often blasted by a little fly, which frequently destroys the plants when they first come up, and very often when they are grown to a moderate size." Not a tobacco farmer in the colonies could ignore these threats to his reputation and livelihood.[3]

Wheat was another vulnerable staple crop that farmers grew throughout Virginia and the Mid-Atlantic region. Several insects—including common flies, weevils, and grubs—nagged farmers who were hoping to meet foreign orders for flour and bread. Landon Carter, a prominent Virginia planter whose fragile ego rested precariously on an annual bumper crop, referred to the pests that pockmarked his growing grain as "the enemy" and, although thoroughly perplexed by their behavior, explained that they "somehow lodged in the grain before it grows hard." After "discovering little eggs and maggots half formed into flies in the grain," Carter honed in on his minuscule subjects, observing them closely, assiduously, several times a day in order to see the conditions under which they eventually emerged. Finally, "in a pleasant evening, after the sun went down, and everything serenely calm, I found the rascals extremely busy amongst my ears, and really very numerous." After carefully placing several specimens in his handkerchief, Carter, "by the magnifiers of my telescope . . . took occasion minutely to examine them." His conclusion was discouraging. Barring extreme measures, like guarding the wheat all day, Carter admitted that "the fly would return again . . . at least every morning and evening, from the casting of the bloom, to the hardening of the grain."[4]

Periodic infestations of wheat ruined many growing seasons, but it was the infamous Hessian fly that changed altogether the prospect of successful wheat production. This insect raised the matter of pest control to a national concern for the first time in American history. The crisis began to unfold in 1778, as the Revolutionary War moved south. It was then that a group of farmers on Long Island observed that their normally flourishing wheat plants were showing unfamiliar deformations. Patches of stalks became badly shriveled. Other plants produced thick, leathery, dark green leaves. Yet others turned brittle and cracked instead of ripening. Although all these symptoms had some precedent in the troubled annals of wheat farming, the unusual extent of the damage placed growers on alert.[5]

Observant Long Islanders could not help but note that Hessian soldiers recently had camped in areas where the devastation to the wheat was most severe. As a result, the name Hessian fly—unfairly or not[6]—entered the American lexicon as quickly as the insect spread to other parts of the young republic. By 1779, not only isolated patches of wheat were being consumed, but entire fields. In the 1780s, the fly wreaked havoc throughout New York, Pennsylvania, and New Jersey. By 1786, one New Jersey farmer wrote about "white Worms which after a few days turn of a Chesnut [sic] Colour—They are deposited by a Fly between the Leaves & the Stalk of the green . . . and are inevitable Death to the Stalks they attack." Thousands of wheat farmers were telling the same story. Five years later, one of those farmers, Thomas Jefferson, called for the establishment of a government committee "to collect materials for forming the natural history of the Hessian fly" in order to find "the best means for its prevention and destruction."[7] In short, the Mid-Atlantic region found itself under siege by an insect that stoked residents' xenophobia while consuming their crops.

Rice, grown throughout South Carolina, was likely the least vulnerable of North America's staple crops to the hazards of an insect attack. As such, it was proposed by Northerners as a viable substitute for wheat. "As far north as Susquehanna," J. B. Bordley insisted, "rice may be tried." In 1781, he recalled, "in a clay loam on upland, in Talbot, Maryland, I grew a garden bed of it, drilled and hoed." The result, he noted, "was good in quality and quantity." By 1801, with the Hessian fly now endemic to agricultural life, the Society for the Promotion of Agriculture, Arts, and Manufactures asked frustrated New Yorkers if "any part of our country [is] adapted to the raising of rice?" Planters, however, were more dubious than writers. "In the dry season," wrote South Carolinian John Drayton, "rice, when growing, is liable to attacks from a small *bug,* equally injurious to it, as the Hessian Fly is said to wheat." While Drayton's comparison of a "small *bug*" with the Hessian fly was most certainly an overstatement, this invasive insect, probably a green leafhopper, could be quite deadly to low country rice crops. "These insects attach themselves to the rice," Drayton continued, "and suck out all the nourishment of the plant." Planters along the coast were usually spared the assaults, but the "hapless inland planter" was much less fortunate. "Patience and hope," Drayton concluded, "are the only sources to which he can apply for consolation."[8]

In a region of North America founded on discipline, order, and steady agricultural production, the insect empire thus posed a dire challenge. The fact that "men," as a Maryland newspaper wrote when the Hessian fly reached the region, "were designed to dwell in the fields" might have been a reassuring reality in the land-rich New World. The fact that the "insect tribes still maintain their ground," however, was the structural weakness behind this design. Rarely did colonial elites admit to losing control of a slave, a servant, a wife, or a child. Rarely, for that matter, would they admit to losing control of their pigs, chickens, and cows. The same cannot be said for insects. When it came to insects, the colonists were quick to throw up their hands and plead helplessness. These beasts were literally out of control. And this scared the men and women trying to make a go of it in the new republic.

Again, Oliver Goldsmith summed up the matter nicely. "These animals," he wrote of insects, "are endued with a degree of strength for their size that at first might exceed credibility." Indeed, "had man an equal degree of strength bulk for bulk, with a louse or flea, the history of Samson would no longer be miraculous. A flea would draw a chain a hundred times heavier than itself; and to compensate for this force, will eat ten times its own size of provision in a single day." Thomas Jefferson, whose agricultural habits bespoke an intense interest in environmental control, agreed, describing the "ravages of insects" in his wheat crop as, quite simply, "not within human controul."[9]

And that was, in essence, the gist of the matter: insects were beyond human control. It is hard to imagine these complaints and admissions being made about any other element of society besides insects. Colonists could define slaves as chattel, servants as labor, and wives as property, and they could (more or less) train them all to behave accordingly. They had no choice, however, but to see insects for what they were: a force of nature to reckon with and, for better or worse, respect. Perhaps Goldsmith best pinpointed the precise nature of this respect: "All other animals are capable of some degree of education; their instincts may be suppressed or altered; the dog may be taught to fetch or carry; the bird to whistle a tune; and the serpent to dance." But the insect was immune to such manipulations. "No arts," he lamented, "can turn it from its instincts." It was, he concluded, "armed with the powers to disturb the peace of an emperor."[10] Not to mention his empire.

"the woods are troublesome"

Although quick to portray themselves as innocent victims, early Americans were partially to blame for the insect attacks they were enduring. As they turned the natural landscape into a bazaar of commodities, the settlers created fertile breeding grounds for the insect pests that plagued their bodies, homes, and crops. In transforming their environment, they were also broadening the impact that insects would have on it. Nowhere was the massive alteration of the North American landscape more evident than in the settlers' systematic effort to deforest the east coast.

In the early nineteenth century, Timothy Dwight, the president of Yale, made note of this development. The rate at which European Americans were consuming the forest especially impressed him. "An Englishman who sees the various fires of his own country sustained by peat and coal only," he wrote, "can not easily form a conception of the quantity of wood, or, if you please, of forest, which is necessary for this purpose." Clear-cutting, Dwight explained, often led to crop planting. But when it did not, he noted, "vigorous shoots sprout from every stump; and, having their nourishment supplied by the roots of the former tree, grow with a thrift and rapidity never seen in the stems derived from the seed." "Almost all the original forests [of New England] had long since been cut down," he continued, replaced with an "exhausting" method of agricultural cultivation. What was once a dense forest had become, in a relatively short time, an "open landscape."

Dwight had presciently seized on a crucial element of environmental change in early America. Perhaps more than any other factor, unprecedented deforestation transformed the American landscape. In so doing, it encouraged the exponential proliferation of insect populations throughout North America. The forest clearing by European settlers, however, was without precedent only in its extent. Before contact, Native Americans had eagerly manipulated the landscape to improve their material lives. Hardly the "noble savages" of myth living in a pristine wilderness, tribes burned extensive patches of grassland, exhausted vast tracts of arable land, and felled stands of old-growth trees to construct elaborate longhouses, build fires to cook and keep warm, and trap game.

But the impact of the Indians' clearing of the forests for the purposes of gathering fuel and hunting should not be overstated. For all their active and self-interested control over their environment, indigenous peoples produced goods not for a distant marketplace, but mainly for themselves and their communities. Trade was local rather than foreign and overtly capitalistic, and the ecosystem reflected the effect of this difference. Native Americans certainly burned expansive tracts of land, but they did so, as one environmental historian writes, while "producing a great deal of ecological security for themselves in the process." Fire, as employed by Native Americans, "played a vital role in the ecology" of North America and even contributed to species diversity while minimizing insect infestation. This distinction between local and market production was critical and, as Dwight well understood, explained why the landscape he was observing appeared as denuded as it did: Anglo-Americans eager to send timber to foreign markets had stripped it with remarkable speed.[11]

Although not unaware of it at the time, Dwight still would have been shocked to learn just how thoroughly the settlers had already altered the Native Americans' practice of land management. At any given time before European settlement, Native Americans had no more than 0.5 percent of the land on the east coast under cultivation—"a minute speck on the landscape," as one scholar puts it. This relatively delicate touch meant that the ecosystem enjoyed a high degree of genetic diversity and natural ecological dynamism, qualities that in and of themselves protected against perennial outbreaks of insect infestations. The Native Americans' interaction with their environment was hardly in *complete* harmony with the ebb and flow of "nature"—which is not possible—but neither did it systematically interfere with the ecosystem's underlying balance and deplete necessary resources. Europeans would do more than intensify what Native Americans were already doing; they would, as Dwight appreciated as early as 1810, fundamentally change it. American colonists, as Alfred Crosby reminds us, "always succeeded in taming whatever portion of temperate North America they wanted within a few decades, and usually a good deal sooner."[12]

Native Americans cleared land to plant and hunt, and, perhaps as a result of these efforts, they certainly confronted invasions of vicious insects. Entomologists have identified more than twenty-five pests native to New York that routinely consumed Indian corn. The Iroquois, moreover, employed precise derogatory terms to describe singularly destructive insects. Analysis of old white-pine boards shows signs of substantial infestation by the pine weevil well before the onset of colonization. Other evidence confirms periodic swarms of red-legged locusts and grasshoppers, as well as examples of Indian groups relocating or repatterning their seasonal migrations to escape infestations of fleas. Native Americans knew, and had many reasons to loathe, invasive pests. But compared with the colonists' experience, they rarely confronted extensive and persistent attacks on their crops, if only because they planted fewer crops in the insects' paths of destruction. Incorporated into ecosystems that remained *relatively* undisturbed by humans for much of their natural history, insects before colonization were generally left to evolve alongside the diverse flora indigenous to eastern North America. Burning practices, moreover, were critical in controlling such insects as fleas, mosquitoes, flies, and ticks. The Native American landscape was no Garden of Eden, and the historian must be very careful not to romanticize the "ecological Indian," but it was a place where basic ecological symbiosis ensured that biological control kept insects in check and, for the most part, out of human goods.[13]

English settlers were astonished by the resources of the American environment. Having hailed from a country where about 10 percent of the land was forested and more than 50 percent of the villages completely lacked woodlands, they understandably saw the timber in their midst as a towering stand of endless wealth. Freed from concerns about diminishing supplies, English settlers who haughtily condemned Indians for not being able "to make use of one fourth part of the Land" cleared forests to plant staple crops, run ironworks and distilleries, construct houses and outbuildings, keep dwellings warm, make staves and potash, graze cattle, build ships and churches, and export timber to hungry European markets. Slashing and burning plots of land, girdling trees, hauling taproots from the soil, and plowing up stubborn surface roots, colonists cleared the forest faster than they could consume its timber. There was nothing willfully wasteful about this clear-cutting. The farmers and their descendants were simply practicing inherited agricultural traditions, albeit on a grander and more exploitative scale (as a result of living in the midst of greater resources) than they had ever done before. Still, no matter where they had acquired their agricultural habits or how intensively they pursued them, colonists were clearing the forest along the east coast at a rate of 0.4 percent a year. And thus they were, in effect, turning the landscape of North America—this vast cornucopia of natural wealth—into a distant image of what it had once been.[14]

This distortion had a direct bearing on insect life. Contemporary evidence confirms a powerful causal link between deforestation and insect infestation. The field of disturbance ecology argues through both historical and contemporary evidence that the rapid removal of old-growth forests, in conjunction with the presence of new crops and nonnative insects, established ideal conditions for insects to become invasive pests. With their traditionally diverse diet undermined, their ecosystem significantly changed, and their basic factors of life altered (factors such as moisture, genetic variability, soil composition, and light), insect species became newly abundant by feeding voraciously on crops they either had not known or had consumed only in moderation as part of a more complex diet. While by no means the sole cause of infestation as it played out from the northern old-growth stands to the southern pine forests, deforestation helped turn native insects from components of the ecosystem into nuisances to be confronted, managed, and, eventually, destroyed. "Simplifying nature had its costs," as Ted Steinberg notes, and insects became a stark manifestation of that maxim on the periphery of British America.[15]

As dramatic as the vegetational changes that settlers wrought were, the ecological responses kept pace. The rapidity of this change is significant in light of a study showing that sudden "canopy openness" in a rain forest had an immediate impact on insect herbivores, "with some populations doubling" and the "abundance of most species increas[ing]." Contemporary studies of mayflies congregating in streams in areas where clearcutting recently occurred reveal the populations increasing by 17.6 times, an explosion that scientists attribute to "changes associated with light, nutrients and temperature following logging." Many of the insect species that proliferated among early American farms were invasive, coming from Europe in the ships, luggage, and hogsheads of the migrants. Ecologists, in turn, have found that the introduction of species to a new environment often intensifies their reaction to forest fragmentation. Nonnative beetles, according to one study, increase dramatically in recently formed edge habitats shortly after deforestation. "Rarer species," the authors explain, "are predicted to be better dispersers and better at persisting." The response of insects to forest disturbance varies according to the intensity of the alteration, isolation from other animals, diversity of the environment, and size of the disruption. Nevertheless, no matter what the specific conditions, the upshot for early America remains the same: many insect species "prefer the productive conditions associated with areas undergoing regeneration from recent disturbance."[16] When Dwight mentioned that "vigorous shoots sprout from every stump," he was unwittingly identifying a critical condition for insect proliferation.

Other ecological consequences associated with deforestation furthered the process of insect proliferation. Plants differ chemically and morphologically, depending on whether they spend more time in the shade or the sun. Many insect species, in a quest to get the most nutritional bang for the buck, have evolved to feed on plants that have endured sustained exposure to natural light and dry conditions. They have done so because plants that grow in the sun for extended periods of time have evolved leaves that are relatively thick and tough as well as high in water and sugar content. These qualities especially appeal to a guild of insects known as crucifer feeders, the most notable member of which is a pest that has driven most home gardeners to fits: the aphid. While crucifer feeders in general can be devastating to a field of crops, it is the aphids, according to entomologist C.J. Hawley, that "are probably enemy number one in temperate zone agriculture."[17] Thus the deforestation practices of early Americans would have directly aided the aphids' ability to perpetuate themselves and the deadly plant viruses they transmit by exposing more plants to longer stints of sunlight while fostering soil erosion that would have dried out the substrate and supported a less diverse array of weeds and grasses.

Additional sunlight was not the only factor involved in fostering infestations of insects. As much as insects are aided by water-stressed plants, they are also kept in check by tightly packed ice and snow. Insects that hibernate in cold climates prepare themselves to handle a certain level of frigidity by reducing the water content in their tissues as freezes commence. Nevertheless, the longer the snow and ice cover the soil, the slimmer the insects' chances of surviving the cold spell. Thus long freezes often were quite beneficial to farmers. The process of deforestation, however, assisted insect breeding in colder climates because it allowed the sun to break through the thinning canopy of trees to warm the ground faster than it had when the canopy was thicker. Insects that otherwise would have perished in the pupae phase now survived to attack plants. In *History of Vermont,* Samuel Williams observed that "the earth is no sooner laid open to the influence of the sun and the winds than . . . the surface of the earth becomes more *warm* and *dry.*" Insects that enjoy earlier exposure to thawed ground and, as a result, easier access to water-stressed foliage were unwelcome reminders of Williams's accurate observation.[18]

More important than the farmers' clearing of forests was their replacing them with fields planted with an array of new, sometimes imported, crops—an agricultural choice that significantly enhanced the prospects of insect infestation. As one ecologist writes, "The pestiferousness of insects in colonial times was in direct proportion to the opportunities that colonists gave natural and imported insects to encounter massed stands of susceptible plants." Orchards planted with apple, pear, peach, and cherry trees abutted extensive kitchen gardens filled with imported vegetables and herbs on land immediately surrounding homesteads. Caterpillars, cankerworms, grasshoppers, fleas, and maggots (all native) wasted no time in adapting to these new "mass stands," which offered them a relatively sudden abundance of nourishment and shelter for a minimal expenditure of energy. The staples of wheat, tobacco, and rice only broadened the menu, dramatically blanketing the east coast before 1830 and, in turn, initiating the systematic rounds of insect infestations that proceeded to plague early farmers. When relatively small and diverse patches of Indian planting grounds yielded to multiacre tracks of cleared and bounded fields, the resulting "botanically and genetically narrowed base of agriculture" created effects that led at least one astute observer, Dr. Benjamin Smith Barton, to explain, "Many of the pernicious insects of the United States seem to be increasing. . . . Some of these insects which originally confined their ravages to the native or wild vegetables have since begun their depredations upon foreign vegetables, which are *often more agreeable to their palates.*"[19]

Grazing cattle further tightened the connection between deforestation and insect proliferation. American farmers who turned their more fertile land over to staple crops routinely dispatched their cattle into semiwooded areas to fatten on whatever the stripped environment had to offer. The cattle consumed roots, grass, weeds, and tree shoots, thereby preventing trees from regenerating while packing the soil into an oxygen-deprived mass that rendered it less useful for future planting and a prime cause of erosion and runoff. The grazing of partially deforested remnants pushed farmers into a dangerous cycle of finding new land for their cattle to graze and, subsequently, destroy. Relocating cattle the following season to yet another well-picked-over patch of forest, which farmers could readily do in a land-rich environment, perpetuated a grazing practice that created acres upon acres of packed soil exposed to direct sunlight. Quality soil was subsequently dried out, washed away, condensed into a clay-like consistency, and rendered unable to absorb rainfall. Runoff poured into creeks and rivers that flooded into pools and ponds. What was left behind supported increasingly impoverished vegetation while the pools became breeding grounds for a variety of insect pests.[20]

Again, the European settlers and their American descendants certainly did not see the relationship between woodland and insect life in such ecological terms. There is really no way they could have. Indeed, in one of the rare examples of a colonist explicitly linking forest clearing and insect invasion, a French Jesuit priest got it completely backward. "I have no doubt," he wrote, "that if this country were cleared, very fertile valleys would be found. The woods are troublesome; they retain the cold, engender the slight frosts, and produce great quantities of vermin, such as grasshoppers, worms, and insects, which are equally destructive in our garden."[21] Although few positive answers were forthcoming, the colonists quickly learned that clearing the country would not protect their gardens from "grasshoppers, worms, and insects." The woods were, ecologically speaking, just fine. It was their rapid destruction that was making things "troublesome." When it came to that destruction, Timothy Dwight, who remarked that the original forests were practically destroyed, seems to have seen it all. But when it came to understanding insects, the connection, as was the case with most early Americans, escaped him.

"my object is rather to elicit from others the result of their observation and experience"

Much as American farmers were unaware of the relationship between human action and insect proliferation, their knowledge about how to confront these populations in a new setting was equally minimal. Lack of formal knowledge, however, should not imply a lack of will to control insect pests. In retrospect, it is tempting to dismiss American farmers as "a disorganized group of amateurs" confronting questions that they were (by later standards) unqualified to answer.[22] But the solutions they pioneered before 1840, before the formalization of professional American science, included a spectrum of useful procedures developed from the ground up.

How useful these procedures were is, of course, a relative assessment. There is no denying that the solutions the planters attempted enjoyed only temporary success at best in removing insect pests from commercial crops. Insects would, even when they took a momentary hiatus, invariably return to plague Americans where they were most vulnerable—their farms. Still, the answers that planters settled on were ultimately consistent with the larger agricultural problem because, even though their schemes never fully eliminated pests, they applied to local conditions and never exposed the human population to extensive harm.[23] Temporary success in the early American context was nothing to scoff at, especially when "failure"—that is, the continued persistence of insects—at least encouraged communication and a familiarity with the environment through ad hoc experimentation, careful observation of natural phenomena, and the option of starting over.

The fact that farmers were initially on their own when it came to controlling insects cannot be overstated. Rather than credentialed experts, farmers were the ones who devised solutions to insect infestations on their own terms, on their own farms, and in their own journals. Without access to chemical insecticides or professional entomologists, planters relied on their systematic interaction with the natural world. With insight garnered from their own observations of what devoured what, what repelled what, and what attracted what, farmers proceeded to manipulate nature to combat nature. Provisional devices prevailed—devices that mirrored planters' extensive agricultural experience and intimate interaction with the wilderness they were subduing. Technological limitations and environmental intimacy, in short, fostered an "untutored ingenuity" that made the early years of American insect control democratic, decentralized, localized, and experimental.[24]

Farmers communicated with one another by word of mouth and in the agricultural press. In contrast to the professionalized scientific assurance that would later dominate formal entomological investigations, articles penned by working farmers revealed an informality characterized by a refreshing willingness to admit ignorance. It was a willingness no doubt buoyed by farmers' acknowledgment that they were first and foremost farmers, not scientists, and thus they were more willing to admit faults when it came to insect control. In his discussion of that "vexatious little depredator"—the aphid—a "Genesee Farmer" had to concede that the pest's winter habits do "not appear to be well understood." He was similarly stumped by the grain worm's "manner of introduction into the [wheat] kernel," noting that the worm, too, "does not yet appear to be fully understood." An expert gardener from New York threw up her hands in frustration over the "vine fretter," admitting that "I have not yet been able to find out where it deposits its eggs" and promising that "as soon as I do, I will make some attempts to destroy them." Samuel Deane, in his entry on "the palmer worm," noted that "the history of this insect is so little known that I will not undertake to say how they may be successfully opposed." A farmer writing in the *Farmer's Cabinet* began his short treatise on injurious insects by asserting that insects were complex animals, and, as a result, he hardly had all the answers. "It is prudent," he added, "to sail with caution when there are breakers ahead."[25] Such humility characterized this ongoing discussion.

And the humility was infectious. Rather than offer definitive answers, farmers who contributed to such journals as the *Farmer's Cabinet, Yankee Farmer, Southern Cultivator,* and *Horticultural Register and Gardener's Magazine* suggested solutions in a spirit of cooperative inquiry. "The Agriculturalist," a columnist for the *Yankee Farmer,* stressed that his article on the grain worm did not so much aim to "describe a remedy," as "to call attention of our readers to [the worm] and excite inquiry [so] that more light may be thrown upon it." Deane explained that, in an effort to control garden fleas, he "once applied some clefts of the stems of green elder to some drills of young cabbages." Although the method proved to be effective (he "could not find that they eat afterwards"), Deane erred on the side of caution, concluding that "as I made this trial but once, I dare not positively assert its efficacy." That job, he implied, was left to other farmers who would be expected to report their results after further experimentation. With the *Southern Cultivator* printing a "remedy for the curlicue, or plum weevil," sent by "Mrs. Kidder, of Boston," it was not uncommon for unexpected boundaries, both gendered and geographical, to be crossed in the collective quest to control pests through a wide range of methods.[26]

In such a manner was entomological information gathered and presented—at once personal, cooperative, haphazard, and unburdened by the imperatives of expertise. Taking on the "peach worm," a farmer writing under the pseudonym "Senex" recounted for readers the details of "my warfare against them." Because "in no system of entomology could I find a description of the insect which has proved so destructive to our peach trees," the author determined "to rely on my own observations for its history and descriptions." After surveying the relevant entomological high points, he concluded with a characteristic note of caution: "I do not presume . . . to think the above the only or best means of abating the evils we suffer from insects; my object is rather to elicit from others the result of their observation and experience." "Jack Planter" addressed his colleagues with the exhortation, "If you think the information [they gathered] will be of service to the readers of your paper, please publish it when most convenient." There were, in light of the task at hand, no single experts. Entomological expertise was instead radically dispersed. It was accrued over time and conveyed through the trial-and-error efforts of hardworking farmers willing to occasionally substitute the pen for the plow.

American farmers may have been humble in their quest for insect control, but they were hardly cowed by the task at hand. As insects attacked farmers' livelihoods with mounting voracity, calls to arms became increasingly impatient. A "Connecticut River Vermont Farmer" insisted on knowing how, given the havoc wreaked by the Hessian fly, planters could possibly still be so "lamentably ignorant on this subject." The indignant Vermonter also believed that he had "a sense of duty" to find answers, and he publicly vowed to do so. "We hope," declared the editors of the *Yankee Farmer* in a similar tone of defiance, "that the attention of farmers will this year be directed particularly to [the wheat worm], or others which may pray on wheat in the field." A writer identified only as "A Farmer" wondered "why should not farmers spend some of their time during these long winter evenings in learning something useful," perhaps by meeting "at a country store" and making "the Hessian Fly a prominent subject of conversation." He knew, after all, that farmers from another town met regularly "in a neighboring wheat field" to trade information on the "habits and character of the said insect." Up north, a "Maine Farmer" began his plea for help by declaring, desperately, "Bugs—bugs—'O! the bugs.'" It was in the midst of these insults, admonishments, and laments that farmers who were actively sharing information in the agricultural press were also turning their cultivated acres into de facto laboratories.[27]

Insight into how east coast farmers applied the homegrown solutions they tested and discussed can be glimpsed in an essay contest sponsored in 1794 by the Massachusetts Society for the Promotion of Agriculture. The cankerworm had been damaging apple orchards in New England for a decade and the society's board finally decided to put out a public call for information, offering $50 to the best report on the cankerworm submitted that year. In April, Ezra Clap sent to board president Oliver Smith an essay that begins with a precise description of the worm's life cycle, followed by the assessment that "it has been said that salt will stop these insects but I have tried it that wont [sic] do." Clap also claimed to have "put a sheep skin with the wool on" near the trees, but had to conclude that "neither will that do." Ending on a note of optimism, he added, "I believe tar is a certain cure if seasonally done." In July, Perez Bradford, another Massachusetts orchard keeper, reported that "if the ground is plowd when the egg is in the earth it turns them head down and places them deeper in itt [sic] and sowing in the summer destroys many of them." Later in July, Benjamin Bassett, who sent his essay to board member James Winthrop, explained that "some years ago the cankerworm did great damage to the little orcharding in this town [Chilmark], but within this few years have almost, if not wholly disappeared." With his "attention much engaged in the subject," Bassett determined that "the most probable remedy was some sort of small vegetation which if known and cultivated around the tree would answer the purpose." He also "found that those orchards adjoining the common and the highways where sheep graze . . . have universally been cleared of the worm." There was nothing particularly authoritative about these responses, other than the fact that they reflected personal experience. Nevertheless, they reveal how actively farmers were investigating solutions on their own terms.[28]

As farmers living on the east coast intensified their discussions about insect problems, the most general solution to infestation proved to be removing the insects manually. Chesapeake Bay planters routinely sent slaves into the fields to "worm tobacco." William Tatham noted that this task "is performed by picking everything of this kind off the respective leaves with the hand, and destroying it with the foot." After listing several possible methods of keeping caterpillars out of fruit trees, Deane admitted that "perhaps the most effectual remedy is the hand, by which the nests may be removed at an early stage." John Abercrombie, an English author of a gardening book that sold well in the United States, agreed with Deane, writing that the trees "should be attended to occasionally, especially . . . young trees, picking off the webs, etc., before [caterpillars] animate considerably." The planter Landon Carter summarized a common English solution to the ravages of the wheat fly, explaining that "the fly may be dislodged whilst it crawls on the ear, for they are so tender that a very little force will destroy them." "Senex" concluded that the best approach to diminishing cutworms was "frequent superficial hoeing." Picking worms and insect nests from crops with bare hands or an old hoe hardly granted to the farmer a sense of environmental dominance. But it was at least a step beyond Thomas Jefferson's assertion that insect ravages were "not within human control."[29]

Farmers certainly relied on forms of manual control besides the hand or hoe. Locating lines of armyworms and then placing a pile of burning hay in front of them would sometimes lure the worms directly into the inferno. To keep worms out of corn, farmers might dig small furrows or postholes at points where lines of worms were observed to cross. If all went according to plan, the worms would crawl in, get trapped, and cook in the afternoon heat. A field surrounded by tar-soaked twine supposedly had a better chance of avoiding an attack of chinch bugs than did an unprotected one, as did crops bordered by "boards set on edge and tarred on the top edge." The *Southern Cultivator* asked its readers: "Have all your trees wren houses in them?" If not, they were to "make them at once, as every one of them will have a tenant, and every one will kill more insects than the farmer could do personally." Deane recommended using domesticated birds to control insects, noting that "if our farms were always plentifully stocked with fowls, and particularly with turkies, these insects would be thinned, as they are fond of them, and eat multitudes." Nevertheless, he explained, adding a phrase that characterized all these techniques, "this can only be a partial remedy."[30]

Another common method that farmers employed to minimize infestations was to attack insects with natural repellents. "Take lighted charcoal in a chafing dish," advised a farmer writing in the *Boston Gazette*, "and throw thereon some pinches of brimstone in powder." After setting the dish "under the branches that are loaded with caterpillars," the farmer then only had to wait for the sulfur vapor ("so mortal to those insects") to "destroy all that are on the tree." The "black worm," according to Deane, succumbed easily to "a decoction of almost any bitter plant, sprinkled on vegetables" (he was especially excited about the potential of hops in this regard). Corn, he added, benefited from "a weak lie of wood ashes" to deter spindle worms. "To drive bugs from vines," the discerning farmer was wise to consider "sifting charcoal dust over the plants" because, according to the *Farmer's Cabinet,* "there is in charcoal some property so obnoxious to these troublesome insects that they fly from it the instant it is applied." To deter the weevil's quest for harvested wheat, one writer recommended, "mix a pint of salt with a barrel of wheat, or put the grain in old salt barrels, and the weevil will not attack it." In a similar vein, "mobs of cloth dipped in salt water and conveyed to the nests on long poles" evidently killed caterpillars crawling on fruit trees. Aphids were bound to defoliate crops unless "suitable applications" of "a wash made by an infusion of tobacco" coated the plants as a preventive measure against this "formidable scourge." Nicholas Culpepper, a Virginia planter and contemporary of Landon Carter, argued that there was "scarce anything more effectual to drive away flies" than "the juice of arsemart"—a prolific weed endowed with repellent qualities. Although nobody did so at the time, one could have filled a hefty volume with these kinds of trial-and-error efforts.[31]

Another common solution was accommodation. Farmers frequently altered their planting and harvesting schedules to work around insect habits. In a letter to Thomas Pinckney of South Carolina, Thomas Jefferson asserted that "the only preservation against the weevil in Virginia" was to thresh wheat immediately after harvest, to which Pinckney responded that Jefferson should also place it "as soon as threshed into cool subterraneous vaults." Mary Griffith echoed a common strategy when she advised farmers to plow their fields in autumn and leave them all winter to frost as a means of "destroying vast numbers of the larvae of insects," explaining that the larvae churned to the top had a better chance of being "frozen to death." To combat grasshoppers, farmers were advised to "cultivate hay crops only on low and moist lands" because the grasshoppers "abide and eat chiefly were the soil is naturally dry." Clover could be planted on high ground and mowed before "the insect has attained its full growth."

"Senex" claimed that one way to avoid the cutworm was "to plant at a distance from any plat or lawn," adding that he had "lost an entire crop of late planted beans" when he failed to follow his own advice. Another farmer, acknowledging the power of the Hessian fly, assured readers that it could still be outwitted. "Sowing late," he insisted, was key, as was "sowing bearded wheat" and "making use of a heavy roller after the chrysalis was formed [and] ploughing up the stubble and thus burying the chrysalis [sic] in the earth."[32]

Lures were another popular method of early American pest control. Gnats might, as the Reverend Henry Melchior Muhlenberg opined, "torment human beings frightfully," but they could be dismissed with a quick "smudge" of cow dung pasted on a nearby board or fence post. Fruit trees infested with "devouring insects" responded well, wrote the gardener Abercrombie, to "some phials filled with sugared water or beer" as long as planters diligently emptied the traps. Quick to admit that the task of protecting young tobacco saplings from the tobacco fly was not an easy one, Tatham nevertheless suggested "a border of mustard seed," having noticed that "the fly prefers mustard, especially white mustard, to any other young plant, and will continue to feed upon that until the tobacco plant waxes strong." The *Yankee Farmer* reported that a French farmer, after placing a pile of sheepskins in his granary, discovered that his grain, much to his delight, was suddenly relieved of its weevil infestation. "Senex" explained that, for melons and cucumbers, "I sow and rake radish or turnep seed on and around each hill; the flies are attracted by these, their more favorite food, from the melons, etc."[33]

The final approach to the insect infestations that early Americans promoted in farm journals, newspapers, and letters was soil manipulation. "A friend," according to a contributor to the *Farmer's Cabinet,* "informs us that he succeeded in destroying cut worms by watering ground infested by them with brine in which hams had been diluted with large portions of water." "In order to kill the eggs of insects which may be in the soil," advised a writer in the *Maine Farmer,* "boil your garden before you plant it." The authority on this tactic was typically elusive as well: "a friend informed us that a neighbor has for a number of years been in the habit of boxing up his beds snugly . . . and then pouring scalding water over every part of it." As a result, "he has never been troubled with grubs." Any hunch that an unidentified friend's neighbor was not the final authority on the matter was confirmed by "R.W.P.," writing from Ohio. He had tried the boiling-water method, too, and reported the unfortunate news that "I poured the boiling water on the peas and let them stand till the ground was cooled . . . [then] to my astonishment the bugs were crawling about as lively as ever!"[34]

It is tempting to dismiss the common admission of failure by "R.W.P." as yet another piece of evidence that this "disorganized group of amateurs" was running in circles. But America's preindustrial farmers would have vehemently disagreed with such an assessment. Driven by a healthy sense of experimentation and held in check by a cautious humility, they were not attempting to *conquer* the insect empire. They wanted to interact with it, develop a better understanding of its processes, and, ultimately, manage it on the local level. On this score, they generally succeeded. Drawing selectively on traditional methods of insect control, American farmers collected and collated their own data and shared and sparred over their findings in popular agricultural publications. What emerged from the cacophony was an increasingly informed, if not altogether harmonious, discussion that, through an open-minded emphasis on trial-and-error experimentation, effectively turned farmers into the de facto authorities on insect management. "The ingenious have attended to the matter," wrote Deane, and he was not being facetious. All things considered, the "ingenious" farmers had made a genuine effort to strike an acceptable balance with the insect empire. Versions of this balance were varied, contingent, and (if they failed) reversible.

The fact remained: from an early date, Americans initiated a paradoxical relationship with the environment of North America. It was a paradox that no matter how diligently farmers worked to manage their insect problems, they continued to perpetuate them by virtue of their economic activity. This conundrum was not—and is not—an easy one to reconcile. After all, early Americans exploited the landscape not to exploit the landscape, but to improve their way of life. All the injustices of capitalistic economies notwithstanding, the free white men and women who settled and developed the colonies and the early republic did so with remarkably ambitious intentions, which fostered and rewarded the attributes of hard work, discipline, and civilization as they knew it. Their efforts yielded the most successful form of republican democracy ever created. But the land that rewarded merit and opportunity for diligent white men was also land that was, in the process, being punished for its inherent wealth. Farmers offered provisional solutions to this larger environmental problem. It was not long, though, before a fresh generation of experts would spend the middle of the nineteenth century heroically trying to improve on the methods of the ingenious.

CHAPTER 2

"there is no Royal Road to the destruction of bugs":
THE RISE OF THE PROFESSIONALS

American farmers revealed a quiet sort of ingenuity as they worked to minimize the threats posed by insect pests in the early republic. Attacking striped beetles before the dew lifted from their wings, luring ants into empty lobster claws, storing grain in salted barrels and close to sheepskin, and soaking the ground in ham brine were tactics that revealed an intimate connection to local conditions and the natural world. Farmers may not have had "a systematic study of chemistry" in their repertoire, but they were the ones who best knew how to "increase the vigor of vegetation." It was knowledge born of careful observation of the immediate environment. Nevertheless, there were limits to it—severe limits—that would, in turn, provide an opening for the birth and development of a more refined expertise.[1]

"I asked him what he knew about Bugs, he answered 'nothing'"

One indication that the "untutored ingenuity" of common farmers had reached a plateau came in 1822 from a young man who had absolutely nothing to do with agriculture: Nathaniel Hawthorne. When an unknown insect attacked fruit trees in his hometown of Salem, Massachusetts, Hawthorne, who was at the time a student at Bowdoin College, took it upon himself to pen an article about the insect that included information on how to combat it. The young writer sent the piece to the *Palladium*, a journal widely read in Salem. The editors, undoubtedly impressed with the prose, anonymously published what would be the future novelist's first article.

The *Palladium* happened to be read religiously by Hawthorne's uncle, a Salem farmer whose orchard recently had been decimated by the pest that Hawthorne had written about with such passionate, but feigned, authority. Many years later, Hawthorne's cousin William H. Foster recalled how the incident unfolded after Hawthorne returned home from Bowdoin for a visit just after the piece was published:

> Mr. Manning [Hawthorne's uncle] requested me to send and get a number of copies of the paper containing this article, as he wished to distribute them among the few who were interested in fruit culture. After they arrived, the farmers discussed the article and all very highly approved of it. After they had gone I found that Nat [home on vacation] had stepped in a side door to surprise me . . . he said if I would keep dark he would tell me a secret as follows. In the Salem papers sent him from home he had read about this insect that was devouring the foliage on the fruit trees, and having nothing to do one evening, he sat down and wrote the article in the *Palladium,* but did not know, until he came [home], that it had been published. . . . I asked him what he knew about Bugs, he answered "nothing." He wrote it to pass an idle hour from his own imagination.[2]

Hawthorne's prank was by no means a common one, but with insects continuing to be a threat to commercial agriculture, farmers could be forgiven for starting to question one another's voices. They could be forgiven, moreover, for occasionally suspecting that the advice they were following, published in magazines to which any literate man could contribute, was coming from "authorities" whose hands may have been too clean to write a decent word about agricultural pests.

Hawthorne was merely an extreme symptom of a common problem. In dozens of journals, farmers had been sharing ideas and solutions with respect to insect control for nearly three decades. But by the 1830s, expressions of helplessness and calls for outside expertise had become too urgent to ignore. In 1838, the *Baltimore Farmer* reported that, for all the farmers' efforts, "grasshoppers have been more numerous and destructive this season than in any former one within the recollection of farmers."

The same year, "The Agriculturalist" complained that, despite the "millions in dollars" in damage exacted by the relentless grain worm, "very little attention has been paid to the subject by Agricultural Societies and legislative bodies." This appeal for help from outside authorities reflected the author's opinion that "individuals [farmers] have been very remiss in their attention to it."[3]

Nobody could argue that enough ink had not been spilled on the matter. The problem was rather that the "many articles on the subject" were "contradictory," leaving planters and gardeners to conclude that "their favorite theories are unfounded." Efforts "to kill lice on swine," according to the *Southern Cultivator*, had "proven insufficient from experience," leading one writer to chastise the editor: "You appear to invite the farmer to write for the *Cultivator*, but it should be your province to print what in your judgment will be profitable to your subscribers." A reader of the *Yankee Farmer*, referring to the Hessian fly, explained, "We would respectfully suggest whether it would not be well for our Legislature, indeed for the Legislatures of each of the wheat growing states, to offer a premium for the discovery of a preventative remedy for this destructive insect."[4]

Many commercializing farmers started to express their frustration that after several decades of homespun experimentation, too many complaints echoed those of a hundred years earlier. Given the demands of commercial farming, progress on the insect front seemed too slow. "The [grain worm]," wrote the *Yankee Farmer*, "has done damage in this country to the amount of millions of dollars . . . we have before us many articles on this subject but it is not yet thoroughly understood." In the late eighteenth century, William Tatham had been lamenting that, when it came to tobacco flies, "no certain remedy against them has as yet been discovered." By 1853, the *Richmond Whig* was still calling it "the worm that never dies." On a visit to Connecticut in 1838, the ornery "Connecticut River Vermont Farmer" walked into a wheat field, drew a head of grain between his thumb and forefinger, and watched in disgust as "half a teaspoon full of their crushed carcasses" fell to the ground. It was with disturbing persistence that grasshoppers, according to the *Baltimore Farmer*, attacked corn, apple trees, and "almost every green thing." For all the fire-and-brimstone literature that had been published on managing grasshoppers, the prescient farmer could not help but notice that, after they devour crops, the grasshoppers "go, almost every one, to a place of plenty." Achieving a balance with the insect world, in short, was one thing. Maintaining it in a nation where farmers were starting to practice a radical form of agricultural capitalism, however, was another. Given that the paradox of environmental dominance and commercial expansion intensified dramatically throughout the nineteenth century, it proved to be a balance destined to tip, pushed by earnest men who knew insects in their own time and place, but, in the grand scheme of national insect control, knew about as much as a precocious college kid who had a way with words and "an idle hour" to kill.[5]

"I have endeavored to treat the subject in a plain familiar manner"

Thaddeus William Harris thus came on the scene as though on cue. Not only were farmers starting to seek outside advice, but the trend toward applicability was becoming popular throughout early American science. Scientists who had isolated themselves from common problems now worked to serve the interests of the American people *and* the fragile American nation. The founders of the American Academy of Arts and Sciences announced their mission as one that would "enrich and aggrandize these confederated states" and "advance the interest, honor, dignity, and happiness of a free, independent, and virtuous people." Academy members were encouraged to pursue "useful experiments and improvements, whereby the interests and happiness of the rising empire may be essentially advanced." Simeon De Witt explained to the Society for the Promotion of Agriculture, Arts, and Sciences that "our business generally is to collect all the improvements within our reach that may be made in agriculture, in manufactures and in the arts of whatsoever kind they may be." These endeavors were to be placed "in a severely practical context" and geared to enrich republican virtue. Harris was therefore on the cutting edge when he explained that "agriculture and horticulture, when aided by science, tend greatly to improve the condition of any people, and . . . form the basis of our prosperity."[6]

To achieve these goals, however, Harris needed time, as well as access to a wider scientific community. "You have never, and can never know," he wrote to an English entomologist in 1829, "what it is to be alone in your pursuits, to want the sympathy and aid and counsel of kindred spirits." To alleviate his intellectual isolation, Harris finally bowed to his true passion and stopped practicing medicine in 1831. Assuming that another

THADDEUS WILLIAM HARRIS

Thaddeus William Harris was the first economic entomologist in the United States to direct his work explicitly toward the insects that attack cultivated crops. His book *A Report of the Insects of Massachusetts Injurious to Vegetation* set a standard for entomological writing that insisted that entomological work should be accessible to "the noble cultivators of the soil."

job in another setting might allow him more time to find a few "kindred spirits," he accepted the post of head librarian at Harvard, a position that his father had held from 1790 to 1792 and for which his inherent organizational skills prepared him well. Not only did Harris quickly master the position, but he gained access to the insects collected by Harvard's Natural History Society, all of which he exactingly identified and catalogued. He was also able to offer lectures twice a week on topics related to natural history, a task he performed with memorable charisma. ("Dr. Harris was so simple and eager," recalled an admiring student, "his tall spare form and thin face took on such a glow and freshness . . . that it was enough to make one love this study of Natural Science.") The job was not an academic appointment—after all, there was only one such job in entomology in the entire country at the time (at Harvard)—but serving as Harvard's librarian enabled Harris to support his family of eleven children while nurturing his love of all things insect-related.[7]

Harris's defining work, *A Report of the Insects of Massachusetts Injurious to Vegetation,* brought a more authoritative tone of expertise and precision to investigations of insect pests. Gone were the incessant qualifiers and caveats. Farmers were now duly instructed to mow marsh grass early in the summer to prevent "the most destructive species" from coming to maturity, which they did "during the latter part of July." They were to allow turkeys "to go at large during the summer" because the birds could "derive the whole of their sustenance from these insects." They were even told to send their children into the fields under the assumption that "an active boy can collect from six to seven kilograms of insect eggs in one day." Even though limited to east coast farms, these instructions were delivered as imperatives, not suggestions. The book also drew a distinction between naturalists and scientists. "The naturalist," he explained, "seldom has it in his power to put in practice the various remedies which his knowledge or experience may suggest." The same would not be true for the new breed of young entomologists that Harris would single-handedly inspire. The "art" of the naturalists' work would, under the influence of Harris, gradually yield to the science of metamorphosis.[8]

Harris was not shy about delivering complex entomological concepts to common farmers. "I have felt it my duty," he wrote to George Emerson, chair of the Commissioners on the Zoological and Botanical Survey of Massachusetts, "to endeavor to make [the book] useful and acceptable to those persons whose honorable employment is the cultivation of the soil." Harris's achievement in doing so was magnified by his influence on other entomologists, many of whom went on to become state entomologists, leaders of entomological societies, and government employees. Most notable was Asa Fitch, a workaholic who became the first state-appointed entomologist, of New York, in 1854. Fitch, who met Harris in 1845, also abandoned a medical career to pursue entomology. Known to his Salem, New York, neighbors as "the bug catcher," he published fourteen pathbreaking reports between 1855 and 1872 that systematically investigated the insects injurious to crops grown in New York. Fitch purloined sleep to pursue his passion, surviving on fewer than five hours a night. "I am," he once wrote, "quite indifferent whether I sleep or remain awake, but prefer the latter." He used his time not only to investigate the behavior of dangerous insects, but to correspond with European entomologists about comparable infestations, lobby the state legislature for more funding, and polish his prose to keep it readable for farmers. "I have," he explained, "endeavored to treat the subject [of insects] in a plain familiar manner."

Other economic entomologists hacked their own path into the young profession. Closely cut in the Harris mold was Joseph Albert Lintner, a close

ASA FITCH

Asa Fitch, an admirer of Thaddeus William Harris, was the first state entomologist of New York and the first entomologist to hold that position in any state. He was a tireless investigator of insect infestations and, as his diary makes clear, slept only a few hours every night.

associate of Fitch's. Similarly concerned with applied entomology, Lintner helped Fitch undertake an entomological survey of New York in the 1850s before moving on to become head of entomology for the New York State Museum. He extensively explored injurious insects in the pages of the *Rural New Yorker, Entomological Contributions,* and the *Country Gentleman*—all publications for working farmers. A colleague of Lintner's noted that his work was "replete with valuable and practical information."

Another entomologist who came of age under the broad umbrella of Harris's influence was Townend Glover. Born in Brazil and raised in England, Glover migrated to the United States in 1836, at age twenty-five. As a fruit farmer in New Rochelle, New York, he developed an abiding interest in insect behavior, so much so that he abandoned his farm to study and paint insect specimens. His concerns took him to New Orleans, Florida, Venezuela, and British Guiana. At some point, he contracted malaria, but survived. His tests for pest control included trying to apply

ASA FITCH'S "BUGHOUSE"
Located on Asa Fitch's property in Salem, New York, this was his unofficial office, the hideout where he studied insects collected from the farms of his fellow New Yorkers. Like many early economic entomologists, Fitch was an active agriculturalist.

alligator blood to a citrus scale insect—just one mark of his experimental curiosity. However eccentric his tactics, though, Glover, like his colleagues, never strayed far from farmers. In 1863, he became the first chief entomologist of the United States under the Department of Agriculture. Later entomologists would look back on the early years of economic entomology and denounce these pioneers as composing a quirky cohort so small as to be insignificant, but the truth is that they were laying the foundation on which the profession would develop an identity.[9]

Aware that their endeavor was unique, Harris and his "students" sought out one another and corresponded actively. The nature of their letters confirms the broader transition from naturalist to applied scientist. For example, in a letter to Fitch, Harris wrote extensively about "gall insects." Twenty years earlier, entomologists likely would have shared information about the size, shape, and quality of the specimen. As a leading proponent of economic entomology, however, Harris was particularly keen on illuminating life-cycle behaviors that might be of use to New York wheat farmers. "If the weather be sufficiently moist," wrote Harris, they "descend to the ground and burrow under the surface, where they remain in a dormant state until spring, when they take the chrysalis form in the earth, and emerge from their retreats as flies in June and July." Wary farmers were to be especially vigilant because "late broods of the maggots are sometimes harvested with the grain, and carried to the barn." Again stressing the applicable bits of information, Harris continued, "This is very likely to happen if the maggots have not come to their full growth, or if their descent to the earth has been retarded by dry weather." If the maggots did not enter the chaff, the solution was basic: "It is highly important that the chaff, dust, and refuse straw . . . should be immediately burnt, so as to destroy these dormant maggots and prevent their change to flies." As mundane as these details might seem, they marked a distinct shift in the way entomologists were conceptualizing their work under the influence of Harris's book.[10]

The first generation of economic entomologists did more than talk and write to one another, however. They enthusiastically solicited the help of the agricultural community, so enthusiastically that Fitch, the state entomologist of New York, remarked that a budding entomologist needed "as many limbs as a centipede" to keep up with his agricultural commitments. Entomologists brought significant agricultural experience of their own to their scientific investigations. Both Fitch and Glover owned and managed large fruit farms in New York. William Saunders, a Canadian entomologist who worked closely with the Americans, operated a large farm in Ontario. Benjamin D. Walsh, one of the most published and opinionated of the early economic entomologists, along with Charles V. Riley, perhaps the most influential entomologist of the late nineteenth century, had long experience in the field. Albert Cook, a Michigan entomologist, moved to California to oversee an orchard. As had the editors of agricultural journals and newspapers, the first generation of entomologists had dirtied their hands as working farmers and, as a result, understood the scientific importance of observing natural phenomena over long periods of time. This agricultural background helped ensure, as W. Conner Sorenson explains, that "there was no educational barrier between expert and client." Indeed, both entomologists and farmers ultimately based their knowledge of insects on evidence gleaned from direct observation, or at least on the earthy insights of other agriculturalists whose noses were kept close to the ground, attuned to the rhythms of agriculture.[11]

Of course, each group had its own role to play in the project of fighting insects. What ultimately distinguished the entomologists from the farmers was their unmatched knowledge of insects per se. Insect life cycles, developmental phases, reproductive habits, and feeding patterns—all this information could, if properly harnessed and disseminated, significantly help farmers manage their pest problems. Making the knowledge available, however, was the key challenge. With this educational mission in mind, economic entomologists working between 1840 and 1870 evolved from an informal "entomological fraternity" into a full-fledged body of experts. Although they lacked the centralizing authority of a powerful bureaucratic agency, these men formed entomological societies, directed state agricultural experiment stations, staffed university science departments, established their own journals, and held professional conferences. In essence, they became a decentralized team of "experts," working locally to manage insect pests with the power of scientific knowledge.

What cannot be overstated is that they professionalized without losing touch with farmers. When A. S. Packard, Jr., noted in 1873 that "he who studies the habit and structure of one insect is a true benefactor to agriculture," the profession already had organized itself around private and public institutions that provided at least some measure of prestige, professionalism, and payment. The Entomological Division of the American Association for the Advancement of Science; American Entomological Society; Ohio Agricultural and Mechanical College; Maine State Agricultural College; Massachusetts Agricultural College; office of state entomologist in New York, Illinois, Ohio, Massachusetts, Missouri, and elsewhere—these far-flung institutions and others provided the collective whetstone on which American entomologists honed their pest-control strategies before publishing their results in the agricultural press for the benefit of farmers. It would be easy to interpret this modest institutional development as proof that entomologists were retreating from fields of grain to fields of knowledge, becoming insular rather than open-minded in orientation. In point of fact, economic entomologists were doing nothing of the sort.[12]

Before the work of Harris and the emergence of economic entomology, farmers and editors published fragmented solutions to pest problems that—despite their roots in daily experience and observation—generally lacked intellectual cohesion and scientific precision. To use the phrase of another historian writing in another context, farmers in early-nineteenth-century America were engaged in the "chaos of experimentation."[13] Professional entomologists imposed order on this "chaos", in several ways, each of which helped root the national quest for insect control in a flexible set of localized tenets. In doing so, Harris and his dedicated followers placed pest control on a path toward strategies of insect management that would be responsive to the myriad agricultural systems that had evolved along the east coast in the seventeenth and eighteenth centuries. Between 1830 and 1890, economic entomologists would battle the insect empire with a range of scientific weapons that, while never perfect, helped systematize the process of insect control while remaining responsive to local concerns. It was due in part to their intimate connection to the experiences of farmers that their pioneering experiment had a fighting chance to make a difference.

His direct advice, in addition to his willingness to place his work in a public context, quickly solidified Thaddeus William Harris's reputation as the most influential economic entomologist in the United States. There was nothing generic about his accomplishment. After all, the professional model he built placed at the center of American entomology the prerequisites for an ethic of pest control that would deal with insects in a scientific and systematic manner. This foundation, in turn, provided a basis to guide a diverse farming community into an agricultural future marked by responsible, locally based, and effective insect management that preserved the spirit of cooperation that prevailed before the arrival of the economic entomologists.

"the greatest quantities can be most readily destroyed by the simplest means"

One primary way in which economic entomologists altered pest-control strategies was in their explicit emphasis on the life cycles of insects. Asa Fitch suggested in an annual report that a fresh perspective was needed: "[I]n this country, where so little accurate knowledge of our insects is diffused among the population . . . an indication of the external appearance and habits of the each species is a great desideratum." Fitch captured the nature of the entomologists' mission when he promised to study only "those insects which are injurious" in a way that would allow "those who are suffering from these pests to devise the most suitable and effectual modes for combating them." The idea was fairly simple. As the editors of the *Prairie Farmer* put it, the nation's most destructive insects "might be easily remembered, and their history and habits become familiar to every young farmer of good observation." This change would prove to be a defining feature of what economic entomologists were attempting to accomplish. The cycles of insect life became critical to almost everything they wrote.[14]

Farmers and entomologists conversed comfortably over this matter. Farmers were obviously well aware that different insects appeared in their crops at different times of the year, but before the involvement of entomologists, they had only a vague understanding of the complete sequence of transformations that insects underwent before they damaged crops. Knowing where and how an organism bred and laid its eggs, under what conditions it changed into larval and pupal phases, when it became an adult and what it ate when, and the behavioral characteristics of each phase—all this information was critical to achieving greater precision with existing insect-control tactics. As the entomologist William LeBaron put it, "There is a period in the lives of most . . . noxious insects . . . when some one or other of the common remedies . . . is effective." The *Practical Entomologist* echoed this theme, editorializing that "the transformations of each species . . . will be faithfully recorded for [the farmers'] information by Entomologists whose time is devoted to this imperfectly understood subject." Only then, it concluded, would farmers "be enabled from the information thus obtained to determine at what period in the insect's life the greatest quantities can be most readily destroyed by the simplest means." Knowledge of insect life cycles thereby allowed farmers to pinpoint when and how the available methods were to be used.[15]

Economic entomologists studiously avoided the academic discussions that had so enthralled earlier generations of naturalists and, instead, worked to get their data on life cycles into the agricultural mainstream. Even for these nineteenth-century scientists, the medium was the message. Ambitious economic entomologists had begun to publish books for wider audiences by the 1850s. Their works reflected the spirit of Thaddeus William Harris's approach, an approach confirmed by his promise in *Report of the Insects of Massachusetts* that "this report is designed for the use of

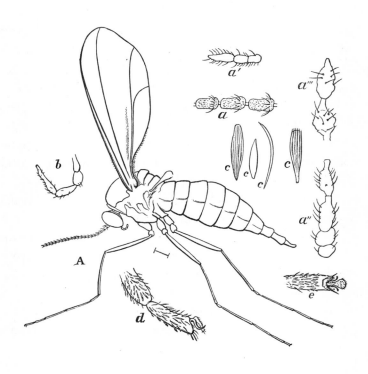
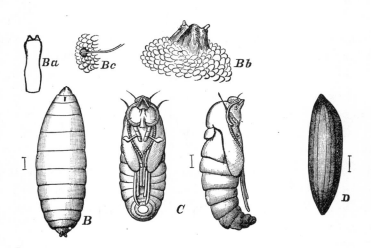

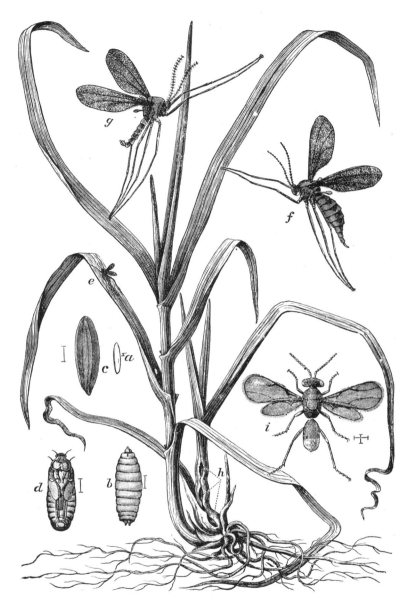

LIFE CYCLE OF THE HESSIAN FLY

Early economic entomologists worked to educate farmers about insect life cycles in order to help them better time their control strategies. These diagrams of the stages in the life cycle of the Hessian fly are examples of such an effort.

persons who may not have elementary and other works on this branch of natural history at their command."

One of the most ambitious of these budding entomologists was Mary Treat, who (to support herself after her divorce) wrote popular manuals on insects and insect control that reflected the spirit of Harris's book, but was even more overt about appealing to a lay audience—frequently publishing her findings, for example, in *Harper's New Monthly Magazine*. In the preface to her *Injurious Insects of the Farm and Garden,* Treat's publisher complained that older entomological works "are written with reference to the identification of the species" rather than with information on how the species may be controlled. Treat's book, by contrast, provided "an account of the most destructive insects and the present knowledge of the methods of preventing their ravages." As the preface made clear, however, little of what followed was going to make much sense if the reader remained ignorant of the basics of insect biology. For several pages, Treat therefore provided an elementary overview of the primary metamorphoses in an insect's life. Her concern about the entomological knowledge of her farming audience was well founded. A correspondent to the *Practical Entomologist* had, just a few years earlier, explained that "it is lamentable to see how wide spread is the ignorance in regard to [the study of insects], even among those whose interest it is to possess a knowledge of insects and whose labors are affected by their operations." Treat fully appreciated how important it was to overcome such ignorance, and she left a trail of evidence that proves her dedication to the cause of educating her readers about basic entomological principles.[16]

Treat's work highlighted remedies that capitalized on the life cycles of insects. With respect to the asparagus beetle, she instructed the reader to follow "correct principles" and, just before the beetles laid their eggs in the early spring, to remove all the asparagus seedlings that had sprung up from the previous year's seed. "Thus," she ended, "the mother beetle is forced to lay her eggs upon the large shoots from the old stools; and as these are cut and sent to market every few days, there are no eggs left to hatch out into larvae for the second brood of beetles." The potential usefulness of this knowledge was undeniable for all committed entomologists and farmers. Fitch, for example, wrote passionately about the ravages caused by the striped flea beetle on New York's cabbage and radish crops. In 1867, however, he was thrilled to report that "the larva state and transformations of this genus of insects has been discovered." Armed with the information that they "remain an egg ten days, a maggot six, and a pupa fourteen days," farmers now had "remedies against this insect . . . that we can scarcely desire anything further in the premises." Through the perpetuation of such "correct principles," Treat, Fitch, and their cohorts furthered Harris's legacy, which insisted that the entomologist empower the farmer with applicable ecological tools based on specialized knowledge about insect metamorphosis.[17]

The increasing emphasis on life cycles, as well as the need to elaborate their complexity, eventually encouraged entomologists to publish their own journals, proceedings, and state-funded reports. Critical to the success of this endeavor was that the scientists systematically solicit farmers' observations on insects life cycles before publishing their findings. This ongoing bottom–up approach to insect control was especially evident in such journals as the *Practical Entomologist* and, later, the *North American Entomologist*. Benjamin D. Walsh, editor of the *Practical Entomologist,* suggested the extent of this connection in the journal's first published article. "There is a new and very destructive enemy of the Potato," he wrote of the potato beetle, before going on to cite the firsthand observations and advice of five farmers who had direct experience with the pest. Stephen A. Forbes, as state entomologist of Illinois, frequently credited his "entomological assistants"—local farmers—for providing the evidence for his reports. These "local observers for the office" allowed Forbes to provide information on insects that he never would have had the chance to study himself. The Committee on Improvement of Lands in Plymouth, Massachusetts, expressed its expectation that "experts" should see their work from the New England farmers' perspective when it wrote, "We think many advantages would result if the committee on produce were required to view crops in the fields to examine themselves." Drawing on the insights of local farmers thus remained a standard approach for economic entomologists as they published their own journals. It was an approach that reflected Walsh's humble observation that "I am perpetually meeting with Farmers and Mechanics who know a great deal more about the Natural History of Insects than I do myself."[18]

This important symbiotic relationship only heightened the entomologists' ability to bring life-cycle data to farmers. An article in the *Michigan Farmer* on the curculio referred to four entomologists in the course of its analysis and, drawing on the finding that the beetles "oviposite in the young fruit" in June, concluded that this discovery "may lead us to most important results." Reporting on the "wheat aphis" in 1864, the *New England Farmer* cited "Prof. [Townend] Glover" and "Dr. [Asa] Fitch" as the leading entomological authorities behind the theory that "unwinged females" deposit eggs "in the autumn on late sown wheat, where they remain all winter, and hatch the following spring." The *Southern Planter* printed a summary of Fitch's *Second Report on the Noxious, Beneficial, and Other Insects of the State of New York*, deeming itself "very much indebted" to the entomologist for his work and encouraging southern states to "undertake a similar enterprise." When a farmer had the nerve to assert his opinion that "our entomologists are woefully ignorant of the tactics of this band of guerrillas [curculios]," he found himself besieged with scornful rejoinders. The *Prairie Farmer*, for one, responded with a seething editorial against slandering entomological expertise: "This is a mistake. Entomologists know, and most fruit men know . . . that the curculio hibernates in the perfect or beetle state, and its natural history is as well known as that of a cow."[19] Through patience and prolific publication, American entomologists worked with farmers to ensure that they indeed knew as much about an insect's life cycle as they about any domesticated animal's life cycle. It was a critical transfer of knowledge.

"little friends that have come to the farmers' rescue"

A second way in which economic entomologists streamlined the task of insect control was through the rationalization of an amorphous method that farmers had been practicing haphazardly for centuries: biological control. "It is well known," wrote Asa Fitch in 1859, "that certain insects have been created apparently for the sole purpose of preying upon other insects." It all boiled down to the fact that, as Fitch concluded, "we have received the evil [from abroad] without the remedy." This signal point joined life-cycle analyses in becoming central to the work of early economic entomology.[20]

One motivation to intensify the entomological focus on biological control after 1850 was the increasing popularity of bogus elixirs marketed to farmers. "Nothing is more certain," wrote editor Benjamin D. Walsh, "than that there is no Royal Road to the destruction of bugs; and the only way we can fight them satisfactorily, is by carefully studying out the habits of each species." Walsh, who would become enemy number one of chemical pesticides, was hardly alone in his opinion. The editors of the *Practical Entomologist*, the leading journal for agricultural entomology, damned the rise of these "washes and decoctions" as "as useless in application as they are ridiculous in composition." They continued, "If the work of destroying insects is to be accomplished satisfactorily, we feel confident that it will have to be the result of no chemical preparations, but of simple means, directed by a knowledge of the history and habits of the depredators." As a result, "enquiring Agriculturalists who read this Bulletin must not expect to find any particular brew . . . as specific for any one or all of our insect enemies." Reiterating the power of traditional biological control over untested patented products, an entomologist advising the Essex Agricultural Society in Massachusetts explained, "It is gratifying to know that the same Hand that sends the bane sends the antidote. These insects not only have numberless enemies among the birds and other insects, but they all have their parasites by whose energies they are swept away." In the mid-nineteenth century, many entomologists were hoping to act in unison with farmers on the potential of such a concrete and observable reality.[21]

Entomologists therefore worked diligently to raise the status of biological control from a popular folk remedy to a legitimate aspect of insect management. They began their effort with birds. Every farmer had seen birds eat grubs, and most had capitalized to some extent on the implications of that observation. "I wish to add my testimony," Mary Treat wrote in the introduction to *Injurious Insects of the Farm and Garden,* "in favor of the various birds that visit our gardens and orchards in the capacity of helpers, as they feed upon some of the most noxious insects that we have to contend with." She noted that purple martins take "rose bugs from the grape vines," orioles "pull the bag worm from his case," and catbirds eat "the unsavory pear slug." The *Southern Planter* reported in 1857 that "the chickens seemed to be unceasing in their labours of destruction of the bugs," so much so that one Virginia farmer who introduced chickens into his fields "in one week's time could not find a bug." The conclusion he drew was as pithy as it was earnest: "farmers, try it." Fitch wrote in a widely distributed report that "by raising a brood of chickens annually in our gardens . . . we can readily prevent these striped beetles from becoming multiplied and injurious." Charles L. Flint, in his annual report as secretary of the Massachusetts Board of Agriculture, recounted that "last winter, I made a bird-house, and the tenements were taken right up. . . . I did not see a slug on my asparagus all summer." As for the birds, "I am going to build them some more tenements, and rent them free."[22]

Economic ornithology—the field that studies bird–insect relationships—occupied the minds of all legitimate economic entomologists, many of whom worked directly with ornithologists to arrive at sound data. Formal efforts began in the 1860s. In 1861, Wilson Flagg declared it his intention "to make a vindication of the feathered race," as "almost every species is indispensable to our agricultural prosperity." Flagg urged recreational hunters to hold their fire because "the gunner who destroys ten birds in the spring, secures the preservation of so many millions of injurious insects to ravage our crops, and to destroy the trees of our forests and orchards." By the 1870s, studies had become more in-depth. In 1874, A. S. Packard, Jr., reported on the progress of a New York entomologist who had sliced open the stomachs of more than 3,000 birds in order to catalogue their consumption of insects. Throughout that decade, entomologists routinely took up like-minded experiments. Stephen A. Forbes investigated the relationship between insect and bird populations in an apple orchard in Illinois and concluded that "birds of the most varied character . . . were either attracted or detained here by the bountiful supply of insect food, and were freely feeding upon the species most abundant." An entomologist from New England wrote that "I have frequently seen where the woodpeckers have taken out the borers; and as they destroy no berries or other fruit, but are entirely carnivorous. . . . I think it well to protect them." Never before had so many scientists and farmers agreed that birds should be employed to the farmers' advantage.[23]

Predaceous insects also became best friends of the farmer as biological control gradually matured into a realistic option in the quest to manage injurious bugs. As early as 1864, Flint noted of plant lice that "were it not that their insect enemies tend to reduce their numbers materially, and keep in check this vast army of suckers, then we might reasonably apprehend the speedy destruction of all our crops." A farmer living near Columbus, South Carolina, wrote in 1857 to the *Southern Planter* about "the operations of the numerous qualities of ants . . . that were lying in wait for the grass-worm," doing little more than expressing the "natural desire for a fresh supply of food." Building on Fitch's work on predaceous insects, the editors of the *Southern Planter* noted, "We have heard from two distinct sources that the ants have been observed in large numbers feeding greedily on the chinch bugs." In 1869, a Massachusetts report drew attention to "a species of Coccophagus" that "destroy[ed] a considerable number of bark lice." By 1871, Packard was declaring that "insects are a most powerful agency in nature." Of course, this declaration was nothing new. Again, though, the task for economic entomologists was now to formalize and standardize this knowledge with an eye toward making it a pragmatic option for insect control on farms. Entomologists began to achieve this goal by spreading the word through familiar channels—prominent scientific journals, agricultural newspapers, government reports, and textbooks.[24]

The knowledge that insects were their own worst enemies soon came to matter a great deal in the quest to control them. The aphids had been "rapidly blighting the grain," but even then, an editorial in the *Michigan Farmer* explained, "when the lice were countless in numbers, and when the winged forms were easily spreading to the oat-fields, the hand of deliverance was discerned in the comparatively few but wondrously prolific enemies of the lice, which had already sounded a halt in the march of destruction." Summarizing the destructive power of grasshoppers, one entomologist remarked that "fortunately there are a considerable number of species of animals"— the blister beetle and ground beetle—"that depend to a greater or less extent upon grasshoppers for substance. . . . All combine in keeping the pests in check." Walsh spoke favorably about the future of biological control, predicting that "our grandchildren will perhaps be the first to reap the benefit of a plan which we ourselves might, just as well as not, adopt at the present day." Charles V. Riley, the federal entomologist whose opinion mattered as much as anyone's, agreed: "[T]here are some instances which there can be no doubt whatever as to the good which would flow from the introduction of beneficial species." Packard added, "It is quite as essential for one to know what insects are beneficial to agriculture as what are injurious."[25]

Closely related to predacious insects in the quest for biological control were the parasitic ones, which lay their eggs inside the pest. The eggs proceed to hatch, and the larvae consume their host. As an article in *Science* described a parasite of the plant louse, "As soon as the parasite has devoured the viscera of the louse, it uses the skin or crust as a sort of cocoon." What emerges from the tiny cocoon is a "tiny black fly," which "is by far the most important of these little friends that have come to the farmers' rescue and saved the wheat, barley, and oat crops." Identifying and writing extensively about the "most important group of parasitic insects" in the early twentieth century, Clarence M. Weed noted that "they are primary parasites of injurious insects" and that, when they deposit their eggs, "the host is doomed." The parasites, "like the fox in the fable, will gnaw away its vitals." Riley explained that parasitic insects are "fatal to their hosts" and, as a result, "quite beneficial to the agriculturalist." In 1869, the Connecticut Board of Agriculture noted that beetles can be injurious to vegetation, but should also be valued because "this great order includes many beneficial carnivorous insects."[26] While entomologists at the time did not know it, parasitic larvae start their attack by strategically feeding on their hosts' least vital organs in order to ensure that the hosts die as slowly as possible. Clearly, they had passed a lot of time perfecting their technique. First-generation entomologists and farmers simply wanted to capitalize on it.

"I only throw it out as a suggestion"

The focused dedication among entomologists to life cycles and biological control should not imply that they were dampening the spirit of experimentation that had prevailed among farmers before their arrival. To the contrary, a third distinguishing feature of their work was an eager willingness to explore a wide spectrum of solutions to the ravages of insects, ranging from the ancient to the unknown, which they did in the spirit of the *Prairie Farmer*'s axiom that "trial is the only real test of measures." Because their profession was highly decentralized in organization, and thus deeply attuned to farmers' specific needs and methods of experimentation, entomologists proved eager to suggest that farmers explore any and all techniques that showed promise. They remained sensitive to the commonly articulated assumption that, as the *Southern Planter* put it, "many intelligent farmers look with distrust on all recommendations issuing from any other source other than that of practical experience."[27] The entomologists' openness to amateur fieldwork preserved the "chaos of experimentation" and the intensity of local emphasis while allowing them to evaluate proposed methods with relative objectivity and to preserve the option of pulling back. A full catalogue of these procedures would be virtually endless, but perhaps the best way to appreciate this inherent flexibility would be to look briefly at how journals and entomological reports presented ideas on chemical experimentation, cultural control, and manual efforts—all tactics of pest management that farmers and entomologists tested as economic entomology matured into an established profession.

The use of natural insecticides had a long history in the annals of pest control. Since the late eighteenth century, American farmers had used pyrethrum, a repellant made from a vegetable powder (ground from petals of two kinds of chrysanthemum flowers and mixed with ashes or flour and water), to control pests. Hellebore, an extraction from a species of herbaceous flowering plant, was another popular vegetable poison that was, according to one set of instructions, "diluted with five to ten parts of flour and dusted on plants through a muslin bag." Tobacco had been refined into "nicotine dust" and used routinely in the eighteenth century for its natural insecticidal qualities. Common items—including soap, potash, and sulfur powder—also had served as acceptable insecticides well before the advent of economic entomology. This long-standing interest in chemical application only intensified after the publication of Harris's *Report of the Insects of Massachusetts,* making its way into state agricultural reports and entomologist-edited farm journals. Some of the solutions were familiar, as was the advice of "showering or syringing the bushes with liquor made by mixing with water the juice expressed from tobacco." Others were relatively novel, such as the Massachusetts Board of Agriculture's claim that "the universal search for an *insect elixir mortis,* of easy application and speedy effect, bids fair to be rewarded in the sea of petroleum or coal oil." Whether farmers read about age-old remedies, such as salting the ground to control for worms, or less familiar tactics, such as applying to their crops "whale oil soap or a weak solution of carbolic acid," they never for a moment believed that these proposed solutions were guaranteed to alleviate their insect problems. Still, the spirit of experimentation prevailed, a characteristic of economic entomology encapsulated in Jabez Fisher's qualification for his advocacy of an early arsenic-based compound: "I do not know that there is anything in it, I only throw it out as a suggestion."[28] Entomologists and farmers were generally willing to try it out.

But not always. Chemicals, despite their popularity, inspired deep skepticism among farmers, especially those in the South, when the remedies came from afar. An article in the *Southern Planter* illuminated this prevailing doubt in 1856 when it remarked that "another cause of the reaction is found in the fact that there exists a class of pseudo-scientific professors

PYRETHRUM CINERARIAEFOLIUM

This species of chrysanthemum was used by many farmers as a natural repellant. The flower petals were ground, cut with ashes or flour, and dusted over crops as a vegetable poison.

whose aim it is to take advantage of the willingness of farmers to believe that the revelations of science may be made directly available to them." Farmers were not fearful of the economic entomologists, but of "self-styled professors" all too eager "to write a prescription for the whole plantation." These doubts, which also directed attention to biological control, predisposed farmers to the myriad methods of cultural control. Since the dawn of agriculture, farmers had been inter-cropping, sowing hybrids, planting early or late, growing lure crops, and manipulating the soil to their benefit. With the establishment of experimental farms in several states, a direct result of the Morrill Land Grant Act (1862), scientists and farmers were able to test and systematize both old and new methods. Previous field testing may have been essential to the "chaos of experimentation" that farmers valued, but, as the Georgia State Agricultural Society declared in 1871, experimenters "compose a confused and unorganized mob." Agricultural experiment stations and agricultural colleges rationalized and funded experimentation while honoring a plea made by Joseph S. Howe, also in 1871: "Could we obtain a record of the experience and practice of [farmers] the result would be invaluable!"[29]

Cultural methods were far more familiar to most farmers than chemical controls and, because they required less knowledge of hard science than other tactics, were the one area where farmers had the upper hand over entomologists. It is therefore not surprising to find in the pages of the *Southern Planter* advice conveyed by "an old Negro" whose method of keeping lice off hogs was to smear tar on the rubbing posts of the pigpen. "I have kept it up," testified one farmer, "and have never had lice upon my hogs." Entomologists listened intently to farmers such as "Mr. Whitney, of Franklin Grove, Lee Co.," when he declared in reference to the bark louse, "I now propose heading down the tops of every tree in my orchard affected by them." Speaking of the Hessian fly, William R. Schuyler wrote to the *Michigan Farmer* to express his opinion that "on the sandy loams of Michigan, *when rightly tilled,* seeding may be safely be [sic] deferred until after the middle of September." The result would be that "a sufficiently vigorous growth in the fall will be insured, while the crop will be far less exposed to the spring and fall attacks." Acknowledging that "the farmer's life is a never ending conflict with weeds and insects," Howe nevertheless had faith that "by constant and timely cultivation," the problem could be managed.[30]

Perhaps the best indication of the entomologists' flexibility comes from their willingness to sanction the most basic and timeworn technique in the farmers' arsenal: manual removal. While entomologists certainly felt pressure to devise and refine methods of pest control that were more scientifically sophisticated, they also knew that nothing was more direct and tangible than picking insects off crops with the simplest tools in the kit: hands. In an article intended to confirm the virtues of progressive farming as it related to pest control, Howe explained that the farmer might make progress "by carefully watching every morning, and crushing every squash bug with a stone." Another scientist writing for the Massachusetts Board of Agriculture suggested that "it might be expedient, during the proper season, for our city authorities to employ persons to gather and kill every morning the caterpillars which may be found in those public walks where they abound." Whatever the particulars, these cases litter the entomological and agrarian literature after the rise of economic entomology as much as they did before. The critical point to appreciate is that entomologists were open-minded enough to realize that, for all their scientific progress on the biology of insect control, sometimes, as one of them put it, "the only remedy . . . is a hand-to-hand fight."[31] At the least, they were prepared to throw some punches.

"the good of creation requires that they should be destroyed"

The fourth and final way that economic entomologists and their work reshaped the quest to control insect pests was in the overall attitude they gradually disseminated about insects per se. The measured reverence that many Americans once held for insects, as well as the mystery that surrounded them, diminished substantially with the rise of economic entomology. It succumbed to a variety of forces, but three primary if disparate factors stand out: the secularization of nature, the quantification of insect destruction, and the rise of the home as a sacred space. Each reason deserves a closer look.

Knowledge demystified power. The breeding experiments and the determination of insect life cycles undermined the spiritual heft that insects once had. It was not so much that Darwinism was taking over during the latter half of the nineteenth century, as that farmers and entomologists were developing a genuine understanding of insect behavior and, to a certain extent, effectively acting on that knowledge in advantageous ways. As C. H. Fernald, writing in *Science,* put it: "I cannot rid myself of the conviction that in economic entomology God helps those most who help themselves."[32] Like the early Americans who took pride in their domesticated animals, Americans in the late nineteenth century took pride in their more detailed sense of insect habits. As a result, they felt themselves potentially more in control of them than they had ever been. Not a farmer on the face of the earth would have felt about insects the way he felt about his milk cows; he would never tame insects the way he tamed his horses. But he would not have been as awed by helplessness as he had been throughout the eighteenth century.

Entomologists therefore actively embraced a new, more muscular rhetoric. They did so in part because, while many animals were declining in number, insects were doing something quite contrary to the crush of economic expansion: they were proliferating. With industrialization transforming the American economy as thoroughly as it was, an ethic of wildlife conservation developed a substantial following. By the latter half of the nineteenth century, private organizations such as the Wildlife Refuge System joined federal programs such as the National Wildlife Preservation System to regulate practices that conservationists deemed a threat to the nation's natural heritage. This emerging preservationist mentality ultimately evolved from what many Americans viewed as the disheartening environmental consequences of economic growth. "However injurious wild animals may be to man," wrote Robert Kennicott in 1856, "he should not forget that he himself is very often the cause of their undue destructiveness."[33]

Although his comment applied perfectly well to insects, Kennicott was speaking of only mammalian quadrupeds. Insects, because so many actually *thrived* under the impact of human development, never made it onto the list of animals deserving of protection—a list that carried a certain amount of emotional weight with a portion of the American public. The ultimate cause behind the decline in wolves, deer, beavers, alligators, grizzly bears, mountain lions, and other increasingly exotic animals was primarily the loss of habitat as a result of land development and farm improvement. The proliferation of insects, however, ran against the grain of this scenario, and thus highlighted insects as the nail that had to be hammered down. By not falling under the umbrella of a conservation ethic, in short, insects stood out to be destroyed. Entomologists soon found themselves to be the only zoologists who worked to kill the animals they studied.[34]

The explicitly economic devastation that insects were causing offered another reason for the demise of their once relatively positive reputation. It was rare for eighteenth- and early-nineteenth-century commentators to speak of infestations of insects in direct monetary terms. Farmers may have complained that their livelihoods were under attack, but they tended not to conceptualize the threat posed by insects in precise financial terms. For them, the problem was more personal than national, more immediate than long term. If they did speak in broader terms, it was along the generalized lines of John Hull's remark that "the canker worm hath for fower years devoured most of the apples of Boston." Economic entomologists, however, were not content with "most of the apples." Perhaps because their work hinged so directly on the economic consequences of insect infestations, they were quick to situate the problem in a larger, often more dramatic, and certainly more quantifiable framework. Scientists across the disciplinary spectrum were coming to embrace what A. Hunter Dupree has called "the measuring behavior of Americans." But entomologists, in conjunction with agricultural scientists, took special care to ensure that their knowledge was "duly classified and arranged." As one agricultural writer put in 1886, just as agricultural experiment stations were focusing on the study of insects, "Accurate observation of what one is doing is first necessary before progress can be made in amending it." It was in this measuring spirit that the insect problems plaguing farmers became quantified and reports such as this one routinely made the rounds in popular newspapers: "While investigating the probable amount of damage done to the wheat crop by insects, professor [Furnald], President of the Association of Economic Entomologists, estimates that 10 percent of the total production is lost. In dollars it amounts to over $38,000,000." American farmers had always aimed to kill insects. But with their devastation cast in such stark economic terms, they were quick to ratchet up the rhetoric of insect destruction. While nobody was yet embracing an ethic of extermination, the language clearly was moving in that direction.[35]

That language, of course, had always applied to the home. As long as humans have lived in shelters, they have had to endure invasions of insects. But in nineteenth-century America, the home became an especially sacred place—not so much integrated into the natural environment as divorced from it. Evolving from a hovel to a haven, the household became a comforting refuge where many Americans embraced and displayed their middle-class status, a status that often hinged on modest urbanization and distance from the farm. The home, in other words, evolved from a place of work to a place of refinement, and there was nothing refined about invasive insects. It was under the influence of this larger cultural change that the prolific literature on pest control in the home reached a fevered pitch. Bedbugs, roaches, flies, moths, beetles, and ants became the victims of a torrent of not only harsh control strategies, but also rhetorical invective. The begrudging admiration that farmers and entomologists had once reserved for the creatures that caused them considerable pain dissolved in the shrill condemnations of those who viewed pests as feckless threats to the charms of domesticity.

Authors of domestic manuals made their points without equivocation. In *The Housekeeper's Receipt Book,* S.A. Oddy explained, "Civilization and the arts having made the desert to blossom as the rose, have also delivered us from the power of ravenous beasts; but we are still liable to be attacked by a more numerous though less powerful host of enemies, who commit their depredations on the animal and vegetable kingdom, and thereby destroy many of the comforts of human life." With the comforts of human life at stake, the most aggressive means were clearly in order. She continued, "It does not become us to be prodigal of life in any form, nor wantonly to seek its destruction; but where any species of animal becomes really noxious, the good of creation requires that they should be destroyed." When a house is filled with fleas, according to Clarissa Packard, author of *Recollections of a Housekeeper,* one may as well live "in a wigwam." The reference to primitive savagery, however prejudicial, reflected an underlying concern that bug-infested homes may very well compromise national stature. As William Cobbett wrote, "There never yet was, and never will be, a nation permanently great, consisting, for the greater part, of wretched and miserable families. In every view of the matter, therefore it is desirable that the families of which a nation consists should be happily off; and this depends, in great degree, upon the management of their concerns." Few concerns were as domestically pressing, he explained, than keeping the home rid of those "nasty things"—insects.[36]

By the late nineteenth and early twentieth centuries, these opinions about insects and the household had become pervasive enough for entomologists to start writing about household pests in similarly disdainful terms in their own books and articles. A classic example is Clarence Weed's *Insects and Insecticides*. Despite the parasitism that prevailed in the natural world, nothing was redeemable about insects in the home. Cockroaches "destroy provisions of every kind," and were best compared to "immigrants from without." Insects were now known not for their discoverer, but for the object of their attacks—for example, clothes moths, bedbugs, and carpet beetles. A clothes moth in Weed's depiction was something like a thief, "working her way into dark corners and deep into the folds of garments." Ants, usually spoken of with special reverence as a result of their superior social organization, were now, in the context of the home, "the most annoying kinds of pests . . . getting into and running over everything." From their nests, they "sally forth and overrun the house, devouring or carrying off particles of food of all descriptions, getting into everything in sight, and often becoming an intolerable nuisance." There was no choice but to have "a successful fight with these [and other] insects."[37]

If fighting with insects was what farmers and housewives wanted to do, economic entomologists, who knew that eradication was impossible, were still eager to engage the battle. Having defined their profession around the pragmatic concerns of farmers, entomologists had, by 1870, a great deal to look back on and celebrate. Following the lead of Thaddeus William Harris, and working in the spirit of his book *A Report of the Insects of Massachusetts Injurious to Vegetation,* they had collectively transformed a hodgepodge of fragmented information scattered throughout dozens of agricultural journals into a coherent and loosely consolidated body of accessible and locally applicable knowledge. Professionalizing without pontificating, entomologists shored up their foundation by disseminating information about insect life cycles; promoting the potential of biological, chemical, cultural, and manual control; and—with the help of number crunchers and the cult of domesticity—lifting the veil of mystery and the attitude of admiration from the insect world.

Harris died from pleurisy in 1856 at the age of sixty. The developments he inspired appeared to him in only the vaguest of forms. In a letter sent to the geologist Ebenezer Emmons in 1845, he wondered if "my investigations pursued at intervals of leisure during some 25 years" had been "spent in vain."[38] But, as A. R. Grote recalled in 1889, Harris "ran the first furrow, and his successors have but widened the field of practical and economic entomology." Those who did the widening of Harris's first well-placed furrow created a field of knowledge based on the early principles of integrated pest control. With this knowledge, they felt themselves prepared to tackle the insect problems that had plagued American farmers since the seventeenth century. Their plans were generally responsible, conservative, safe, linked to the past but not averse to change, and mired in the persistent power of observation. These qualities were in many ways the opposite of what would come to define the insect wars after 1900. But to fully understand how the quest to control insects ran in a direction that Harris could never have predicted, and most certainly would never have endorsed, we must turn to the frontier—the region that sustained America's ongoing insect paradox.

CHAPTER 3

"Let us conquer space":
BREAKING THE PLAINS AND FIGHTING THE INSECTS

It is unlikely that Solon Robinson knew, or had even heard of, Thaddeus William Harris. Had they met, however, they would have found that they had little in common. Robinson was an adventurer; Harris, a librarian and scientist. Robinson was an explorer; Harris rarely left Cambridge. Perhaps most notably, Harris worked under the assumption that the essence of American agriculture was embodied in east coast farms and plantations. But Robinson, deeply familiar with the Old Northwest, envisioned a scale of farming and ranching that would soon transform agriculture into a kind of commercial enterprise that rendered Harris's work irrelevant. In other words, as Robinson led the charge west, America's insect paradox—which Harris was hoping to manage—came to demand a more concentrated and intensified response.

"there are more bugs than wheat"

Solon Robinson's rhetoric of expansion presaged the rapid and largely human-inspired transformation of the grassland prairie into cash-crop farmland. While the extent of the transition can easily be overstated, it was true that midwestern pioneers stripped much of the land of its native grasses and—as their forebears had done back east after clearing old-growth forests—replaced them with large stands of merchantable crops, mostly corn and wheat, but also tubers, vegetables, sugar beets, and orchard fruits.

Although not all-encompassing, the alteration was one of the most significant human-driven environmental changes in American history.

What made the pioneers qualitatively different from the Native American cultures of the Great Plains was the extent of their economic activity and intensity of their market orientation. Indeed, the rationale behind the transition from grassland to farmland responded, as had the transition from woodland to farmland along the east coast in the eighteenth century, directly to commercial forces. It was (or so it was thought) more profitable for farmers to practice monocultural agriculture and ship loads of a single crop to urban markets than it was to practice diversified agriculture and serve local needs. The lush native grasses, given these imperatives, had to be either exploited or removed. For a time, they were exploited—mostly to graze cattle. But soon this option was swamped by a market logic that dictated removal. As with deforestation in the East, this transformation happened quickly.[1]

Removal had serious consequences. For one, it led to the concentration of millions of head of livestock into feeding enclosures. Farmers who pursued this option reasoned that it was cheaper to fatten livestock with corn than to allow them free range across diminishing supplies of prairie grasses. Hence the golden sheaves' rise to prominence as the native grasses, which composed what one observer called "earth oceans," retreated to "the margins of cultivation." Around this transformation of the land, an agricultural philosophy that rewarded specialization over diversity, quantity over quality, and short-term economic gain over long-term ecological stability further cohered into an agricultural way of life. To be sure, this intensive market activity, and subsequent environmental consequences, did not suddenly supplant a culture that was in harmonious balance with a stable ecosystem. The overlapping populations of buffalo, horses, Indians, and Europeans marking the economy of the Great Plains, according to one analysis, "resulted in a loss of two thirds of [the Indians'] former plant lore."[2] It was, all things considered, a multifaceted effort. Nevertheless, the confinement of livestock in combination with the monoculture of grain represented a new practice of agriculture that would proceed well into the twentieth century without significant opposition.

But there was a hitch. As pioneers like Robinson "broke" prairies and introduced foreign crops, they nurtured more than just thriving fields of grain. Insect infestations gripped the Midwest during the second half of the nineteenth century. Whereas Robinson wrote home about the "indescribable delightful feeling" that the prairie inspired, other pioneers—including a farmer named E. Snider of Highland, Kansas—described a different kind of development. As Snider put it, locusts "acting more like the advance skirmishes of an advancing army . . . commenced coming thicker and faster, and they again were followed by vast columns, or bodies looking like clouds in the atmosphere . . . rattling and pattering on the houses, and against the windows, falling in the fields, on the prairies and in the waters."[3] What he was witnessing would become for midwestern settlers a horrific reality of life. Insects joined other environmental hurdles to check the progress of Manifest Destiny in ways the promotional literature never mentioned. Despite the portrayals of the Midwest as a land of unfettered opportunity, American pioneers had inherited a deep, often biblical, sense that the progress of civilization was bound to confront seemingly insurmountable challenges. These pests fit the bill. While the settlers certainly did not expect the insects to greet them as aggressively as they did, in no way were they going to abandon their goals as a result.

Farmers and ranchers in the Midwest furthered the insect paradox while being well aware of the consequences. The region offered extensive evidence of the success of animal confinement (penned cattle) and, at the same time, equally extensive evidence of the failure of crop domestication (insect outbreaks). Farmers now knew that this connection was no coincidence. Economic entomologists, involved as they were in formulating strategies of pest management, made attempts to offer solutions. They encouraged farmers to loosen monocultural practices in favor of agricultural diversification. Cereal crops had to be interspersed with mixed plantings of peas, beans, sorghum, broom corn, tomatoes, and sweet potatoes if pest outbreaks were to be managed—or so farmers were warned. Wilson Flagg, of the Essex Agricultural Society in Massachusetts, recommended "resorting to the original wild potato, and the restocking the country by seeds procured by this source." Asa Fitch reiterated that "a large portion of the insects which now infest our fruit trees originally subsisted on the native forest of this country." As early as 1852, the *Southern Cultivator* was advising farmers to diversify their products: "It has always appeared to me surprising that the planters of the South should persist in the cultivation of large crops of cotton to the manifest injury of their land." An article in an 1870 issue of the *American Farmer* chastised farmers for "lamenting their light crops of wheat" when they knew that "more of a mixed husbandry will save the West." And as late as 1885, the president's address to the Entomological Society of Washington reiterated the entomologists' collective conviction that "certain [insect] individuals of a species, which has hitherto fed in obscurity on some wild plant, may take to feeding on a cultivated plant, and with the change of habit undergo in the course of a few years a sufficient change of character to be counted a new species." Eventually, "the species finally becomes a pest" and, as such, "necessarily attracts the attention of the farmer."[4] Cultural cures, in other words, were widely available as monoculture spread west.

Pioneer farmers, for their part, rarely paid heed. Wedded to an agricultural system that thrived on the maximization of a single crop, they were inclined to seek the "Royal Road to the destruction of bugs" rather than take steps toward moderation and balance. Any motivation to embrace diversified agriculture, moreover, would have been further undermined by factors generally beyond their control: the increasingly efficient transportation systems, the rise of primed urban markets, the relative ease of entering those markets, the availability of government subsidies, and the emergence of specialized industries capable of meeting needs that more diversified family farms once met on their own. The entire infrastructure of midwestern agriculture, in short, supported cash cropping. Add to these factors access to easy credit buoyed by a boom in farm prices, and it makes sense that pioneers introduced extensive farming to midwestern land. "Most of us crossed the Mississippi or Missouri with no money," wrote a state official, and were thus inspired by the "haste to get rich." Although closely tied to the landscape, the typical pioneer was, like his east coast ancestors, more concerned with transforming the land than achieving ecological balance with it.[5]

The arrival of settlers on the Great Plains between 1840 and 1870 thus perpetuated and intensified the insect paradox. At the very time when Thaddeus William Harris, Asa Fitch, and other young entomologists were turning their earnest attention to agricultural concerns, those agricultural concerns were expanding—one might even say exploding—into an entirely new phase of development. Not only were American farmers moving west, and not only were they transforming the environment as they did so, but they were farming on a scale that made even the most ambitious tobacco, rice, or wheat farm back east appear modest in scale and scope.[6]

The destruction of the prairie grasslands, and their replacement with monocultured staple crops, fundamentally altered both the insect population and the strategies of insect control. They did so, moreover, at the precise historical moment when entomologists were establishing the very institutions and strategies designed to work cooperatively with American farmers to manage the insect problems that had plagued them since the seventeenth century. The farming operations in the Midwest were becoming too large, too monocultural, too vulnerable, and too commercialized to incorporate the insect-control strategies designed for farms in the Northeast, which were smaller and more self-sufficient. The problems that Harris and his acolytes worked to *manage*, in short, were problems that midwestern farmers wanted permanently *solved*. It was an unfortunate and irreconcilable clash of expectations.

"I'm for freedom. That's why I'm going out on the prairies"

"Prairie breaking," wrote an Iowa pioneer in 1869, "is going on in every direction." All beasts of burden that happened to be available were put to use, "from two horses up to the heavy team of six yoke of oxen." These teams pulled furrows at a depth of 2 inches in order to uproot the native sod. With the help of flax sown a couple of months earlier to steal nutrients from unmoored grass, the layer of entangled switchgrass and sloughgrass rotted, leaving the broken fields primed for the mass cultivation of wheat or corn. For many pioneers, the experience sparked mixed emotions. Cow Vandemark, Herbert Quick's fictional frontiersman, observed that "breaking prairie was the most beautiful, the most epochal, and the most hopeful, and, as I look back at it, in one way the most pathetic thing man ever did, for in it, one of the loveliest things ever created began to come to its predestined end." "Predestined" is the key word. After all, he relished "the thrilling sound as the knife went through the roots." It was, for Vandemark and thousands of his nonfictional prairie-breaking cohorts, ultimately the sound of progress, the sound of control, and the sound that perhaps reminded the breaker why he was going to the trouble of transforming the landscape in the first place. As Vandemark put it: "I'm for freedom. That's why I'm going out on the prairies."[7]

The disruptive effects of sod breaking were further intensified by grazing. Before pioneers shifted cattle into feedlots (around 1850), they allowed them to roam across the plains. Conceivably, this approach could have led to a more sustainable use of the land, one not entirely inconsistent with the system the Native Americans practiced as they maintained a patchwork of plains, pastures, croplands, and savannas. A wide range of native mammals, after all, had been grazing for centuries. Settlers soon discovered, however, that cattle did not easily digest the native fare (with the exception of big and little bluestem, which they quickly depleted), nor did the native fare hold up well under the relentless pounding that the beasts inevitably delivered. Before long, pioneers who had yet to turn the plains over to grain were importing "cool-season grasses" that proved more palatable to cattle, had a longer growing season, and were sturdier under hooves. These imports, while more trample resistant, drove out native species while laying the groundwork for invasions of timothy grass, red clover, sweet clover, and thatches of foreign weeds.

As this transition took place, the soil of the plains deteriorated. It was soil, moreover, that had once been the most productive in the world. Native grasses had been especially effective at directing photosynthate material to their roots. When these rich roots decayed, they infused the soil with dense concentrations of carbon and nitrogen that nurtured the growth of even healthier and more diversified grasses. Such was not the case with the replacement crops, and, as a result, "the destruction of North America's mid-latitude grasslands took less than a century." Despite the speed of this devastation, farmers at the time were often unable to appreciate its full significance. For example, writing in 1837, a pioneer from Maine explained that "the deterioration [of the soil] is so imperceptible to the cultivator . . . that his farm becomes worn out before he is aware of it." By 1925, however, one ecologist surveying the transformation of the prairie in the nineteenth century wrote, "The disappearance of a major natural unit of vegetation from the face of the earth is an event worthy of cause and consideration by any nation"—especially when "civilized man" was "destroying a masterpiece of nature."[8] The masterpiece he was speaking about was the soil of the Great Plains.

Imported grasses may have diminished the damaging impact of grazing, but overgrazing inevitably became a common condition of commercial agriculture. Ranchers tended to graze cattle in limited (and fenced) areas along waterways and nearby sources of hay, a practice that—unlike with the roaming buffalo—concentrated destructive power in areas that would otherwise have offered especially fertile and well-placed land for more sustainable mixed husbandry. Not only were the native grasses quickly trampled and uprooted, but the soil itself was degraded. Studies of overgrazed land have shown a 50 percent reduction in soil moisture, a 75 percent reduction in organic matter, and a 50 percent reduction in nitrogen content. The problem did not end with soil quality. Compacted soil increases runoff, water and wind erosion, and methane levels in the atmosphere. Soil that becomes too dense cannot support even the toughest root systems because water infiltration is inhibited or completely blocked. Effectively, such soil develops the properties of concrete—a phenomenon that sent ranchers and farmers farther and farther west to seek new land rather than intensify production on old plots.[9]

These observations about soil and grasses are not recent discoveries. Writing in reference to Nevada, the Department of the Interior reported in 1880 that "pastures were now crowded, particularly during the winter season, and in many sections were deteriorating from overstocking. . . . For several seasons the pasturage has suffered from overgrazing." In 1898, the same department argued that "a long discussion on the effects of overgrazing" was due, noting ominously that "the killing of natural vegetation through trampling and over grazing has only barely been begun." In his book *The Fight for Conservation,* published in 1910, Gifford Pinchot, former director of the Forest Service of the Department of Agriculture, summarized a generation of ecological wisdom when he wrote that "the great cattle and sheep ranges of the West, because of overgrazing, are capable in an average year, of carrying but half the stock they once could support." While many of these complaints echoed across the Far West, they reflected legacies passed on by the prairie breakers of the Midwest—with their furrows and plows and teams of oxen. But, again, with so much land in front of them, even the more progressive farmers and ranchers had little incentive to adjust their practices to more efficient models of production.[10]

Agricultural change responded to the incentive of short-term profit. Indeed, by the time many western ranchers were lamenting the ills of overgrazing, their midwestern counterparts had altered the way they raised cattle. Instead of grazing them on native and imported grasses, they started to feed them corn grown on what had once been grasslands. This transition stood at the heart of the frontier experience. As more settlers populated land west of the Mississippi, property values soared and, as they did, midwestern farmers who were now competing with western ranchers decided to stop "wasting" land and plant it with staple crops. This decision consequently enabled prairie farmers to purchase calves from western ranchers, pen them in confined feedlots, and fatten them on home-grown corn. This system made further market sense given that the Civil War had cut off grain markets in the South, which made it cheaper, in terms of freight charges, to ship cattle instead of grain. Whatever the ultimate reason for converting cows to corn feed, farmers were quick to praise the transition. "I esteem this a very important production in this part of the country," wrote one farmer about corn, "it ripens early enough and produces a good sized ear with deep kernels, tolerably easy masticated when fed in the shuck to cattle, and easily shelled when fed to hogs." An Illinois farmer said of his corn that he would "not sell a bushel, but feed it all to the stock." This bovine diet became customary enough for the Department of Agriculture to publish remedies for farmers who accidentally fed "smutty corn to cattle."[11]

Growing corn and other staple crops had the ecological impact of worsening soil that had already been nutritionally depleted by grazing. To be sure, many farmers expressed enthusiasm for, if not complete reliance on, the "cornification" of the Midwest. "This grain," wrote an Illinois farmer, "is the indigent farmer's main dependence, for without it, I do not see if he could live and support his stock." This mentality, however, failed to consider the decline in soil fertility that corn, wheat, and other crops caused when farmers refused to intersperse them with clover and alfalfa or to nurture them with a careful system of manure application and field rotation. "By plowing and tilling the soil," writes one authority on soil conservation, farmers "disturb the soil a million times more effectively than animals." Long-term plot studies undertaken throughout the Midwest in the 1930s showed that continuous cropping depleted supplies of nitrogen and organic matter to the point where the soil's structure eventually underwent a fundamental change. Basically, with so much organic matter sucked from it, the soil collapsed on itself as though cows were still trampling it. "There is no portion of the globe," wrote the *Ohio Cultivator*, "that is being exhausted of its fertility by injurious cultivation so rapidly as the Mississippi Valley." Exaggerated or not, the comment confirms one government official's disapproving remark that the farmer's mantra was "corn, corn, corn, forty years in succession, and then move to the far west."[12]

The nutritional depletion of the soil, combined with the removal of the natural protective cover that native vegetation once provided, fostered patterns of severe erosion. The earliest national studies of soil erosion, also undertaken in the dust bowl days of 1930s, showed that 50 million acres of land—most of it in the South and the Great Plains—were rendered useless by extensive cultivation of single crops. "Unrestrained soil erosion," wrote an official from the Department of the Interior in 1934, had turned the nation into "an empire of worn-out land." He blamed the deterioration of what he nostalgically called "a virgin paradise" on nineteenth-century pioneers who "found a region so rich in land, timber, grass, game, fish, fur and navigable streams that they early developed . . . a false concept of inexhaustible resources."[13]

The assessment touched on several truths. As more recent studies have shown, soil erosion and the consequent sediment runoff were indeed stark nineteenth-century realities. Investigations in Wisconsin, Michigan, and Illinois reveal that the region "experienced some of the highest rates of erosion in the northern United States." Studies of changing widths of streams that drew on data recorded in the 1830s found "postsettlement alteration of erosion and sedimentation" to be the driving force behind geographical change. With the native flora and fauna so severely disturbed, according to the same Department of the Interior official, "lands which had been thoroughly protected through thousands of years of time by unbroken mantles of vegetation were suddenly exposed over extensive areas to the dash and sweep of torrential rains." The topsoil of the Midwest was "literally swept away," leading to extensive sheet erosion that undermined a landscape that had once achieved a relatively symbiotic relationship with its original inhabitants.[14]

The cattle–corn complex—centered as it was on the cultivation of extensive swaths of land and the aggressive replacement of native flora with imported species—was thus more than an extension of the agricultural history of the colonial and early republican periods. It provided a new model for the agricultural imperialism that gradually conquered the Midwest and Far West. The well-honed pattern of land exploitation, short-term market maximization, and extensive cultivation—not to mention the general rejection of "clean agriculture"—supported the production of several crops, few of them native, that covered the Midwest and Far West in the last quarter of the nineteenth century. In addition to corn, wheat, and cotton, there were fruit orchards (apple, pear, plum) stretching from Idaho to California, sugar beets growing from Kansas to Colorado to Montana, citrus groves dominating Nevada and California, and potato fields extending from Minnesota to Colorado. All these changes to the land were made in earnest and in the name of such respectable pursuits as commerce, entrepreneurialism, rugged individualism, Manifest Destiny, and God. In less than a century, however, they helped alter the ecology of the United States in a way that exposed it to attacks by insects.

When Solon Robinson encountered his first prairie and deemed it "a rich mine of wealth," he spoke for generations of American entrepreneurs who scoured the landscape for a chance to improve their lives. These young men and women were behaving in an entirely customary manner. It was a manner that thrived on abundant natural resources, a compulsion to exploit those resources, a willingness to migrate, and the overwhelming confidence that, howling as the wilderness might be, it was there to be conquered. Moving rivers; driving cattle; mining the earth; changing the soil; depleting the flora (and, in turn, the fauna); opening the land to the forces of erosion; building homes, fences, stables, and marketplaces—all reflected the confidence that American pioneers had nurtured since the settlement era. The alteration of the Great Plains was central to the idea of American progress, and it certainly reaped short-term economic rewards. But it also led to long-term ecological instability. There was no greater manifestation of that instability than the insects that invaded the fields, crops, farms, and ranches that the pioneers had built and tended.[15] Indeed, millions of insects would now take advantage of new conditions to destroy crops on a scale that Thaddeus William Harris and his founding brothers could never have imagined.

"just as Sherman marched to the sea"

The first pest to grab the attention of midwestern farmers was the Colorado potato beetle, which had been identified in 1824 by the American naturalist Thomas Say. Say was the great-nephew of another famous American naturalist, William Bartram, and the founder of the Natural Academy of Sciences in Philadelphia. Economic entomologists and agriculturalists who came of age under the influence of Thaddeus William Harris considered Say an important naturalist. Still, they certainly would have found his description of the Colorado potato beetle lacking in a few critical details. As Say described it, *Doryphora decemlineata* lived primarily in the Rocky Mountain region, where it fed on a wild species of potato called *Solanun rostratum;* was 0.5 inch long; was light yellow with ten dark lines running lengthwise down its carapace; and had pink wings and a dark brown head. As every post–Civil War farmer with access to a local newspaper knew, however, the Colorado potato beetle was so much more than a prettily colored bug that flitted from one wild potato patch to the next. It was, as early as the 1860s, a voraciously destructive insect that, "just as Sherman marched to the sea," charged easterly, against the tide of human settlement, to infest potato fields from lower Colorado to upstate New York.[16]

By the end of the nineteenth century, Americans knew more than they wanted to know about the creature that Say had identified. They knew that it had changed its diet from wild nightshades to cultivated potatoes and followed the crop as though magnetized by it. By 1859, the Colorado potato beetle had extended its reach beyond Colorado to Omaha, Nebraska. Two years later, it was destroying potato fields in Iowa, spending the next three years infesting potato patches across the entire state. Writing to the *Prairie Farmer,* an Iowa planter warned readers: "[T]hey made their appearance upon the vines . . . devouring them as fast as they grew." Swarms of beetles crossed the Mississippi River in late 1864 and proceeded to infest western Illinois. In 1865, Benjamin D. Walsh weighed in on "the new invader from the West," predicting that the beetle would "travel onward to the Atlantic, establishing a permanent colony wherever it goes."[17]

Walsh was correct. After "marching through Illinois," the Colorado potato beetle hit Wisconsin and, by 1868, was settled in western Indiana. The next year, the *Ohio Farmer* announced what farmers throughout the state feared: "[T]he insects soon traversed the State of Illinois and reached the shores of Lake Michigan, where they might have met a watery grave, but, unfortunately their course was only deflected southward." By the summer of 1870, the Detroit River "was literally swarming with the beetles, and they were crossing Lake Erie on ships, chips, staves, boards, or any other floating object which presented itself." By 1874, the Chesapeake Bay region, the Mid-Atlantic states, and New England were infested. "They appear to have an irresistible tendency to travel East," a Brooklyn resident wrote to Charles V. Riley, "and are only stopped by the waves of the Atlantic Ocean." It was with good reason that Mary Treat, the popularizing entomologist, wrote of the beetle, "At the time of its discovery, neither Mr. Say nor any of his associates could have had the remotest idea that this insect would at some future day become one of the greatest pests that ever afflicted the farms and gardens of this country."[18]

It did not take long for entomologists to quantify the beetles' destruction in monetary terms. "Various estimates of the cash loss to the country by the ravages of the beetle have been made," wrote the *New York Times*, "all of which unite in placing the decrease in production alone at $8,000,000." Other estimates put losses at $10 million to $12 million. Making matters more troubling, the *Times* continued, was that "the Colorado beetle is noted for its permanency, and rarely abandons localities until it has ravaged them for several seasons in succession." So many farmers had temporarily abandoned the potato due to the ravages of the beetle that "many a family had to forgo the luxury of a product which a few years before had been one of the cheapest on the farm, and so abundant as to enter largely into the feed of all kinds of stock."[19]

For all the market havoc the beetle wreaked, however, observers of its damage could not escape the conclusion that its spread was neither a freak accident nor a natural disaster. To the contrary, as Riley pointed out in 1876, it was "civilization . . . and settlement on the plains" that provided "the means of bringing [the beetle]." The devastation was unquestionably severe, but it would not have been possible "without man's direct assistance or carriage." Looking back on the era, E. Porter Felt, writing about the Colorado potato beetle, noted, "Man has disturbed the balance of nature by planting large areas to single crops." Humans were to blame, for it was they who "provided conditions favorable to the development of the pest," rendering themselves "well nigh helpless."[20]

"the San Jose scale was at last in the east"

John Henry Comstock, an entomologist at Cornell University who was doing research on scale insects in California, discovered a strange specimen in the summer of 1880. Sensing its destructive nature, he named it *Aspidiotus perniciousus* in his first published account of the San Jose scale: "I think it is the most pernicious scale insect known in this country; certainly I never saw a species this abundant as this is in certain orchards I have visited." Perhaps unfairly, the scale became known as the *San Jose* scale because it turned out to have arrived from a commercial orchard in San Jose, California. The

San Jose scale is smaller than the head of a pin, and it spends most of its life hidden underneath a waxy covering that protects it from predators. There is, however, nothing microscopic or subtle about its destructive power. The San Jose scale attacks nearly every cultivated fruit tree—including apple, pear, quince, apricot, plum, and cherry—as well as currant and even gooseberry shrubs. It destroys by sucking the sap from the trees' bark, wood, leaves, and fruit. "Young branches thickly infested by this species," wrote entomologist Daniel William Coquillet, "soon present a gnarled, knotted, and stunted appearance, and if everything is favorable to the rapid increase of these scales the tree is finally killed by them." Commercial fruit growers were soon maligning the tiny insect as "one of the worst fruit tree pests known," an insect that "is with us as a permanent resident." Writing fifty years after the discovery of the scale, E. Porter Felt remarked that it "was so destructive that in the judgment of many progressive horticulturalists apple growing as an industry was threatened."[21]

The San Jose scale did not debut with a bang. Unlike the Colorado potato beetle, it is not particularly skilled in the art of relocation. As John Smith, the state entomologist of New Jersey, observed in 1897, "the scale has little power of its own to get from tree to tree." The *San Jose Mercury News* mentioned in 1888, almost in passing, "having noticed the progress of the San Jose scale at Los Angeles," and by 1891 the *Tacoma Daily News* reported "the importation of oranges to Washington [State]" as the cause of an outbreak in the Pacific Northwest. Entomologists were so underwhelmed by the scale's modest itinerancy that they were theorizing that it moved up and down the west coast on the talons of hawks. In 1893, however, the scale's reputation for lethargy came to a halt. A professor of agriculture at the University of Virginia discovered some strange spots on a shipment of pear trees that had come to his lab from a nursery in New Jersey. The professor sent the trees to an assistant entomologist at the Department of Agriculture who, by his own account, "jumped from my chair in excitement on recognition of the fact that the San Jose scale was at last in the east." Orchard keepers were less excited, but they joined the entomologist in wondering how the relatively sedentary scale had crossed the country with all the subtlety of a lightning bolt.[22]

The answer turned out to involve markets and modern transportation. Just as the Colorado potato beetle hopped the occasional ride on a ship or train to move from one potato field to the next, the San Jose scale sat tight on young trees grown in California but destined for nurseries, and thus orchards, along the east coast. The first transcontinental specimens infected with the San Jose scale were plum trees sent to researchers at eastern agricultural experiment stations that hoped to use them, ironically, to hybridize trees resistant to the plum curculio, another common orchard pest. How these scales wound up in two New Jersey nurseries remains unclear; but what is clear is that these nurseries had an active client base. Between 1893 and 1899, the San Jose scale had spread through portions of every eastern and midwestern state.

Local news blurbs trace the story in dramatic detail. In 1895, the *New York Times* predicted that "the fruit growers of New-York will feel disastrous results of the scourge." An article in the *Wheeling Register* noted the arrival in West Virginia of "a dandruff like scale" that it called "one of the worst fruit tree pests known . . . a menace to the fruit industry of the state." The following year, North Carolina was hit, with the Raleigh paper reporting that the state entomologist had "found the San Jose scale in the nursery at Biltmore." By 1900, the San Jose scale had ravaged the southern plains, with the *Dallas Morning News* deeming it the "greatest obstacle to the success of those who are engaged in horticultural pursuits throughout this country." By the turn of the century, Leland O. Howard wrote, "it had occupied the attention of nearly every meeting of farmers and fruit-growers that had been held in the Eastern states."[23]

The driving force behind the demand for fruit trees was Europe's hunger for dried fruit from the United States. Insult was thus added to injury when European governments sought to block the importation of American produce as a result of the well-publicized San Jose scale scare in the United States. Market resistance began in January 1898, when the German government "refused [an American shipment] admission into Germany because the fruit is alleged to be infected with the San Jose scale." With that decision, a line of dominoes collapsed. In March, the British board of agriculture decided to "take steps to prevent the landing in England of a consignment of American apples," followed by the Canadian resolution "prohibiting the importation of nursery stock from the United States," followed by the French decree in December "forbidding the admittance into France of fruit and plants from the United States." Smith put the matter in sobering perspective when he calculated that he would have to inspect 600 trees a day for 800 days in a row before he could be sure that an annual stock of trees was safe for export. He naturally argued against undertaking such a task, instead suggesting that "the willfully neglectful individuals could be made responsible for the consequences of their neglect."[24]

But as T. D. A. Cockerell reminded his readers, *anyone* who cultivated the soil was ultimately "responsible for the consequences of their neglect." The cultivated environment, wrote Cockerell, was in "a condition of unstable equilibrium owing to the remarkable changes lately brought about by the hand of man." The nature of those changes may have reflected "the advent of civilized man," but they also meant that "the land now covered with abundant crops will . . . support a far greater number of insects than formerly." Perhaps more troubling than the number of bugs was that "their character is largely different"—"kinds which formerly were scarce or limited to a small area now increase and become widespread, while others lose their ground or become extinct." While the prospect of extinction was sure to seem "a melancholy thing to entomologists," it was the "horticulturalists and farmer," Cockerell argued, who "will doubtless feel much more melancholy when contemplating the reverse state of affairs—of an insect once rare increasing enormously." Due to the environmental changes wrought by American farmers, ranchers, and miners—that is, due to the "hand of man"—the San Jose scale had joined the Colorado potato beetle in making the second half of the nineteenth century a time of unprecedented insect infestation.[25]

"the ravages of the chinch bug had been very destructive"

The Colorado potato beetle and the San Jose scale were joined by the chinch bug. "No greater national calamity," wrote the *Milwaukee Sentinel*, "could befall the country than that the corn crop should fail to yield the usual abundance." The year was 1874, and the reference to the diminished corn crop was due directly to a pest that is native to North America but had become a severe force of destruction. An adult chinch bug is less than 0.25 inch long, is clothed in a grayish down not visible to the naked eye, and has honey-yellow antennae and white wings. The *Sentinel* article cared very little, however, about how the insect looked. Most people already knew. The article thus continued, "It is not to be disguised that the ravages of the chinch bug are reported from so many quarters and over so wide an extent of country in the corn region." The cinch bug had become, after all, nothing less than "a resident of the West."[26]

If so, it was an especially ill-mannered tenant. Thousands of reports pouring from agricultural journals across the nation had been saying so since the 1850s. Virginia began consistently lashing out against the chinch bug in 1854; Connecticut, in 1860; Illinois, by 1864; and the rest of the Midwest, by 1870. "The Western states might have congratulated themselves that they were free," wrote the *New York Times*, "but they have in the chinch-bug an even greater pest than the [wheat] midge." Indeed, by 1890 nearly every region of the United States—from Maine to southern California—was reporting massive outbreaks of the chinch bug, which, according to *Science,* moved in "a wave like propagation." By the turn of the twentieth century, C. L. Marlatt of the Department of Agriculture explained, "the chinch bug is certainly responsible for as great annual losses to farm crops as any other injurious species of insect known." Charles V. Riley, chief entomologist of the United States, had said as much twenty years earlier, noting that "few if any insects attracted more general attention . . . or did more serious injury . . . than did the chinch bug."[27]

Although always present, the chinch bug had never been as injurious as it eventually became. Corn crops in North Carolina fell under the first reported attack in 1797. Scattered reports followed every few years thereafter. In 1821, a newspaper in Boston reported that "the wheat crop is expected at all events to be shorter than the last" due to the chinch bug. Residents of Alexandria, Virginia, reported in 1823 that chinch bugs were "flying in our city." The same year, another Boston newspaper had to note that "we are sorry to learn that this bug (so called from the fetid smell when mashed) has made its appearance among the wheat." By the 1830s, there were occasional sightings in the North and, more commonly, the lower South. "In North Carolina," wrote one report, "the ravages of the chinch bug had been very destructive." These occurrences, though, were comparatively few and far between. They in no way foreshadowed the panicked assessments that characterized reports of predation by chinch bugs during the latter half of the century. Few, indeed, could have predicted that the pest would cause more than $22 million in damage in Illinois alone on the way toward achieving its Department of Agriculture–defined status as one of the nation's most destructive insects.[28]

As chinch bugs became more aggressive, entomologists began to delve deeper into their history, life cycle, and habits. By the 1870s, it had become clear precisely how the chinch bug rose to power as quickly as it did. Most entomologists agreed that the chinch bug had more or less always lived from the east coast to eastern Colorado, that it subsisted on a diverse diet of several native grasses, and that it was kept in check both by a diversified ecosystem and by a fungus that thrived especially well in edge habitats. They also agreed that the chinch bug propagated primarily through wheat. The explosion of wheat and corn cultivation—the bulk of it concentrated in the Midwest—provided the chinch bug's path to infamy, because the insect could seize on wheat, increase in numbers, and then attack corn. Much as the Colorado potato beetle switched from wild nightshades to cultivated potatoes, the chinch bug switched from wild grasses to cultivated grains. Because the ecological changes across the Midwest had been dramatic, so was the result of the chinch bug's infestation. Within a decade of breaking the plains, farmers watched the chinch bug cover Minnesota, Iowa, Illinois, Nebraska, Kansas, and Oklahoma. Wheat was the starting point, a way to get a foothold in a region. Once entrenched, the chinch bug went on to consume corn, oats, barley, rye, sorghum, and a variety of imported grasses until it could get back to wheat.[29]

Agricultural reports soon read like doomsday screeds. In Illinois, to cite a typical complaint, an agricultural reporter wrote in 1889 that "this season the pest seems to have appeared simultaneously over a far wider extent and in a far greater number of places." Every county in 1888 had filed an update with the state entomologist on the chinch bug's progress. Here, too, as a sample shows, the assessments were bleak:

> Carroll—Nearly destroyed some fields of spring wheat
> Coles—Chinch bugs have made their appearance in considerable numbers
> Cumberland—Have been here throughout the year. All over the county to some extent. Some fields of corn already ruined
> Gallatin—Noticed since about the first of June. One farm was overrun by them.
> Hardin—Increasing here very fast. Probably ten times as many as ever before.[30]

Yet again, only the most willfully myopic farmer could have missed the lesson of these reports. "The fact of its attacking cultivated plants," wrote the state entomologist of Ohio, "does not by any means necessarily imply a recent introduction." This was a point that farmers had to grasp. "All or nearly all native insects," continued the entomologist, "adapt themselves to cultivated plants only when forced to do so by the encroachments of the latter upon their natural food plants, and I think we can show that *Blissus leucopterus* is not an exception." Soon entomologists were pleading with farmers to quit grains altogether. "Chinch bugs originate almost exclusively in spring wheat or barley," wrote William LeBaron, the state entomologist of Illinois, "and we have it in our power . . . of getting rid of these destructive insects, and keeping clear of them, by abandoning the raising of these two kinds of grain." At least a few farmers took the advice. Writing from Bureau County, Illinois, one farmer explained, "There has been very little wheat sown here for many years past, and farmers think that this accounts for the absence of the chinch bug." Samuel Frost of McDonough County had a similar experience: "We have no chinch bugs in our county. There is very little spring wheat raised here, which may account for their absence." It certainly did. But midwestern farmers, like their colleagues east and west, proved to be more intent on building a better bug than reforming agricultural practices that were clearly going to die hard.[31]

"Again and again have fields of young wheat become infested"

"The wheat plants have been attacked," an agricultural report noted, "by the Hessian fly in a way that threatens the destruction of the crop." This news came in 1879, a full century after this small dust-colored fly began its systematic attacks on North American wheat. In the two decades after its discovery, the Hessian fly rapidly followed and destroyed wheat crops from New York to Pennsylvania to Ohio to North Carolina. Not unlike the chinch bug, the Hessian fly, after an outburst of destruction, seemed to retreat into relative quietude. Maine filed a report in 1823 that the Hessian fly was spotted; Michigan noted its appearance in 1837; Wisconsin, Indiana, and Illinois hosted the fly in 1844; Iowa and Minnesota confronted it in 1860. Surveying the situation in 1860, one might reasonably have concluded that corrective measures undertaken in the East—notably, planting bearded wheat, early sowing, and implementing a variety of rotation schemes—had kept the pest in check. It was, for the most part, considered to be history.[32]

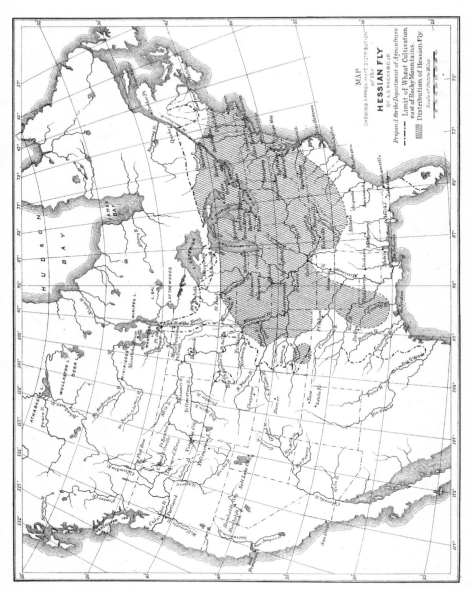

DISTRIBUTION OF THE HESSIAN FLY

The Hessian fly is a pest normally associated with the Northeast during the Revolutionary era. As this map shows, however, it continued to spread westward in the nineteenth century, fueled by wheat monoculture and the expansion of commercial agriculture across the Midwest

By the 1870s, however, as wheat became more of a staple crop across the plains and elsewhere, it became clear that entomologists had not wasted their time in continuing to study the Hessian fly's life cycle. As the pest found new opportunities to breed, reports of attacks by Hessian flies quickly became more urgent and frequent. An article published in 1878 about New York reminded the nation that while the Hessian fly spread, it did not leave its adopted home base, but rather remained entrenched *and* spread: "Reports from central New York [indicate] that the Hessian fly is seriously damaging the wheat crop in that part of the state." When the fly spread, moreover, its travel plans were ambitious. "The Hessian fly is at work on the wheat," the *Dallas Weekly Herald* reported in 1877, in a county-by-county assessment of the wheat crop. In 1879, an Ohio newspaper noted that "southern Indiana and some places in Kentucky" were under siege by the Hessian fly. An assessment of the Hessian fly in Illinois in 1881 confirmed a 59 percent reduction in wheat production over a year as a result of the pest, which "destroyed most of the wheat that was not winter killed." By 1883, it was Pennsylvania farmers who realized that "their main crops will be seriously endangered should the fly multiply." By 1889, it was evident that things had not improved in Illinois, as "the Hessian fly is destroying the wheat crop in central Illinois . . . whole fields of have been destroyed." The following decade was even worse.[33]

To stress the reasons behind the outbreak of the Hessian fly, entomologist F. M. Webster conveyed a lesson he had learned from a small experiment he had conducted in the garden behind his Washington, D. C., townhouse. It was a lesson that, for farmers, was either hidden in plain sight or, more likely, purposefully ignored: "I have had no difficulty attracting Hessian flies to a small plat of fall wheat sown in my garden in the midst of a city with 30,000 inhabitants, with no wheat fields within a mile. Again and again have fields of young wheat become infested along the border nearest to infested stubble fields situated long distances away." Once again, the point was hard to miss: human decisions were directly responsible for the devastations that these insects were causing. Webster, commenting on the difficulty of identifying the Hessian fly, wrote that "a failure to observe the Hessian fly until long after it became established in a State or Province, cannot be taken as proof of . . . stupidity on the part of the wheat grower."[34] While his opinion seemed reasonable enough, the same could not be said for the wheat growers' choice to rely on wheat as a staple crop in the first place. Few farmers, however, were willing to make the necessary changes. These were entrepreneurs, after all, not environmentalists, not ecologists, and their mission was, with Horace Greeley trumpeting the call to "go West, young man," beyond reproach. Unfortunately, the Hessian fly went as well.

"the grasses did not wave"

Sarcasm was generally not a popular sentiment among rock-ribbed pioneers developing the American West. But on July 29, 1874, readers of the *Daily Constitution* got a healthy dose of it from an anonymous contributor who began, "What can be pleasanter than the life of a Missouri farmer?" His answer was anything but sincere: "At daylight he gets up and examines the holes around his corn hills for cut worms, then he smashes coddling moth larvae with a how handle until breakfast." After battling potato beetles and "pour[ing] boiling water on chintz bugs in the corn and wheat fields," he attends "a brief session of family devotion at the shrine of the night flying coleoptera." When evening comes, "all the folks retire and sleep soundly till Aurora reddens the east and the grasshoppers tinkle against the panes and summon them to the labors of another day." It was a fair account, except for one small detail. The grasshoppers that tinkled Missouri windowpanes in 1874 had congregated to the point that they were being called another name altogether. This sarcastic writer was referring to a pest that there was really nothing to joke about—locusts.[35]

Swarms of locusts dwelling in the Rocky Mountains reached outbreak proportions, migrated in search of food, found it, and denuded the prairies of the Midwest from 1873 to 1877. "Homesteaders came from the east," writes one biologist, "locusts came from the west, and they met in the Great Plains." In 1875, the swarm that migrated across North America may have contained more than 3 trillion locusts. As entomologist Jeffrey Lockwood puts it, "The swarm outweighed a man to the same degree that the biomass of a whale exceeds that of a mouse." Charles V. Riley, the state entomologist of Missouri at the time of the outbreak, wrote that the largest of the swarms "covered a swath equal to the combined areas of Connecticut, Delaware, Maine, Maryland, Massachusetts, New Hampshire, New Jersey, New York, Pennsylvania, Rhode Island, and Vermont." Swarms achieved

FLIGHTS OF THE ROCKY MOUNTAIN LOCUST

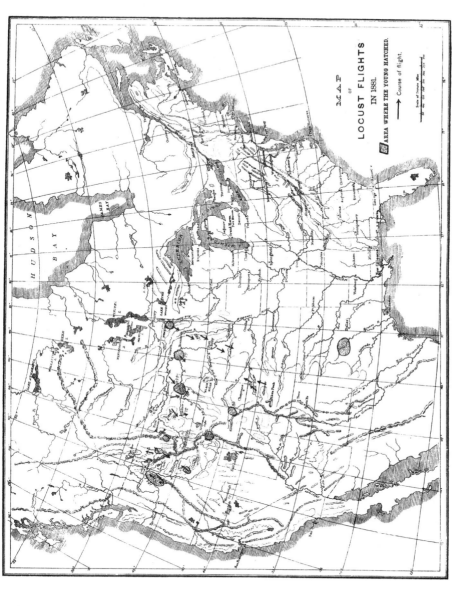

Entomologists did their best to track the Rocky Mountain locust as it moved across the West in the 1870s. The arrows on this map show the directions in which swarms of locusts were moving as well as the areas of their most intense concentration.

such enormous numbers by taking advantage of drought conditions to multiply even more rapidly than normally. They fed productively on weakened crops. "The grasshoppers ate and ate; they devoured everything from barley and buckwheat to spruce and tobacco," writes Charles Bomar. They "even ate blankets that women put over the crops to protect them." The locust, not surprisingly, became, according to the United States Entomological Commission, "the single greatest impediment to the settlement of the country between Mississippi and the Rocky Mountains."[36]

Yet again, the settlers were the insects' greatest assistant and, yet again, they knew it. "Wheat and grasshoppers could not grow on the same land," wrote one homesteader, "and grasshoppers already had the first claim." Locusts might very well have blanketed the prairie grassland had it been conserved, but they thrived more readily on the grain crops that settlers made abundantly available. "Only one family in 10," reported another settler, "had enough provision to last for the winter" because of the crop devastations. Evolutionary pressure had traditionally helped native plants resist locust attacks. Not so with the newer cultivars. The Entomological Commission, led by Riley, which had been assigned to deal with the locust outbreaks, urged planters to practice diversified agriculture. Noting that range grass escaped damage, it pushed a combination of broom corn, tomatoes, sweet potatoes, and beans, and even suggested quitting wheat altogether. Another option was not to plant at all—that is, strategically place large fallows across the plains. "However," writes Lockwood, "the imposition of such crop free zones never materialized as no government had the courage to order farmers not to plant after they'd already been decimated by the swarms."[37]

God was invoked—but clouds of locusts continued to darken not only the landscape but the entire prospect of pioneering new land. How could families persist when, as Laura Ingalls Wilder wrote, "the grasshoppers were walking shoulder to shoulder and end to end, so crowded that the ground seemed to be moving"? Where was the hope when "the whole day long the grasshoppers walked west . . . walked without stopping"? How could anyone do anything but give up when locust swarms "walked steadily over the house . . . over the stable . . . into Plum Creek . . . until the dead grasshoppers choked the creek and filled the water and live grasshoppers walked across them"? What were farmers to do when the nation's brigadier general went to the secretary of war because "several citizens have besought me to send rations to the famine stricken families of southern and western Nebraska"?[38]

The locusts answered these questions on their own. It was not until 1990 that entomologists figured out why the Rocky Mountain locust made its last appearance in 1902. Much as settlers and their agricultural practices initially nurtured the locust, the same practices—through what can only be described as an accident of nature—drove the locust to extinction. As we have seen, farmers persisted with traditional agricultural habits. This time, and this time only, they beat the insects. At the very moment when migrants began to extend their aggressive plowing and grazing practices into the West, the Rocky Mountain locust was concentrating its breeding grounds in narrower and denser plots. The settler who remarked that "wheat and grasshoppers could not grow on the same land" was absolutely right, but not in the way he intended. Wheat, corn, livestock, and other environmentally transformative ventures killed the locust.[39] In short, when it came to the locusts, the pioneers got lucky.

But scientists, entomologists included, are not fond of luck. With American farmers taking agricultural practices to extremes that fostered massive infestations of insects, more aggressive and authoritative measures were in order, and fortune—no matter how well it worked for the locust—was not an option. Entomologists and farmers were remarkably grounded in the relevant scientific principles. Thanks to the work of early entomologists they were well aware that stripped lands, overgrazed pastures, and monocultured crops enabled the potato beetles, scales, chinch bugs, Hessian flies, and locusts to do their damage. Farmers, who were still intimately tied to the land much as their forebears had been, knew as well as anyone that the pests they encountered were linked to precise agricultural decisions. Their field experience repeatedly confirmed this connection. Entomologists, still working in the powerful orbit of Thaddeus William Harris, Asa Fitch, and the other prime movers in economic entomology, not only observed this relationship but—due to their careful study of insect life cycles—understood the reasons for it. Why this knowledge was not put to use to end these harmful practices at this devastating moment remains one of the most elusive questions in American agricultural history.

That is, until one recalls the power and logic of commercial agriculture in the United States. Expansion, profit, and the short-term maximization of resources were the touchstones of this endeavor. It had always been that way, and, as far as anyone knew, it would continue to be so. Insect pests might be on the rise, and they might make for a tragic year here and there for hardworking farmers, but they were, it was thought, the necessary step back that many farmers had to take before bounding forward. In light of Senator John Calhoun's famous dictum—"Let us conquer space"—farmers were loath to sacrifice the chance of short-term profits for radical and long-term agricultural reforms designed to reduce insect infestations. And in light of the imperative that economic entomologists solicitously serve the interests of farmers, many entomologists were loath to advise their charges—their reason d'être—to the contrary. They would, as they had done from the start, work with farmers on the farmers' terms. The insect pests that capitalized on the expansive agricultural developments of the mid- and late nineteenth century thus did not create an impasse. They created the need for a more organized bureaucratic response to their infestations. "It should not be surprising," wrote John T. Schlebecker, "to find that the hearty individualism of the settler crumpled before the plagues of insects." Nor should it have been surprising, he continued, "that he should demand that the government do something."[40] The government, for better or worse, would be called on to manage the insect paradox.

CHAPTER 4

"a great schemer":
CHARLES V. RILEY AND THE BROKEN PROMISES OF EARLY INSECTICIDES

As American farmers were experiencing unprecedented outbreaks of insect attacks throughout the second half of the nineteenth century, the nation's foremost government entomologists—Charles V. Riley and Leland O. Howard—drank a lot of beer. "In those early days," Howard later recalled, "entomology and beer went together." Riley and Howard, who respectively served as director and deputy director of the Division of Entomology in the Department of Agriculture, had formed the Entomological Society of Washington in 1884. At their weekly meetings, they and a few other entomologists hashed out the entomological news of the day. Contrary to Howard's whimsical remark that the biggest choice the group faced was whether "to have light beer or dark," the society confronted serious choices that would shape the field of entomology for generations to come. Foremost among them was how to streamline insect-control strategies to make them useful to a national constituency of farmers. No two men were better placed to make these difficult decisions than Riley and Howard.[1]

These men were of their time. Economic entomology was standing on a relatively firm foundation after the Civil War. Having experimented with cultural, biological, and chemical controls, entomologists were accustomed to working closely with farmers in order to determine the best combination of methods appropriate for specific problems in specific locations. Second-generation entomologists generally placed the highest value on working with farmers in a local context. Following the legacy of

Thaddeus William Harris, they were well accustomed to confronting and negotiating a problem's local contingencies. The formation of the Division of Entomology in 1878 certainly posed a possible threat to the provincial, flexible, and open-minded approach to pest control that the earlier generation of entomologists had worked to nurture, but such a fear proved to be unfounded. Indeed, under the guidance of Riley, the division preserved the discipline's founding qualities while adding greater prestige, institutional organization, and funding to the larger entomological mission. The "chaos of experimentation" that farmers had relied on throughout the colonial and early American periods, and that economic entomologists had effectively preserved after 1840, persevered through the rest of the nineteenth century. Post–Civil War economic entomologists, despite the press of institutional centralization, would instinctively favor the tradition of decentralization, resisting the impulse to seek a one-size-fits-all solution to the insect wars. One of the methods that the profession stressed under the leadership of Riley—biological control—promised to be economic entomology's guiding light into the twentieth century, and thus the primary strategy to manage the insect paradox.

"Professor Riley"

Charles Valentine Riley was born in London on September 18, 1843. At the age of seventeen, he moved to the United States and found work on an Illinois ranch. A co-worker recalled Riley expressing an "intense love of rural life." "There was not a sick animal of the three hundred on the place that he did not understand and help," G. B. Goode wrote, adding that Riley "kept a lot of bees . . . and spent his Sundays reading, sketching, and studying insects." Riley's work ethic bordered on the obsessive, so much so that after three years of working on the farm he suffered a bout of exhaustion that drove him to seek less physically demanding work in Chicago. "Had his health not failed him," wrote Goode in 1890, "my opinion is that he would be a farmer today."[2]

In Chicago, Riley wrote articles on invasive insect species for the *Prairie Farmer* and, after a year of freelancing, was asked to edit the paper's entomological page. This position provided him with an opportunity to read extensively about insect behavior and control, particularly with respect to agricultural practices. His articles reflected the tenacity of his research, standing out for their readability, reliability, keen attention to local conditions, and self-confidence. (In a note submitted to the *Canadian Entomologist*, he said of fellow entomologist Samuel Scudder's description of an insect, "I doubt whether any of these characteristics are of as much value as those I have given.") Pieces he wrote on the Colorado potato beetle and May beetle were circulated nationally and, in time, praised by other entomologists. When the iconoclastic entomologist Benjamin D. Walsh read Riley's work, he invited the ambitious young writer to co-edit a journal that he was launching: *American Entomologist*. Riley, only twenty-six, jumped at the chance, one that brought him into intimate contact with farmers, insects, control measures, and a fraternity of seasoned economic entomologists that included Asa Fitch, A. S. Packard Jr., and Townend Glover. These men wore Thaddeus William Harris's legacy on their sleeves. Riley, for his part, felt right at home.[3]

Working with Walsh and writing for the *American Entomologist* steeped Riley in the fundamentals of economic entomology. He delved into the intricacies of insect life cycles, worked frequently in the field, explored a wide range of control options, got earful upon earful of frank opinions from Walsh, and came to appreciate the idea that insects should be managed but not eliminated. The impressive quality and range of Riley's work, as well as his growing reputation and network of professional connections, resulted in his appointment as the state entomologist of Missouri in 1869. A week after his appointment became official, his friend and mentor Benjamin Walsh was killed by a train as he crossed the tracks, absentmindedly reading his mail.

Riley carried Walsh's torch to Missouri. Perched atop this mantle of government authority, modest though it may have been, he published nine annual reports in which he reaffirmed the cardinal principle that effective insect control derives from an intricate understanding of insect behavior in a local setting. These documents set a new standard, gradually changing the nature of entomological writing. Although they offered little in the way of novel biological discoveries, their signature accomplishment was to make accessibility a top priority. When Leland Howard later admitted that Riley had "set new standards in the treatment of insect pests," he was speaking specifically of his work's "readability and attractiveness," an assessment confirmed by another entomologist who recalled that "Riley intended his reports primarily for the farmer and every effort was made to keep style and language within the understanding of his readers." To further disseminate his work, Riley taught at the University of Missouri. He wowed his students by drawing different insects on the chalkboard with both hands at the same time. They affectionately called their ambidextrous teacher, a man without formal education, "Professor Riley." The name stuck.[4]

In Missouri, the field became Riley's preferred laboratory. It was there that he was able to tap his encyclopedic knowledge of control tactics to achieve a number of accomplishments. Most notable was his research into the phylloxera, or grape root louse. Building on Walsh's maxim that "one of the most effectual means of controlling noxious insects is . . . the artificial propagation of such cannibal species as . . . prey on them," Riley initially introduced the larvae of the syrphys fly into louse-infected soil. "Wonderful indeed," he wrote, "is the instinct which teaches this blind larva to penetrate the soil in search of its prey." When this method revealed its limitations, he then experimented with mites (*Tyroglyphus phylloxerae*) that "prey extensively on this root-inhabiting type." Results were promising enough for French officials to import the mites in order to save the country's phylloxera-ravaged vineyards. While the mites did indeed have a positive impact on the louse, it was ultimately Riley's suggestion that the French plant wild stocks of American grape varieties that eventually struck gold, or at least earned Riley the honor of a gold medal from France on behalf of the nation's jubilant vintners. Riley's studies of the Hessian fly, cabbage worm, and plum curculio—in addition to "the insect-catching attributes of various plants"—enhanced his status as one of the nation's most creative entomologists while reiterating the very basic premise of economic entomology that mere "descriptive work . . . has little other result than to confuse and perplex."[5] What mattered, as Harris had taught a generation earlier, was application. Riley was indeed working in Harris's furrow.

All these accomplishments, however, were dress rehearsals for Riley's next performance. In 1877, Congress appointed Riley to head the first United States Entomological Commission. His assignment was to study the habits and depredations of the Rocky Mountain locust, a pest that was decimating crops throughout the West when Riley was working in Missouri. The position provided Riley the chance to formalize what amounted to a set of entomological commandments. According to these imperatives,

CHARLES VALENTINE RILEY
An impassioned entomologist who was able to draw illustrations of insects with both hands at once, Charles Valentine Riley served as the second director of the Division of Entomology. As a young man, he worked so hard on an Illinois farm that he came close to a nervous breakdown.

economic entomologists had to explore all options, their efforts had to involve farmers in the field, their results had to be accessible to the public, and their methods had to respect local variation. Riley and his commissioners explored arsenical insecticides, natural enemies, and cultural methods in equal measure, choosing to tactfully downplay the advice of petrified politicians that farmers pray, fast, and publicly atone for their sins so that God would ease their suffering. Their contact with farmers was extensive. Edmund P. Russell, in fact, has noted that the commissioners "had little to add to the list of remedial measures already developed by the citizenry." Their communication with the general public, through two enormous but easily readable annual reports, was more than dutiful. They were, many noted, masterpieces of entomological research.[6]

After an investigation into numerous methods of insect control, including a range of chemical options, the most promising of the commissioners' tactics turned out to be cultural and biological. Through intensive documentation of the relationship between local weather conditions and locust hatching, Riley and his team were able to predict with respectable accuracy the area that was ripe for an outbreak and the length of time it would last. The predictions were reliable enough for Riley to call on farmers to diversify crops, burn land, dredge ditches, and destroy locust eggs through plowing. (They were also predictable enough for pastors to call on congregations to fast on the day before Riley said an outbreak would end!) As for biological methods, Riley and his commissioners published encouraging reports on the locusts' "inveterate enemies" that "carry on their good work most effectually." Focusing on blister beetles, which eat locust eggs, Riley argued that once the locust became "unduly multiplied," its "natural enemies [would] invariably multiply in increasing proportion, until, in their turn, they get the upper hand, and bring about a natural balance." In perpetuating the virtues of cultural and biological control, Riley's opinion about chemical solutions may have diminished. He wrote in the *First Annual Report . . . Relating to the Rocky Mountain Locust*, "We have never had much faith in the application to the plant or the insect of any chemical mixture, fluid, or powder, as a means of destroying the locusts." Riley and his commissioners never found a magic bullet to end locust plagues (economic entomologists of Harris's generation did not think in such terms), but they performed well enough for the *American Naturalist* to conclude correctly that "this locust will never be so destructive as in the past, and due credit has been given to . . . the U.S. Entomological Commission."[7]

Riley's career finally came to rest in Washington, D.C., where Riley succeeded Townend Glover in 1878 as the United States entomologist in the Department of Agriculture. Working in the service of the nation as a whole challenged Riley's adherence to radical localism, forcing him to hone his approach in ways that favored uniform applicability over provisional effectiveness. This was not easy for Riley, and he became a prickly man for whom to work. No longer was he dealing with the insect pests of Missouri or the Great Plains, nor could he meet with every farmer who had an interesting story to tell. Riley proved to be an acerbic administrator, perhaps frustrated by the pressure to devise national solutions for thousands of specific problems that he could not micromanage.[8] That said, the reports he wrote as the federal entomologist suggest that he was doing his best to apply his skills to the national context. The characteristic attention to all available forms of control—cultural, biological, and chemical—was clearly evident, as were the essential qualities of clarity, communication with farmers, relative flexibility, firm faith in the power of biological principles, and enduring belief in insect management rather than extermination. The commandments, in other words, were still followed, if a bit less religiously.

What did change in Riley's approach to combating insects at a national level was a subtle but very important shift in emphasis. Between 1878 and 1894, Riley, who had previously refused to overtly favor one form of control over another, chose to finally place extra emphasis on biological methods. This emphasis, while important, must be kept in perspective. Riley did not "choose" biological control over other methods; he simply gave it unprecedented scientific attention. Four years before his tenure came to an end, Riley delivered a speech to the American Association of Economic Entomologists that could just as easily have been presented by Harris fifty years earlier. "Our work," he rhapsodized, "is elevating in its sympathies for the struggle and suffering of others."[9] Riley poured his passions and ambitions into the unmistakable mold of an old-school economic entomologist. But now the mission had a twist, and central to it was the traditional idea that every predacious insect has an enemy. Riley dedicated himself and the division to exploiting this deceptively simple concept.

"fully one half of our worst Insect Foes are not native"

The origins of Charles Riley's faith in biological control derived not only from his farming experience, but also from his relationship with Benjamin D. Walsh. Walsh lived as he died: with a bang. He was a brash bête noire who turned down divinity school, as he remembered it, to avoid being around hypocrites. He instead chose to run a 300-acre farm where he developed notions of biological control that he would later push as the most sensible solution to the nation's pest outbreaks. Walsh always valued his active fieldwork, once disparaging the investigations of a university-trained entomologist with the remark that "if the Professor will only open his eyes the next time he walks out into the fields . . . " As Riley's iconoclastic co-editor of the *American Entomologist*, Walsh was one of the nation's first entomologists to publicly promote biological control as the most favorable solution to insect infestations. "Accident has furnished us with the bane," he explained in reference to the nation's most destructive insect pests, and "science must furnish us with the remedy."[10]

BENJAMIN DANN WALSH

"Everything about him was walshian," wrote his student Charles V. Riley. Benjamin Dann Walsh was a brash economic entomologist and staunch advocate of biological control of insect pests. He lambasted chemical insecticides from the moment they appeared on the agricultural market, calling them "Hellbroths!"

The challenge, of course, was heightened by the fact that many injurious pests had come from abroad while their enemies had remained behind. "It is a remarkable fact," Walsh wrote in 1866, "that fully one half of our worst Insect Foes are not native American citizens but have been introduced here from Europe." Because the United States "had only three or four parasites to check [their] increase," one "common sense remedy" was "to import the European parasites that in their own country prey upon the wheat midge, the Hessian fly, and the other imported insects that afflict the North American farmer." As Walsh saw it, the United States was also choked with flora from the Old World, and those crops were too often exposed to both indigenous and foreign insects prepared to consume them. The implication could not have been more basic: "[I]f we wish to fight effectually against those noxious insects which have been introduced among us from Europe, we must fight them by the instrumentality of the strong and energetic foes that make war upon them in their own country." Thus Walsh's idea of proper science was, he admitted, "as old as the hills." Nonetheless, farmers had no choice but to "assist nature whenever they have thwarted and controlled her, if they wish to appease her wrath." As Walsh repeatedly reiterated this point in the pages of his journal,

Riley—who was thirty-five years Walsh's junior—took careful notes from the charismatic master.[11]

Another influence on Riley's emerging advocacy for biological control was Charles Darwin. Walsh had been a classmate of Darwin's at Cambridge, and both Riley and Walsh were quick to see that it was possible to simultaneously believe in the theory of the balance of nature while accepting the logic of natural selection. Their interest in biological control only heightened their appreciation for Darwin's revolutionary theory. Darwin, wrote Riley, "brought light out of darkness" through his investigations into insect life. With natural selection, as opposed to special creation, now "recognized as a law," the balance that nature achieved had to reflect natural processes rather than divine ordination. Leland Howard later captured the essence of this balance when he referred to "the constant struggles between species in the apparent effort to preserve a just balance." It was this Darwinian version of balance that Walsh had in mind when he wrote that "whenever man, by his artificial arrangements, violates great natural laws . . . he pays the penalty affixed to his offense." Interfering with biological processes, Walsh and Riley insisted, led to imbalance, and in the Darwinian scheme of things it was incumbent upon humans to "restore the natural equilibrium." Biological and cultural controls were much more conducive to this demand than chemical applications. Any doubts that Riley had about the seamless relationship between biological control and Darwin's theory would have been assuaged by a fan letter he got in 1871: "I received some little time ago your Report on Noxious Insects, and I have now read the whole with the greatest interest. . . . Pray accept my cordial thanks for the instruction and interest which I have received." The letter, in a scratchy scrawl, was from Mr. Darwin himself.[12]

Riley's appointment to head the Division of Entomology provided him with a national platform to lobby for serious attention to be paid to biological control, but he refused to neglect other options as he was pursuing his favored method. Under his tenure, the Division of Entomology developed, tested, and frequently promoted a wide array of insecticidal methods, often with Riley's enthusiastic approval. Riley's general approach to the division's research on insecticides was to fund it and acknowledge success when it was well deserved, but to do so with several caveats and persistent skepticism. For example, in a report to the Philosophical Society of Washington presented in 1884, Riley praised the ability of kerosene to occasionally defeat grape phylloxera, noting that "the discovery is of great importance." Then, in the next breath, he condemned pyrethrum as an insecticide that, "contrary to popular belief," has "a peculiar and toxic effect on higher animals as well as lower forms of life." In 1887, at a meeting with New York farmers, Riley advised them "to kill the [bark] beetle and kill its young" with "some form of arsenic," but then reminded them that "there is no elm free from its attacks" and that "every one of those insects has a number of enemies and parasites" that chemicals could easily destroy. After praising the power of Paris green as an effective insecticide, he followed with the reminder that it "injures the plant" and is subject to the vagaries of the weather.[13]

There was, in essence, nothing doctrinaire about Riley's preference for biological control. Couching his general concerns about insecticides in cautious rather than oppositional terms, he urged entomologists to "recognize . . . the necessity of drawing our inspiration more directly from the vital manifestations of nature." Such manifestations, as the earliest entomologists well understood, were generally not to be found in a scientific laboratory while mixing chemicals to kill pests. Nevertheless, the point remains: while Riley was publishing papers on such topics as predators of the cotton worm and parasites of the Hessian fly, he was simultaneously pursuing the chemical option.[14]

Within a decade of his tenure, however, the balance slightly tipped. Riley's emphasis on the biological option was officially confirmed by an uncompromised victory. In 1886, an insect called the cottony-cushion (or fluted) scale, which had been imported accidentally from Australia, proliferated to the point of undermining the California citrus industry. At the behest of Riley, the Department of Agriculture sent Albert Koebele, an insect enthusiast, to Australia in 1888 to research the cottony-cushion scale's natural predators. Koebele quickly discovered a ladybug, known as the Vedalia beetle, that consumed the scale with impressive voracity. In January 1889, he sent three boxes containing 129 beetles to California, where state entomologists bred them for experimentation in local citrus fields. In June, the entomologists turned loose 10,555 Vedalia beetles. In so doing, they transformed the scale from a destructive insect into one that "ceased to be a major pest." The positive results were hard to overstate. Shipments of oranges from California tripled within a year. "We fully expect," editorialized *Insect Life*, "to learn of the increase and rapid spread of this new introduction, as well as of the other predaceous species which have been introduced." The Vedalia story, according to Edward O. Essig, an early-twentieth-century entomologist, "furnished to the world the first demonstration of effective natural control." Even the *New York Times* weighed in, labeling the Vedalia experiment "a complete success."[15]

The buzz was contagious. By the late 1880s and early 1890s, biological control was enjoying unprecedented popularity. "The subject of parasitism is of intense interest," Riley reported to the Entomological Society of Washington in 1893. There was now ample evidence to buttress the enthusiasm. F. M. Webster, the state entomologist of Ohio, wrote in a study of the Hessian fly that "it is proper to say here that the pest suffers much from the attacks of several minute parasites, which attack and destroy it in both the egg and larval or maggot stage." A survey conducted by the Department of Agriculture noted how "the boll worm was scarce during the past season," with the surmised reason being "a common species of soldier bug [that] was found devouring a large full-grown boll worm." The larva of this soldier bug was observed to "puncture the [bollworm] eggs and suck their contents." A test larva placed "on a branch of cotton with some newly hatched boll worms" proved that the worms indeed had fallen "victim to its beak." While Koebele was solving the cottony-cushion scale problem, other entomologists in California were reporting that "the larva of a little moth . . . is also known to feed on the eggs" of the scale. Yet others were encouraged by a scale-eating chalcid fly, explaining that "the probability is that this little friend was introduced from Australia with its host." Two of Riley's last reports for the Division of Entomology crowned the achievements of biological control with an overview of its many accomplishments. His final assessment was that "there can be no doubt whatever as to the good that would flow from the introduction of beneficial species."[16]

It looked as though Walsh (who often was dismissed as an ornery polemicist) and Riley (who had cautiously cast his lot with biology) would be rewarded for their advocacy of biological control of insects. "There is no question," Riley wrote about the fluted scale, "but that it is very desirable to introduce . . . enemies and parasites as can be introduced." Reports of a parasite of the Hessian fly inspired Riley in 1890 to import "from England a foreign species of these parasites, some of which have, by his instruction, been turned loose in the fields in the vicinity of Columbus [Ohio]."

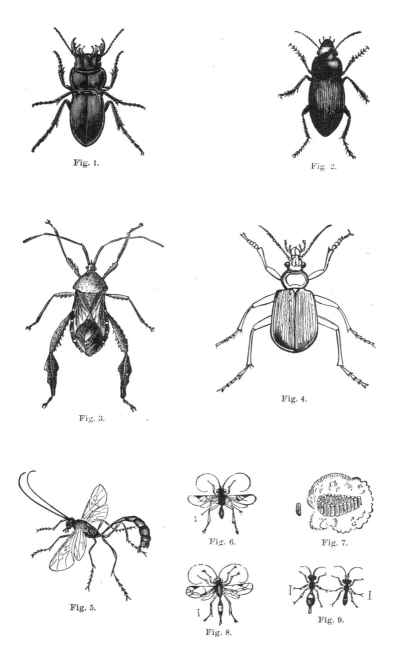

BIOLOGICAL CONTROL

Part of the mission of the Division of Entomology was to publish reports and prints that helped farmers identify the insects that were consuming their crops as well as the enemies of those insects. These diagrams show the (*left*) life cycle and (*right*) enemies and parasites of the armyworm.

A bulletin published by the Department of Agriculture on insects that infest wheat spilled considerable ink on biological control, noting that "the Hessian fly was subject to many predators in Europe" that were "effective . . . in limiting damage." Of the wheat plant louse, the bulletin noted, "fortunately this species has many natural enemies, including various insect-feeding beetles and flies and also true internal parasites." An update on the bollworm in 1894 placed "special stress and importance" on "the egg parasite," reiterating how "its value cannot be overstated." Studies of ants, moreover, "make it absolutely certain that at that season they frequently capture a boll worm." A "small Capsid found abundantly on corn silks" was promised to be treated more thoroughly in a future report. Again, the attention that entomologists were paying to biological control was by no means exclusive. They were, in the midst of this excitement over biological control, still pursuing chemical and cultural options. The clear emphasis on biological methods, however, was proof that Riley, as head of the Division of Entomology, had ensured that biological control would survive the turn of the century.[17]

"a few drops of potent benzene"

Charles Riley's preference for biological control developed while scientists at the state and federal level were conducting intensive research into chemical insecticides. In the 1870s, several states established experimental farms where they tested a variety of control measures, including chemical ones. Riley, who had been a state entomologist himself and remained a firm believer that the best entomology was local, proved eager to cooperate with these farms; in fact, he viewed the experimental farms as the means through which national entomological investigations could remain both comprehensive and bureaucratically decentralized. The Hatch Act, passed with overwhelming congressional support in 1887, tightened the connections between state and federal entomologists as they worked to evaluate the quality of the various insecticides that had been trickling into the agricultural market by the 1870s. Specifically, the act established and funded agricultural experiment stations (at $15,000 a year) in connection with state agricultural colleges founded under the Morrill Land Grant College Act (1862). Under a more authoritarian leader, the act might easily have segregated economic entomologists into local and federal agents who pursued competitive agendas and rarely communicated. Riley, however, worked diligently to foster good relations among entomologists who were now spread across the county, variously employed by the Department of Agriculture, agricultural experiment stations, and agricultural colleges. While respecting regional differences and provincial orientations, he also transcended them by establishing the American Association of Economic Entomologists in 1889, broadening the scope of the Entomological Society of Washington, and sending dozens of federal fieldworkers across the country to keep tabs on local matters. As economic entomologists of all stripes communicated about insect control, they returned repeatedly to an idea that was coming to tempt entomologists and farmers caught in the crossfire of commercial agriculture and insect outbreaks. That is, as entomologists began to see themselves as crusaders for a national cause, they began to take seriously the proposition that chemical insecticides could be the modern scientist's preferred way to fight insects.[18]

It was a heady proposition. Whether it was misguided, however, was as yet anyone's guess. Assessments of the most common insecticides that farmers began to use in the 1870s were circulated by word of mouth, in newspaper articles, and in agricultural magazines. If a farmer reported success, others jumped on the bandwagon and a variation of the "solution" usually made the rounds. Unlike their twentieth-century counterparts, the nineteenth-century insecticides derived from plants and minerals. Sometimes chemical companies synthesized and sold them as washes and dusts; sometimes farmers bought the ingredients and manufactured the solutions themselves. Whatever their origin, and whatever their avenue of introduction, dozens of insecticides quickly entered the agricultural lexicon, and a few stood out for their popularity.

A short list would include the following:

- Paris green, a bright-green arsenite derived from copper dye and conventionally used as paint and wallpaper pigment, was being applied to kill insects as early as 1868, supposedly after a farmer who was painting his window shutters noticed that beetles in his tomato plants died after paint splattered the plants.
- London purple, an arsenite of lime and derived from aniline dye, began to make the agricultural rounds in 1878 after Hemingway and Company of London sent three sample barrels of it to the

Iowa Agricultural Experiment Station, encouraging it to spray the solution on potato plants. It was soon being used in much the same way as Paris green, although as a finer powder it suspended better in solution.

- Kerosene emulsion—1 pound of commercial soap and 1 pound of water heated to the boiling point, plus 2 gallons of pure kerosene oil (normally used as lamp fuel)—yielded upon cooling a "gelatinous mass" that "at any time may be diluted in water to the desired strength and used."
- Carbolic acid emulsion came from adding "one part crude carbolic acid to from five to seven parts of [a] soap solution."

Other early popular insecticides included arsenite of lead, white arsenic, sulfate of copper and lime, bisulfide of carbon, and a mixture of lime, salt, and sulfur.[19] As agricultural experiment stations turned their attention to these insecticides, they already were becoming part of a typical agricultural arsenal as sporadic reports of success sparked the interest of insect-plagued farmers.

Farmers turned to chemicals in part because Paris green, London purple, kerosene emulsions, and other mixtures were hardly foreign to their experience. The insecticides reflected a history of broad experimentation that encouraged farmers and entomologists to explore a variety of options, chemical ones included. Foremost among them, as we have seen, were natural insecticides—pyrethrum, hellebore, tobacco dust, slacked lime, and sulfur being the most common. Although more sophisticated in terms of production and application, the newer arsenical sprays and kerosene emulsions that started to gain traction in the 1870s were still the legitimate heirs of a familiar tradition, and thus were insecticides that farmers and entomologists were inclined to accept. Farmers generally tended to be reticent on the matter of incorporating chemicals into their agricultural operations, but the silence was, in its own way, telling. When Joel Gillingham of Salem, Massachusetts, began to purchase spray pumps in 1892, or when Philip Martino began to use Paris green in the early twentieth century, their extensive account books leave no evidence that there was anything unusual about their purchase and use of these methods of control.[20]

Plus, in certain circumstances, the chemical insecticides appeared to do something rather profound: they worked. "Hot alum water," according to an article in a New Hampshire newspaper, "will destroy red and black ants, cockroaches, spiders, chintz bugs, and all the crawling pests which infect our houses." The *Farmer's Cabinet* promised in 1862 that "scattering chloride of lime" on wooden surfaces would eliminate "all kinds of flies." A newspaper in Macon, Georgia, working from the assumption that "our houses and gardens are more or less infested," advocated in 1861 "a few drops of potent benzene," noting that "the bodies of insects killed by it become so rigid that their wings, legs, &c., will break rather than bend if touched."[21] In many respects, the measure of success, as well as the extent of the skepticism, was lower because the potential outcome—a pest-free farm or garden—was so appealing. Farmers must have doubted the effectiveness and safety of these products, but the risk was, it seems, well worth it.

Skeptics may have frowned on these reports of domestic eradication made by mere newspapermen, but they had a harder time dismissing the positive assessments of professional scientists, many of whom responded favorably to the prospects of insecticides. With the Department of Agriculture reporting in 1882 that Paris green "was successfully used for the control of the cotton worm," with a former California horticulture officer declaring in 1883 that London purple wiped out cutworms in California, with a professor at Michigan Agricultural College reporting in 1877 that a mixture of carbolic acid and whale oil eliminated orange tree scales, and with Riley himself occasionally confirming the power of chemical insecticides, one could forgive farmers and their advocates for thinking that—with the advent of Paris green, London purple, and kerosene—they had discovered a cache of magic bullets.[22]

"a hodge-podge of ingredients such as enters the witches' caldron in Macbeth*"*

One danger, of course, was that these bullets could backfire. Legitimate insecticides competed with hundreds of quack remedies, a problem that plagued the Division of Entomology and agricultural experiment stations as they pursued chemical research. Advertisements published in agricultural newspapers and bulletins promised the world to ailing farmers in the form of elixirs and potions designed to kill any and all insects. "Hellbroths!" Benjamin Walsh fumed, while a contemporary scholar has gone on to deem them "worse than worthless." "The most absurd claims" proliferated, charged another federal entomologist, based on the assumption that "farmers seemed to be especially easy to fool." Evidently some were. Hapless planters who took a chance on Slug Shot, Bug Death, Black Death, Smith's Vermin Exterminator, or Grape Dust too frequently saw a year's work disintegrate under the influence of a bad decision. While many of these "insecticides" carried a patent, none were submitted for registration, a step that would have required the manufacturer to list their ingredients. Revealing ingredients was exactly what a man like Benjamin Best, manufacturer of Best Patent Fruit Tree and Vine Invigorator, aimed to avoid. He could therefore promote his potion as "the most useful combination of ingredients ever offered to any people," causing Walsh to huff that it was actually "a hodge-podge of ingredients such as enters the witches' caldron in *Macbeth*." The farmer, caught in the crossfire, often did not know Shakespeare from rank hucksterism, and his crops suffered the consequences.[23]

As predisposed as entomologists and farmers may have been to experiment with chemical solutions, what ultimately mattered most was the insecticides' level of effectiveness and safety. Would they systematically work as a long-term solution to the insect wars? State agricultural experiment stations and the Department of Agriculture may have been under political pressure to seek solutions to insect infestations, but, to their credit, their reports struggled honestly and patiently with questions about the efficacy of insecticides. Given the nation's ultimate embrace of insecticides, however, what stands out so conspicuously in the reports of hundreds of investigations is the considerable evidence found against these new products. When it came to basic matters of safety, application, effectiveness, and practicality, the insecticides tested by agricultural experiment stations proved to be *problematic on every score*. This is not to suggest that the insecticides produced in the late nineteenth century were entirely without merit. As Charles Riley was the first to admit, there was little doubt that pro-insecticide propaganda masked kernels of truth, and that insecticides occasionally saved many American farmers much heartache and numerous foreclosures—at least for the time being. Under highly controlled circumstances, and measured by short-term objectives, petroleum- and arsenic-based insecticides certainly raised some optimistic eyebrows and inspired a few deserved encomiums. But the trend that reports published by agricultural experiment stations reveal more consistently than any other is that,

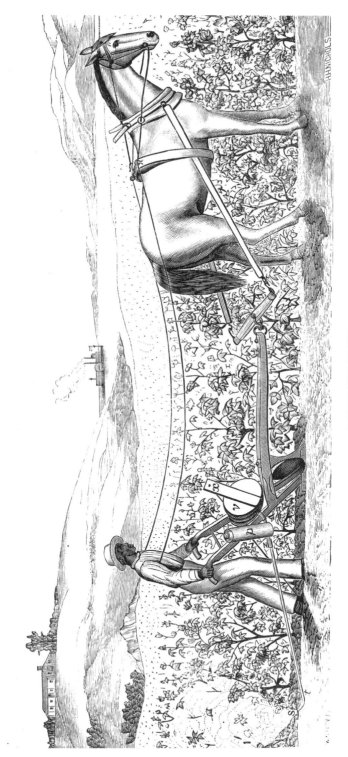

PLOW BELLOWS SPRAYER

Spraying insecticides proved to be a dangerous task, according to early tests. This sprayer, designed to blow insecticide mist behind the plow through a bellows, was suggested as a safe way to avoid direct exposure.

for all their promise and supposed safety, the new insecticides were seriously flawed. It was true that the diverse range of American farms in the latter half of the nineteenth century favored the pursuit of a one-size-fits-all solution. One can certainly sympathize with farmers and entomologists who were seeking a comprehensive end to insect infestations. But on closer inspection of the evidence, it appears that no matter how hard economic entomologists looked for good news, it was always a stretch to find it. Legitimate chemical solutions were certainly more than a "hodgepodge," but as far as Riley was concerned, the Division of Entomology had sound justifications for sticking by its commitment to biological control.

Safety was an obvious concern in these reports.[24] Enthusiastic as he was about the compounds he studied, C. H. Tyler Townsend had no choice but to reveal the "supposed danger to man and animals from spraying with the arsenites." There was, he explained with some regret in 1892, "prejudice in the minds of some persons against the use of arsenical sprays for injurious insects." His appeal to the contrary was somewhat compromised by his claim that "the only means by which fruits or vegetables of any kind can convey the poison to the consumer is through the very small quantity of arsenic which may be left upon the outside of the edible portions of the plant." This was cold comfort. A year earlier, the London newspaper *Pall Mall Gazette* had published a scathing report about a shipment of American apples, calling them "unsafe on account of the large amount of arsenic which American orchardists spray upon their fruits." F. H. Fowler, a Massachusetts entomologist, pleaded with consumers to "keep all poisons out of reach of children and stock" and to "keep stock out of orchards for a few days after spraying." An article in the magazine *Insect Life* recounted that, as an investigator sprayed London purple onto "worm infested trees," a gust of wind "[made] the work exceedingly disagreeable, one of the men being made sick by having the poison blown in his face." A report from the Kentucky Agricultural Experiment Station mentioned that "the use of arsenical poisons in combating insects on cabbage . . . led to some misgivings regarding the possible dangers of this method to human beings" after several consumers claimed to have been poisoned after eating "dusted cabbages." In another experiment, after a laborer entered a room where livestock had been fumigated for lice, he "suddenly fell forward on his face unconscious." At the end of an endorsement for bisulfide of carbon, John B. Smith, the state entomologist of New Jersey, noted in a near parody of understatement: "A word of warning, however, may be in order." The warning: "Bisulphide of carbon is poisonous. Its vapors kill animals and plants as well as insects. It is exceedingly flammable, and a spark is all that is necessary to cause a flash."[25]

Many reports attempted to manage or even downplay the evident dangers. A report on arsenical insecticides published by the Illinois Horticultural Society noted that "it is well understood by orchardists that the deadly poisons here discussed must be used with a certain caution." Powders were to be thrown with the wind, it warned, and gloves were to be worn at all time. Perhaps Stephen A. Forbes, the author, had read the widely circulated bulletin on lime sulfur from the Virginia Agricultural Experiment Station reporting that "the men found the wash quite unpleasant to their hands and very obnoxious to the face and eyes, and it became necessary for them to wear gloves and cover their faces with veils." Drawing on investigations conducted by experiment stations, Forbes summarized "careful experiments with respect to the poisoning of pasture under the trees sprayed." The findings were not encouraging. Tests done on grass under sprayed trees found that it contained 20 percent of a poisonous dose of arsenic for a cow and 10 percent of that for a horse, figures that increased when one considers, as Forbes did, that "arsenic is a cumulative poison, and a daily feeding for two or three weeks upon grass which had been sprayed with arsenites might have very different consequences than a meal or two." He then acknowledged that "it may be well to note the poisonous nature of these substances for man."[26] Whatever the amount, there was no denying the message that these substances were a threat to more than just the insects they were designed to kill.

But life is full of risks, and a few examples of insecticide poisoning were not going to convince most farmers of the dangers the products posed if they ultimately saved their crops. If safety concerns alone did not dampen popular acceptance of chemical insecticides, however, one might have expected the inherent challenges of their application to have done so. Even if insecticides had been perfectly safe, they would have been compromised by the near impossibility of mixing and spraying them uniformly—critical factors in the effective use of insecticides. Kerosene emulsions, according to several agricultural experiment stations, led to "considerable difficulty" because "the mixtures are only of a temporary nature and not at all stable." Paris green, likewise, "sinks unless constantly stirred." The available

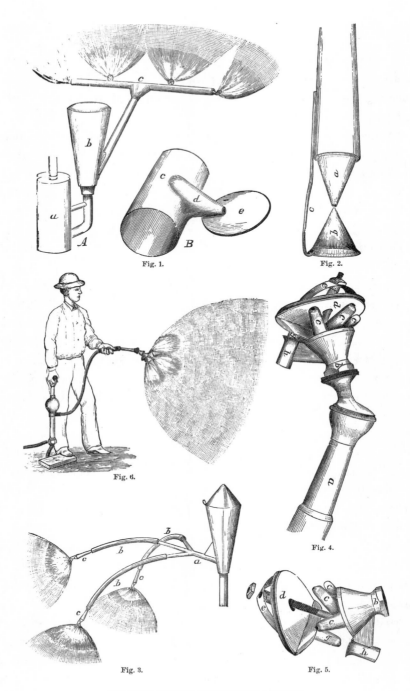

DEFLECTOR NOZZLES AND BROADCAST SPRAYER
Early tests of chemical insecticides revealed that spraying a uniform mist over the proper parts of a crop was a daunting challenge. These nozzles, which farmers assembled themselves, were designed to make the application of chemical insecticides safer and easier. Other designs, such as the broadcast sprayer (*right*), were less safe.

equipment did not help, since "the various spraying machines which have been constructed for the specific purpose of making mechanical mixtures ... have never been quite satisfactory." As emulsions quickly separated in spraying machines, the substance that doused the trees had either so little oil that it was "of no insecticidal value" or so much oil that it could "burn the foliage and otherwise endanger the plants." Stations routinely urged that just enough oil be used "to moisten the surface of the leaves" without producing a "stream of oil which might collect in various places." Considering that this moisture had to land in the form of a fine mist and settle evenly on both sides of the leaves, farmers were quick to shrug off such a requirement as an unnecessary burden. The Ohio Agricultural Experiment Station was theoretically correct in asserting that "the essential points to be regarded in the application of kerosene are the finest possible spray, the completest and thinnest possible coating over the entire surface, and weather conditions favoring rapid evaporation." For good measure, the trees also had to be bone dry prior to application.[27]

As a practical matter these requirements were especially discouraging. No sprayer on the market could emit "an almost impalpable mist," and, even if it could, the mist would dissipate in the slightest breeze. Beyond that problem, kerosene tended to sear leaves in the summer, when the advised evaporation would have been more likely. Coating the underside of leaves was nearly impossible with a sprayer, which emitted a dense spray downward. Plus, who could be sure that the weather would stay dry during the several days needed to spray so carefully and thoroughly? Many a farmer must have ridiculed the suggestion, made by a Professor Webster in Ohio, that kerosene would work best "if applied lightly with a brush."[28]

Preparing these solutions on their own—a task that farmers were strongly advised to undertake with so many bogus products infecting the market[29]—was akin to following an unwieldy recipe. Consider Townsend's directions for making a "Winter Resin Wash" to use against the San Jose scale. The listed ingredients included 30 pounds of resin, 9 pounds of caustic soda, 4.5 pints of fish oil (as a binder), and 100 gallons of water. Townsend instructed:

> Prepare by placing all the ingredients in a large boiler, and pouring over them about 20 gallons of water. Boil briskly for three hours, or until the compound is perfectly soluble in water. The boiler must

now be slowly filled with hot water, care being taken to stir well, until it makes 50 gallons of hot solution. This should be strained through a fine wire sieve or a piece of fine muslin, when it can be diluted with an equal quantity of cold water, as needed for spraying.

Not only was this procedure beyond the reach of the average farmer, but, no matter how closely the directions were followed, problems inevitably ensued. After preparing "the formula advocated by Professor Cook," Herbert Osborn reported in *Insect Life*, "the result was that we had what appeared to be an excellent emulsion." Osborn had mixed a kerosene solution in a glass jar, but his hopes were dashed as he watched "the separation taking place, the white emulsified part rising to the top and the water or soapsuds gradually increasing at the bottom" and hardening into a useless "jelly." One can only imagine the challenges involved in achieving proper mixtures of insecticides to combat a pest such as the red scale, whose control required that an insecticide be applied "every day for three or four months consecutively." These impracticalities of application lent credence to one entomologist's wry claim that "investigations are giving us improvement after improvement, some of which, unfortunately, are no improvement at all."[30]

Related to the difficulties of mixing and applying insecticides were shifting external variables that farmers were asked to consider if they chose to spray with them. Weather and climate, for example, were critical to effective spraying. Kerosene required hot and dry weather, but "caustic soda" needed dampness. "If a solution of caustic soda alone is sprayed on the trees in dry weather," warned a bulletin from the New Mexico Agricultural Experiment Station, "its causticity is lost and its action as an insecticide is at an end." But was this true for trees in arid New Mexico only? Few concrete answers were forthcoming. Some trees reacted adversely to some insecticides when still in bloom, while root vegetables avoided damage only if they were saturated at a very precise stage of growth. Was this true for every plant? Every insecticide? Again, nobody was saying. "All educated men," declared Fowler, "should pronounce vehemently and with one voice against spraying our fruit trees with arsenites till the blossoms have all fallen." But were it only so simple. The problem was that "less damage has been done by the arsenical insecticides when applied in May"—the month when flowers were in full bloom in most climates. Likewise, this requirement butted against the advice that arsenic "should not be used on cabbage after the heads are two-thirds formed." An entomologist in Idaho insisted that, "in all cases the first application be made just after the petals fall and before the calyx closes." Confusing matters further, spraying had to work around the schedule of bees, so as "to avoid poisoning [them] and other useful insects which visit the flowers." If a farmer wanted to use hellebore, he had to know that the application should occur at night, that the hellebore had to be diluted with flour, and that the mixture had to be "dusted on plants through a muslin bag."[31]

The idiosyncrasies, contradictions, and health hazards of mixing and applying insecticides might have been worth the trouble if insecticides had proved to be systematically effective. But, in perhaps the most puzzling finding of all, the results that late-nineteenth-century entomologists reported were, in the most generous assessment, mediocre. The most common negative charge to come from agricultural experiment stations and test farms was that the insecticides under review were, for all intents and purposes, *useless*. After assessing the impact of Paris green, London purple, and white arsenic on apple trees, a team of entomologists in Vermont concluded that "no benefit was derived from the application of the poison." In another report, kerosene was found to have "no insecticidal value." Researchers at the New Jersey Agricultural Experiment Station discovered that an insecticide called salimene caused "no perceptible effect as the scale developed as freely on unsprayed trees." They deemed an arsenic-based insecticide "absolutely ineffective," while dismissing the impact of lime and sulfur as "so small as to be hardly noticeable." Sulfite of soda was judged to be "ineffective," potassium sulfide yielded results "not good enough to justify recommending the mixture," and trees doused with caustic soda "were as bad as ever." A California report concluded after a lengthy study of insecticides for scale insects that "it was better not to spray at all." Pyrethrum had "no effect whatever" on plant lice, according to the Department of Agriculture, while a dusting of pyrethrum on grub worms ended up producing a brood of "healthy pupae." Other descriptions from other reports reveal such conclusions as "a perfect failure," "nonsense," and "worthless."[32]

The discovery that an insecticide had no effect actually came as a relief to many researchers. After all, it was not uncommon for an insecticide to cause immediate and irreparable damage. "All the treated trees," according to the New Jersey Agricultural Experiment Station's evaluation of a mixture of lime, sulfur, and salt, "were as bad or worse than when the work was begun." A study done in California revealed that "those orchards which had not been sprayed were found upon examination to be freer from scale than those which had been sprayed annually." Crude oil and other "distillation products" used as insecticides produced oranges covered in strange white spots, since "the chief injury to vegetable tissue from oils was caused by a penetration of oils into the interior of the plants." In Illinois, researchers found that "London purple is certainly so caustic to the leaves [of peach trees] as to forbid its use under any circumstances." An apple tree treated with London purple was "very badly scorched." "Some injury was done to the tree," reported the Missouri Agricultural Experiment Station, "when the carbon bisulfide was injected into the soil too close to the trunk" in a futile effort to kill woolly aphids. Apple trees in another experiment were "worse affected by the apple worm than the check trees not treated," while a test of London purple was "thoroughly unsatisfactory," with the mixtures "defoliating or at least badly damaging, the trees, and not protecting the fruit." White arsenic in solution "should undoubtedly be abandoned as dangerous, if not destructive, to foliage . . . if its application be followed by rain, it would probably even then take disastrous effect." On and on the reports went.[33]

None of these results surprised the Division of Entomology. After all, a comprehensive report published by the division in 1886 confirmed the ambiguous conclusions that were coming in from dozens of investigations conducted by agricultural experiment stations. Eighty-six experiments testing a variety of insecticides on a plentitude of pests did little to promote the benefits of these chemical applications. In only three experiments—copperas water on cabbage worms, soluble pinoleum on woolly aphids, and kerosene emulsion on caterpillars—were all the larvae killed. Aside from those, only nine of the eighty-six experiments could be classified as successful (with more than 50 percent of the adults or larvae destroyed). Thus fully 85 percent of the experiments were failures, in terms of both being useless and/or causing damage. The remarks after each summary, in many ways, were telling indications of the overall mood regarding insecticides: "all worms have returned to the leaves and are actively feeding"; "this injured both plants, one quite seriously"; "the larvae did not seem to suffer any inconvenience"; "cannot see that any were destroyed"; "three days later, the beetles had returned"; "the ants had returned to work in the old burrows." Surveying the results, Riley concluded that "a pretty strong case is made against the [insecticide] remedy in the reports which now follow." The survey, in short, provided good reasons to question whether or not C. H. Fernald's labeling of the late nineteenth century as "the period of insecticides" was little more than generic propaganda.[34]

"here apparently was my chance"

Charles Riley, who never condemned chemical solutions per se and often praised their potential, nevertheless viewed these test results as further confirmation of his interest in biological control. By 1894, however, the indefatigable economic entomologist had finally had enough. Obsessively workaholic as he had always been, not to mention tired of bureaucratic wrangling, Riley needed a break from the grind of churning out reports that tried to make sense of the nation's insect problems as a whole. Upon retiring from the Division of Entomology, he applied for the position of deputy director of the Department of Agriculture. After that job did not materialize, he sent his résumé for a professorship of entomology at Oxford, located near his birthplace. When that went to another candidate, a dejected Riley immersed himself in the two passions that had drawn him into his career to begin with: painting and collecting insects. It is tempting to think that Riley, with time on his hands to write at a rational pace, was planning a magnum opus on the virtues of biological control, perhaps a paradigm-shifting study like Thaddeus William Harris's *Report of the Insects of Massachusetts Injurious to Vegetation*. His intentions, however, must remain unknown. It would have been hard to top the suddenness of Benjamin Walsh's death, but on September 14, 1895, while riding his bicycle to the Natural History Museum of the Smithsonian Institution to look at insect specimens, Charles Valentine Riley hit a paving stone that catapulted him onto the sidewalk. The impact broke his skull, and Riley, fifty-two years old, never recovered, leaving behind a wife, five children, and a legacy that placed him alongside Harris as a giant in the field of economic entomology.

With Charles Riley's death went biological control's greatest expert and advocate. And with his fall came the rise of Leland Howard, Riley's deputy for almost twenty years. Howard never celebrated the tragic death of his boss, but contemporaries would not have been surprised had he done so. When Howard assumed Riley's former position in 1894, he recalled thinking, "Here apparently was my chance." Howard had chafed under Riley's leadership, taking the position of deputy director of the Division of Entomology in 1878 only because, as he explained, it was "an agreeable way of earning a living until I could go to something bigger and broader." Along the way, he evidently paid some ego-bruising dues. Howard, who had come at entomology not through experimental work in the field but through pure science studied at Cornell University, never forgave Riley for burdening him with mindless office work. "Much to my disappointment," he later wrote, "I found that Professor Riley wished to use me . . . more as a clerk than as a scientific assistant." The two men clearly did not enjoy each other's company. Entomologist E. A. Schwartz, in a letter to H. G. Hubbard, mentioned that "Prof. Riley is said to have returned from Europe. . . . Upon hearing of his return, Mr. Howard has escaped into the mountains of New York." The best authority on Howard's feeling for Riley, however, was Howard himself, who many years later confessed that he thought Riley was little more than "a great schemer."[35]

Howard, however, had schemes of his own to pursue. The relatively conservative and decentralized conventions of nineteenth-century entomology—based as they were on biological control—would experience its last gasp in the early twentieth century, and it would do so largely due to the influence of Howard. Not only did Riley and Howard not get along personally, but their animosity carried over into their theories of pest control. Howard said little about this simmering disagreement in his memoir, quipping in reference to the brew-quaffing members of the Entomological Society of Washington that "a few glasses of beer will make a stupid remark sound witty, but there was no necessity for such stimulus . . . because all the remarks were witty."[36] But they were not witty enough to prevent Howard from blazing a notably different trail into the profession, one that took the Division of Entomology in a new direction.

For reasons that should be clear, Riley had been cautiously skeptical about the looming transition to chemical insecticides. At best, he regarded chemical solutions as one option among many others, occasionally useful, often harmful, hardly a panacea. But Howard, who would control the renamed Bureau of Entomology until 1927, became a firm believer in the power of insecticides to solve the problems caused by insect pests on the national level. The eventual triumph of Leland O. Howard's vision would shape the issue of national pest control for the entire twentieth century. The details of that triumph—one that placed the federal government at the helm of national insect control—constitute a story in and of itself.[37]

CHAPTER 5

"let us spray":
MOSQUITOES, WAR, AND CHEMICALS

As for leisure activity, when Leland O. Howard was not quaffing a beer, he was, of all things, riding his bicycle. Howard's interest in cycling was genuine enough for him to have joined a local club that went on weekend outings around Washington, D.C. The men rode on tall models fashioned with oversize front wheels. "The old high bicycle had made its appearance in the streets," he recalled, "and it fascinated a lot of us." The only thing that may have been higher than Howard's bike at that point in his life was his optimism. Howard's assumption of Charles V. Riley's position as director of the Bureau of Entomology in 1894 was a veritable coup for an ambitious scientist who hoped to use the full force of the federal government to back what would become his lifelong agenda. And now, with Riley killed in a bicycle accident, the chance to comprehensively control insects through federal programs became for Howard a genuine possibility. He would fulfill it beyond his wildest predictions.

"the practical importance of entomology did not appeal to me"

Leland Howard's personal background prepared him to favor chemical solutions as the preferred approach to insect infestations. Whereas his predecessors had entered the profession of entomology because their work on farms nurtured their fascination with insects in an agricultural environment, Howard took a more academic route. As a student attending the best private schools in Ithaca, New York, where his father practiced law,

LELAND OSSIAN HOWARD

A towering figure in American entomology, Leland Ossian Howard was the third director of the Bureau of Entomology. Under his watch, and with his encouragement, the United States became an insecticide nation.

Howard read Helen S. Conant's *Butterfly Hunters* and Thaddeus William Harris's *Report of the Insects of Massachusetts Injurious to Vegetation*. When he entered Cornell University, he took courses in modern languages and history. After unexpected circumstances landed him in the laboratory of John Henry Comstock, however, his early love of natural history came rushing back to the fore. Comstock recently had been appointed the first professor of entomology at Cornell, and he proved eager to take Howard under his wing and introduce him to his library of entomological volumes. Howard, whose fieldwork almost always took place in either a laboratory or a library, consumed the works of Asa Fitch, Benjamin D. Walsh, and Charles Riley while writing a conventional thesis on the morphology of wasps. If the commandments of these founding fathers of economic entomology influenced Howard, it rarely showed. Instead of expressing enthusiasm for agriculture, Howard repeatedly articulated his disappointment in Riley's "emphasis on agriculture." He later admitted that "the practical importance of entomology did not appeal to me. . . . I studied insects simply as a fascinating form of life."[1]

The historical context in which Howard came of age further prepared him to accept insecticides as a means to manage the predation of insects. It was a time when science in the United States was maturing and specializing in a way that status was coming to matter. Entomology—often dismissed as a branch of science full of crusty bug catchers—had always occupied the lower rungs of the professional ladder. Entomologists were conventionally stereotyped as being out of touch with cutting-edge developments in the rarified realm of "pure" science. Conducting experiments with insecticides, however, had the potential to pull entomologists into a more prestigious fold of scientific inquiry and enable them to think in broader scientific terms rather than react inductively to random agricultural problems. In a scientific atmosphere where "abstractness correlated closely with status," distance from the roots of entomology—family farms and agricultural journals—held seductive appeal to a young, formally trained scientist who hoped to make his mark. Howard likely sensed that backing insecticides might lend entomologists an immediate and welcome dose of professional prestige.[2]

Politics were a formative influence on Howard's mind-set as well. The contemporary political environment—defined by a delicate combination of urban progressive and rural populist agendas—served as a welcoming context for the advocates of chemical insecticides. Pragmatism, the philosophical foundation of progressivism, was a philosophy that promoted "truth" as whatever "worked" in a publicly applicable fashion. Progressivism, for its part, demanded that problems addressed by the federal, state, and local governments be amenable to quantification by teams of credentialed "experts." Under tightly controlled conditions, many insecticides delivered their blows with lightning speed, and, predictably, faster concrete results were more politically palatable than delayed ambiguous ones—a political reality that Howard certainly appreciated. Populism complemented progressivism by urging the federal government, in particular, to address the needs of agriculture. The populists' push for reforms, such as lower interest rates and leveraged grain storage, is well known, but the quest for control of pests with insecticides was part of the larger populist agenda as well. These overlapping political trends provided insecticides with a welcoming framework while making it difficult for any responsible entomologist to ignore the potential promise of chemical solutions.[3]

"egg on his face"

Within this broader context, Leland Howard would become the critical pivot in the revolution toward the popular acceptance of chemical insecticides. But this revolution took place at an evolutionary pace. In fact, during Howard's first decade as director of the Bureau of Entomology, it sometimes seemed as though Charles Riley were still at the helm, at least with respect to the methods of insect management that the bureau tested and advocated. Much as Riley had favored biological control but investigated all possible approaches, so Howard dutifully explored cultural and biological methods. The Farmer's Bulletins published by the Department of Agriculture between 1900 and 1904 effectively reflected the bureau's flexibility in applying specific solutions to various problems. Advised remedies for infestations of the tobacco flea beetle were cultural ("the destruction of weeds along the margins of the field") and chemical ("the use of arsenical poisons"). Insects that fed on growing wheat responded to strategies that were biological ("important natural parasites" and "control by fungus diseases"), cultural ("trap crops," "rotation," and "burning stubble"), and chemical ("a very strong oily insecticide"). To be sure, Howard often tipped his hand. He once described manual methods of controlling the elm leaf beetle as "always be[ing] incomplete," while "paris green, London purple, or arsenate of lead" led to "absolute success." But in 1897, Howard, who evidently had come to accept his inextricable links to agriculture, described his work as "directed toward answering the question: How may the farmer and fruit grower avoid damage to their crops by insects?" Many of the reports he published during the first decade of his tenure reveal an earnest and openly experimental approach toward answering this important question. For a while at least, he nurtured his roots in the experimental values of economic entomology.[4]

Two insect pests, however, pushed the limits of Howard's patience and eventually convinced him that more aggressive, uniform, and broadly applicable measures were in order: the gypsy moth and the boll weevil. The moth, which the French-born botanist and amateur entomologist Étienne Léopold Trouvelot accidentally had released in the 1860s, bred in a vacant lot next to Trouvelet's house in Medford, Massachusetts. By 1900, it had exploded into a mass of defoliating caterpillars that were destroying trees throughout New England. Trouvelot was experimenting with the

LELAND HOWARD AND BIOLOGICAL CONTROL
Although Leland Howard would ultimately become a staunch supporter of chemical insecticides, he did not completely dismiss the potential of biological control of insect pests until the 1930s. In 1912, he toured biological control facilities with Paul Marchal, a French entomologist.

cross-fertilization of silkworm breeds when the moths broke free. Howard, now the nation's top professional entomologist, became obsessed with solving the problem.

As a testament to his initial willingness to explore all options, Howard, after extensive consultation with state entomologists in Massachusetts, traveled to Europe, evoking no doubt the example of Albert Koebele, to investigate the moth's natural enemies. What he found daunted him. Although Howard had little knowledge of the moth's ecology, he did discover that it had no fewer than fifty predators. Importing them, releasing them, and waiting for them to do their work would be, he correctly predicted, "extremely complicated." But he persevered in his plans, complementing them with cultural strategies and a modest level of spraying.[5]

This strategy, focused on the importation of natural enemies and burdened by the expectation of quick results, proved to be a public-relations bust. Imported parasites and predators of the moth died after release, disappeared, failed to breed, and—if they did attack the moth—made not a dent in the population. It was not that the strategy of biological control was untenable—Howard would not reach this conclusion until much later in life—but that much more research was required before it could be systematically employed. This time-consuming demand melded poorly with a bureaucratic, Progressive Era mentality that called for rapid outcomes for public problems. Howard began the gypsy moth project with the bold assessment that he needed only three years to get the moth under control. By 1911, he was admitting failure based on inadequate knowledge of his chosen strategy. And by 1933, he was calling his early faith in biological control "nothing less than absurd." The entire experience, in Edmund P. Russell's words, "put egg on his face."[6]

The efforts made by the Bureau of Entomology to control the boll weevil only intensified Howard's sense of helplessness while furthering his interest in chemical options. The boll weevil had entered Texas in 1892, crossing the border with a vengeance and quickly burning a wide swath throughout the Deep South. The bureau confronted the problem in 1894 by sending C. H. Tyler Townsend, an entomologist working in Jamaica, to Texas to study the boll weevil's habits and life cycle. After surveying the situation, Townsend made a strong plea for cultural control. The most efficient way to handle this pest, which was consuming up to 90 percent of the cotton crop, he explained, was to strategically burn fields, allow hogs to forage, rotate crops, and—most controversially—create "no cotton" zones. At the same time, he suggested that the bureau send an agent to Mexico to search for natural enemies of the boll weevil and study ways to make indigenous parasites more efficient. Howard embraced these recommendations wholeheartedly, and he even persuaded the governor of Texas to take action on the plan. However, overwhelmed by protests, the Texas legislature ultimately balked at regulating the cultivation of cotton. The governor, Howard, and the bureau were ultimately left holding a blueprint without a builder willing to carry it through.

Although Texas would be left to its own devices, the Bureau of Entomology continued to pursue methods of cultural control elsewhere. After lobbying Congress for funding to fight the boll weevil (by, in part, showing up for a hearing with a 2-foot-long papier-mâché model of the weevil), Howard went national with his plans. Every southern state soon learned about the benefits of planting cotton that matured early (before weevil

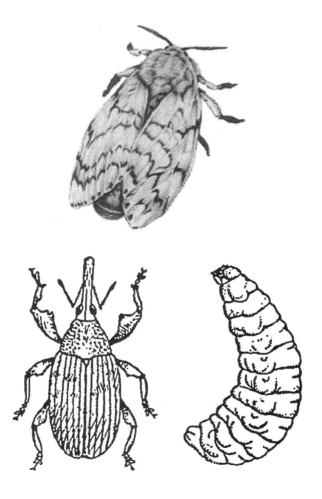

GYPSY MOTH AND BOLL WEEVIL

The (*top*) gypsy moth and (*bottom*) boll weevil—shown in its adult, pupal, and larval stages—daunted Leland Howard. Their resistance to biological and cultural controls (respectively) may have pushed him to consider more seriously the benefits of chemical control.

populations exploded) and the virtues of removing stalks where weevils "over-wintered." But, for reasons more political than scientific, it all came to naught. Whereas the problem with biological control of the gypsy moth was a lack of information, the problem with cultural control of the boll weevil turned out to be an absence of coordination. There was little doubt that these cultural methods would have worked, but farmers had to act comprehensively and in unison, a performance that only loophole-free legislation could reasonably choreograph. Because no southern state was able to marshal enough political will to pass regulatory laws, the massive problem of the boll weevil was, as Frederick W. Malley, the state entomologist of Texas, concluded in 1901, left "to be met and mastered by the planters themselves." In his memoir, Howard would remember thinking that "the action of the world as a whole seemed rather slow."[7] While this perhaps was true, the world also awaited more forceful answers.

"The horizon opened up very greatly"

These highly publicized and embarrassing failures pushed Leland Howard to reevaluate his opinions on the biological and cultural control of insect pests. At the same time, he began to sharpen his support for the application of chemical insecticides. What was once a studied openness to a wide array of solutions, in essence, gradually yielded to a stated preference for more streamlined chemical procedures. In some ways, this reorientation was subtle, as in Howard's decision to focus an unusual amount of the Bureau of Entomology's attention on shade trees. Unlike fields of crops, which may sprawl across political jurisdictions, shade trees usually are consolidated in urban regions with clearly defined municipal governments whose constituents shudder to see their streets denuded of beautiful foliage. This important difference made it all the more feasible for Howard to advocate chemical solutions to insect infestations under the assumption that a centralized commission would oversee their implementation with proper organization and broad public support. Such was the situation when he insisted in 1905 that "the only thoroughly satisfactory safeguard against this insect [elm leaf beetle] consists in spraying the trees with an arsenical solution." Howard was proud that "ten years ago," such a claim "would have been received with ridicule," but was now so conventional that "there is no hesitancy in commending it to more general city use." All that was needed were annual appropriations for cities to buy chemicals and pumps, because, as he had learned in Texas, "it is unreasonable to expect that a private individual" will responsibly undertake a "public duty." Thus it was with a new level of civic enthusiasm, most of it reserved exclusively for chemical control, that Howard advocated, for example, the municipal spraying of maple trees with creosote to protect them against the tussock moth, an insect that, only a few years earlier, he had claimed could be controlled biologically. Parasitism had burned him, but chemicals, if Howard chose the right battles, could salve his wounds.[8]

Howard's subtle transition from acceptance of biological and cultural approaches to advocacy of chemical insecticides had a literary component as well. Aware that a history of American entomology had never been published, Howard took it upon himself to write one. He did so, however, in a way that made the embrace of chemical insecticides seem like the next logical step in the field's preordained progression. There were three elements to his revisionist agenda. First, Howard mocked the "chaos of experimentation" that had characterized the earliest phase in economic entomology, dismissing the first systematic prescriptions for insect control—mostly cultural methods—as being "frequently nonsensical," "seldom based on any accurate knowledge," and "rather amusing" and of offering "practically nothing original." Second, he downplayed the obvious contributions of his predecessor as United States entomologist, Charles Riley. Although effusive when it came to Riley's accomplishments in Missouri, Howard entirely ignored his work with the Division of Entomology—including his work with biological control. In another example of tactical omission, Howard's narrative skimmed from the unschooled "utterances of impractical and ignorant persons" (pre–Thaddeus William Harris) to "a new era in remedies" (present-day Howard) that included Paris green, other arsenical compounds, hydrocyanic acid gas, and kerosene emulsions—all described as profound "discoveries" left to Howard's generation to exploit. Finally, Howard, in direct contradiction to scores of studies that highlighted the dangers of insecticides, declared any concerns about the safety of these compounds to be the pedestrian worries of old timers. At length he quoted, and then debunked, Riley for having warned that Paris green was "eminently dangerous" and that it never should be used until "every other available remedy had been tried." Such caution made little sense, Howard concluded, when Paris green was "the first great start which the new economic entomologist received." This was, in short, history with a current objective in mind: "warfare against injurious insects." Chemical insecticides, which Howard portrayed as flawless, were the only viable weapon to be used.[9] History, after all, said as much.

All things considered, these were mild tactics. The emphasis on shade trees and the revisionist history, however effective, paled next to Howard's exploitation of an insect that had heretofore been of only minimal interest to applied entomologists—the mosquito. The mosquito ultimately revealed Howard's ingenious ability to strategically and temporarily switch the focus of the Bureau of Entomology from agriculture to public health, a reorientation that favored chemical insecticides. What made this move possible was malaria. "The discovery . . . of the carriage of the malaria by certain mosquitoes," according to Howard, "was of the greatest importance to all mankind."[10] He wrote these words after hearing about the accomplishments of a British scientist named Ronald Ross.

Ross was an unlikely candidate to confirm a vector theory of malaria transmission that would go on to save millions of lives. For one, he was clumsy with the microscope, constantly asking bemused colleagues to point out plasmodia that were literally right under his nose. For another, he was, as Howard noted, "more interested in mathematics and in poetry than in medicine or entomology." Ross spent much of his time while serving as a British physician in Secunderabad, India, engaged in such taxing endeavors as golfing and fishing. That he also wrote a romance novel and designed a perpetual-motion machine that never quite got moving further supports the charge of underemployment. Due to this endearing dilettantism, few would have guessed that it would be Ross who, on August 20, 1897, after bungling the dissection of several mosquitoes, would locate in the gut of an *Anopheles* specimen a cluster of pigmented cells that were distinctly foreign in appearance. "These bodies struck me at once," he declared. Ross, a previously unknown scientist, confirmed what many scientists had long thought but had failed to prove: mosquitoes were the vector of malaria parasites among humans. "The little brute," as Ross called the parasite, had finally been cornered.[11]

Howard appreciated the implications of Ross's discovery. "I felt," he recalled, "that I was no longer engaged in work that might be called of comparatively little importance. . . . The horizon opened up very greatly." As a man who initially defined his career around the challenges posed by agricultural pests, Howard could now, as the chief of the Bureau of Entomology, directly apply his knowledge to a threat that was not limited to farmers, but affected humanity at large. Howard had been personally interested in mosquitoes from a young age, and the youthful anecdote that he chose to record in his autobiography is worth noting: "I had not only studied their transformations out of mere curiosity, but I had found that in this

MOSQUITO AND MALARIA

The *Anopheles* mosquito, which was determined to be the vector for malaria in 1897, provided the basis for Leland Howard to link economic entomology with medical entomology and, in turn, push the benefits of chemical control for public health as well as agricultural productivity. Ronald Ross discovered the connection between the bite of the mosquito and the spread of malaria.

aquatic stage they were readily killed by pouring a few drops of kerosene on the surface of the water."[12] And then there was a note from Walter Reed. As Reed—who, as the head of the Yellow Fever Commission, had confirmed the mosquito as the vector of yellow fever—made clear in a letter to Howard written in 1901, kerosene was a viable weapon to use against mosquitoes. "The mosquito theory for the propagation of yellow fever is no longer a 'theory,'" Reed bragged, "but a *well-established fact*. Isn't it enough to make a fellow happy?" He went on, "*Anopheles* and *Culex* [two genera of mosquitoes] are a gay old pair!" After declaring "what havoc they have wrought to our species during the last three centuries!" Reed then got to his point: "But with Howard and Kerosene we are going to knock them out."[13]

It would be a mistake, though, to conclude that Howard intended immediately to focus the bureau's full attention on kerosene. His first published article on mosquitoes, written in 1893, mentioned kerosene as only one possible means of control among many others. Similarly, pamphlets he wrote in 1900 advocated a plethora of mosquito-reducing procedures—some chemical, some not. Most important of all, however, was the work of his colleague John Smith, who developed another approach altogether.[14]

"New Jersey mosquitoes do not bite New Yorkers"

In 1894, the year in which Howard had become director of the Bureau of Entomology, John B. Smith, an attorney turned entomologist, became the state entomologist of New Jersey. Smith would also find his professional life altered by the pioneering work of Ronald Ross. "It is my intention," he wrote with typical understatement after learning of Ross's discovery, "to devote some time during the present season to an investigation of the mosquito question." The "mosquito question," as it turned out, would occupy his career until his premature death in 1912.

Like Charles Riley, with whom he had worked, Smith was by no means categorically averse to insecticides. Indeed, his vision was at once encompassing, experimental, naturalistic, and diverse.[15] This predilection had much to do with his ecological sensibilities. Although he never worked as a farmer, Smith routinely impressed colleagues with his deep knowledge of natural history and ecological interdependence. In his textbook, *Economic Entomology for the Farmer and Fruit-Grower,* Smith wrote that "prevention is always better than cure," insisting that the best method to avoid insect attack is "the rotation of crops." One of his contemporaries, Herbert S. Barber, recalled "being out in the country one full day with Dr. Smith" and noting his being "overjoyed" with the outdoors. "It was," wrote Barber, "one of the prettiest things I ever saw in my life." After Smith accepted a position as an entomologist at Rutgers University in 1889, he immediately contacted scores of local farmers as he undertook studies to combat the horn fly, the elm leaf beetle, the plum curculio, and the asparagus beetle. As had Thaddeus William Harris in the 1840s, Smith—who insisted that "especial attention be paid to the life history and breeding places" of the mosquito—recognized that farmers understood insect behavior as well as anyone, and, to a significant extent, he trusted that his own agrarian-based faith in ecological connectivity would lead him, however circuitously, to effective solutions to insect infestations. In short, there was something of the old-school naturalist in Smith, a quality so evident that when a friend once described him as having "legs which scarcely raised him above the meadow grass," Smith beamed with pride.[16]

When Smith confidently insisted that "mosquitoes were not a necessary affliction," he was not promoting the power of kerosene, but pondering the potential of a more complex answer to the mosquito menace. Smith's proposed approach to controlling mosquitoes tellingly began with a debunking of one of Howard's more basic premises: "[I]t has been believed and so often stated by Dr. Howard that [*Culex sollicitans*] would not breed in water as salty as the sea itself." His own investigations, however, suggested that "the contrary was true." "Collections made along the shore soon proved that this general belief was incorrect," he wrote, as "larvae were found in great abundance in pools and ponds in which the water was fully twenty-

JOHN BERNHARD SMITH

As the state entomologist of New Jersey, John Bernhard Smith pioneered nonchemical methods of insect control, including the draining of swamps and the introduction of mosquito-eating fish.

five percent more salty than sea water." Not only that, but the mosquitoes that bred in the saline swamps departed to infest areas more than 30 miles away, a migratory accomplishment that Howard never noticed.

With these misconceptions clarified, Smith proceeded to outline a plan of action, one that targeted the limited territories in which mosquitoes congregated to reproduce. "There seems to be no very great difficulty in deciding upon the character of the work that should be done to destroy the breeding places of these insects," he wrote, revealing his intention to attack the problem preemptively. And then he laid out what would become another model of mosquito control, which consisted "partly of draining, partly of ditching, and partly of opening ways for the free entrance of tides, and with them small fish that feed on mosquito larvae." Smith aimed primarily to eradicate mosquitoes by eliminating swamps and allowing their natural enemies into their breeding pools. His idea of control was more cultural and biological than chemical.[17]

Smith's multifaceted plans were well received by the scientific community and the general public, even far afield from his base in New Jersey. "While the mosquito has been asleep," explained an article in the *Dallas*

Morning News that summarized the scientific reaction to Smith's proposals, "the scientist has been awake." With "war having been declared against the insect in many parts of the country," it continued, citizens were well advised to stay informed about relevant battle tactics. Acknowledging that "in many instances immediate relief has been obtained by spraying ponds and bogs with kerosene," the article noted that "it has been generally agreed by students of the subject that permanent security can be secured only by draining the breeding places or by stocking mosquito-infested ponds with larvae eating fish." The *Grand Forks Herald* in North Dakota insisted that "it is simply ridiculous that the city of Grand Forks should be seriously troubled by mosquitoes," and then proceeded to summarize Smith's program in three phases: (1) abolish breeding places, (2) screen rain barrels or stock them with fish, and (3) "encourage and support natural enemies in permanent waters." The *Baltimore Sun* reported to its readers who lived in swampy areas around Chesapeake Bay that kerosene was always a viable option, but "the best way to get rid of the nuisance of mosquitoes is to abolish their breeding places." All this advice was straight from John B. Smith's *The Common Mosquitoes of New Jersey*.[18]

Smith's policy of wetlands drainage and natural control intensified significantly between 1900 and the outbreak of World War I. Funding from the state of New Jersey, coupled with steadfast legislative support, allowed Smith to systematize his efforts in a way that made New Jersey a potential model of mosquito control. By 1903, the Newark meadows had been reclaimed. In 1905, Smith received an appropriation of $350,000 from the state legislature to drain the swamps around Elizabeth. Next came South Orange, where Smith worked with the South Orange Improvement Association to drain swamps there. Organizations like the North Jersey Mosquito Extermination League, the Associated Executives of Mosquito Control, and the New Jersey Mosquito Extermination Association found support in enactments such as the County Mosquito Extermination Commission Law to place Smith's pioneering work at the head of New Jersey's agenda, ensuring that drainage was conducted throughout the state as a matter of course. By 1912, state law required that every county undertake "the active work of mosquito control." For the most part, this local and regional work was done with few chemicals.[19]

The results of Smith's well-coordinated efforts were promising. Massachusetts, Rhode Island, Connecticut, New York, and Delaware made serious efforts to follow suit. When the Associated Executives of Mosquito Control began to hold annual conventions in 1912, public-health representatives from several states in the Midwest as well as California returned home with ideas that, if not implemented, were at least considered, disseminated, and taken seriously. Whether or not other states adopted Smith's digging, trenching, draining, and fish-importing methods, the ultimate proof was in the New Jersey swamps, and the evidence there was increasingly impressive. There had been "such a material reduction of the number of mosquitos [sic] in the state," reported the *New York Times*, "that real estate values have increased." A mosquito-infested region had, over the course of a decade, become a place where "living has been to a great degree relieved of the mosquito annoyance." Through a "comprehensive and scientific effort," the *Times* added, "a large measure of relief from the salt marsh has been afforded to about 1,725,000 people." New Jersey's successful effort to reduce the number of mosquitoes in the wake of Ronald Ross's discovery became such a source of state pride that the health commissioner, responding to the charge that Long Island was infested with mosquitoes from New Jersey, could deliver an outrageous rejoinder with a straight face: "New Jersey mosquitoes do not bite New Yorkers." The ludicrousness of such a claim only reinforces the power of John B. Smith's approach to this recently discovered challenge to public health.[20]

"the best and cheapest method of drowning the prospective mosquito crop"

Leland Howard followed these developments closely. Given the transition that he was undergoing, one might have expected him to dismiss the proposals of John Smith outright. This was not the case. Despite his chemical interests, Howard was an open-minded scientist who came by his prejudices for valid reasons, and thus he was by no means uninterested in Smith's strategies. Publishing in 1901 what would become an early classic in the entomological literature, *Mosquitoes: How They Live; How They Carry Disease; How They Are Classified; How They May Be Destroyed*, Howard showed considerable tolerance for Smith's earthmoving methods but, at the same time, began to favor an insecticidal approach as the final solution to the mosquito problem. Like Smith, he had little patience for public nonchalance about the matter. Too often, he griped, "what is everybody's business is nobody's business, and the result is that in many localities everyone submits to the mosquito evil." He could genuinely acknowledge "the indefatigable and successful work of Doctor Smith," but loyalty did not necessarily follow. Kerosene, he intended to show, was everybody's business.

Howard had good reasons for leaning in the kerosene direction. He was, after all, in a position that demanded clear and measurable results, the kind of results that, in the context of political progressivism, could be packaged as universally applicable and easily standardized. As the head of a major division of the federal government during the Progressive Era, Howard was under constant pressure to adopt a method of pest control that was expedient, affordable, relevant to the entire country, and built on a record of past success. In the impassioned pages of his book, he repeatedly stressed these qualities of kerosene as he buttressed the case for its use. Not only was kerosene "abundant and cheap in America," he noted, but considerable progress had been made with it "before the malaria discoveries directed attention . . . to the question of mosquito control." Referring to Howard's book only a month after its publication, A. S. Packard Jr. wrote, "The reclaiming of these extensive districts is too great an undertaking to be accomplished *in a short time*, but small areas of water and a limited extent of marshy land can be freed from mosquitoes by means of kerosene." Here was the rub. Whatever its drawbacks, kerosene was a quick and, at least in the short term over small areas, effective killer. And therefore it was with kerosene that Howard ultimately hoped to act on the very same conviction that drove Smith to elaborate on his own useful, if more dilatory and variable, methods. As Howard put it, "There is no reason why any community should submit to the mosquito plague." With the community wanting, in C. H. Tyler Townsend's words, "a poison to kill them quick," there was no reason for it to wait longer than was necessary to eradicate its mosquitoes.[21]

Much as did Smith's plan, the promise of kerosene developed its followers. Insisting that the "war on the mosquito should be waged without abatement," a newspaper in Columbia, South Carolina, approvingly quoted Howard to promote the virtues of "light crude oil," advising the use of "carnivorous fish" only in circumstances where "kerosene cannot be used." A "Mosquito Bulletin" distributed throughout California instructed homeowners to "catch mosquitoes upon the walls in kerosene cups" as "probably the best" of all solutions. Stressing the critical issues of cost and efficiency, a state health board noted, "It is found that the best and cheapest method of drowning the prospective mosquito crop is to spray the water surfaces with kerosene and crude oil mixed in equal parts." Hoping "to strike the last blow and get rid of the last surviving mosquito," an entomologist in Texas who supported Howard's plan impugned mosquitoes as being "as dangerous as rattlesnakes" before dumping oil into every local pond and cistern. In his laudatory review of Howard's book, Packard concluded, "It is refreshing to read of the immense inroads made by fishes upon the larvae, by dragon flies and by birds . . . but it will afford the reader still more satisfaction to know how easily these dangerous pests can be exterminated by the use of so simple a remedy as petroleum." With expediency, efficiency, and comparatively modest cost, Howard and kerosene assured Americans that they were well within their bounds to declare, as one newspaper did, that "The Anopheles must go!"[22]

"the greatest misconception"

John Smith was by no means ready to roll over under Leland Howard's influence. He had, in fact, a history of professional combativeness. Earlier battles over entomological matters had become intense enough for one of his frequent opponents, Harrison G. Dyar, to go so far as to bestow the species name *corpulentis* on an insect that he had discovered in honor of the chronically overweight Smith. To this entomological insult, Smith retaliated by naming his next discovery *dyaria*. But for all his feistiness, Smith eventually found Howard's influence to be insurmountable. While Smith was toiling away in New Jersey, Howard was working in the halls of Congress to pass a federal law that would regulate insecticides. In 1910, with Howard's strong support, Congress passed the Insecticide Act, one of the first pieces of federal legislation to confront the physical well-being of American citizens. The act passed with considerable backing from prominent entomologists who, under the leadership of E. Dwight Sanderson, the state entomologist of New Hampshire, had worked through the Standing Committee on Proprietary Insecticides of the Association of Economic Entomologists to ensure that, when it came to insecticides, "one may know with what he is dealing." The nation's earliest commercial producers of chemical insecticides—companies like Sherwin Williams, Sharp and Dohme, Marshall Oil, Zenner Disinfectant, Parke & Davis, and

Ansbacher—backed the standards enthusiastically. They welcomed the law's ability to remove quack patent insecticides from an increasingly competitive market. Plus, largely due to Sanderson's savvy politicking, the companies were able to play a direct role in shaping the legislation's most critical provisions. (As the bill circulated through Congress, Sanderson accurately predicted "a widespread sentiment . . . in favor of such legislation" among the corporate class.) It was all good news for Howard because federal legislation, whatever its strengths and weaknesses, lent a critical imprimatur to commercial insecticides, the companies that made them, and, by extension, Howard's plans to fight his battles chemically. Another confirmation of the insecticidal approach came in 1912, when Howard set up a field station in Mound, Louisiana, for the express purpose of investigating chemical control and placed at its head C. L. Van Dine, who recently had applied the kerosene solution to mosquitoes in Hawaii. There was little that Smith, who would die two years later of a rare disease, could do to counter these developments.[23]

A glance ahead to the 1920s and 1930s reveals how thoroughly Howard's vision won the day. A popular entomology text published in 1920 confirmed what had become conventional wisdom: the "effort at control [of mosquitoes] . . . is accomplished by pouring oil on the water." A manual advising housewives how to keep a pest-free home stated without qualification that "one of the best methods is to pour oil upon the surface of the water." A textbook from 1930 instructed students that "water may be oiled every two weeks" to control for mosquitoes. And finally, an article on controlling mosquitoes in hydroelectric plants advised that kerosene be "sprayed over the surface of the water either alone by air pressure or by mixed with water in pumps which throw a stream of the mixture as far as one hundred feet."[24]

It was with considerable dominance that kerosene became standardized as a mosquito killer. Nevertheless, the story of Howard's chemical triumph does not end with mosquitoes. In 1914, the world went to war. After the United States joined the Allies in 1917, they fought more than just the Central Powers. They also fought the insect pests that fought them. And if there ever was a specific geopolitical context that had complete disdain for gradualist, multifaceted, and ecologically sensitive approaches to insect control, war was it. After the war, cultural and biological approaches to insect control—applied to both mosquitoes and common agricultural pests in general—moved to the periphery of scientific research. Howard, once again, would brilliantly exploit a nonagricultural event—World War I—to pave the way for an approach to pest control that would effectively dominate twentieth-century agriculture.

"warfare as real and as systematic as the conflict of militarism and democracy"

When the Allies settled into trenches in Europe in their fight to make the world safe for democracy, American entomologists were called on to make the war effort safe from infestation. As he had with mosquitoes, Leland Howard was quick to realize the impact that the war might have on the future of both entomology and chemical insecticides. "[W]ar conditions," he wrote in 1919, "have intensified the work of the entomologists and have enabled them to make the importance of their work felt as almost never before." Whereas entomologists were once dismissed as men "trained to count the spots on a mosquito wing," they could now, as scientists working for national security, be valued as "competent investigators whose advice and help meant everything in the warfare against insect life." Not only did Howard write to the surgeon general of the army promoting a plethora of "men well trained in applied entomology who could be used to advantage and who were anxious to serve," but he effectively laid out the precise ways in which they could contribute. The status of professional entomologists could only go up, he predicted, as they confronted the insect pests that endangered soldiers, protected the food that fed them, and cleared the trenches in which they camped.[25]

What made Howard's efforts all the more significant was that war was overtly hostile to naturalistic methods of pest control. According to a historian of science, this was a time when "the need for immediate remedies took precedence" over everything else, especially ecologically driven ideas of cultural and biological control of insects. Although Howard could never have predicted that a compound called DDT would replace kerosene as the preferred killer of mosquitoes, he must have known that, if he could send the Bureau of Entomology to war, an even further onslaught of powerful chemicals to fight the insect menace was bound to follow.[26]

Howard's inspiration to link war and entomology began with the British zoologist A. E. Shipley. Howard recalled being influenced by his articles published in the *British Medical Journal* from 1907 to 1912. The pieces were later collected in Shipley's book *The Minor Horrors of War*. "I deal with certain little Invertebrata," Shipley wrote in the introduction, "animals which work in darkness and in stealth . . . little animals which in times of War may make or unmake an army corps." Lively chapters on the bedbug, louse, housefly, flour moth, mite, and tick condemned these pests as the causes of "minor discomforts of an army," "frequenter[s] of human habitations," creatures that "infest human sores," and vectors of disease. Repeating the old adage that "an army marches on its stomach," Shipley obsessed about pests that threatened "our soldiers' food," spent page after instructive page on the ways to combat insects like the "biscuit moth," and claimed of the fly that "the poisoning of the soldiers' food-supply is its chief role in war." Not a single possible niche that insects might exploit among men at war—maggots in wounds, blowflies in ears, lice in soldiers' clothing—escaped his attention. The book was, in its own way, a tour de force of applied entomology. Perhaps most importantly, Shipley's reliance on benzene, carbolic acid, sulfur baths, chloride of lime, creosote, and alkaline soaks in his fight against these enemies revealed his commitment to chemical efficiency. Howard quickly became an acolyte.[27]

Anticipating the impact that the war would have on crop production in the United States, Howard presciently coordinated a national information network on war-related insect infestations. Bringing together entomologists from state agencies, agricultural experiment stations, and universities, as well as demonstration agents, he created, by his own estimation, "as a far as possible almost a census of insect damage and prospects, so that the earliest possible information should be gained as to any alarming increase in numbers of any given pest." Consolidation was critical. If and when the war machine said "jump," he wanted to be sure that American entomologists jumped ably and in unison. He insisted that the disparate information he collected be "received at a common point (Washington) and distributed where it should be of the most good." These preparations paid off. When Congress turned its attention to the nation's insect enemies after declaring war in April 1917, entomologists suddenly found themselves thrust to the center of a wartime bureaucratic maze. Ample funding materialized, as it has a way of doing during war, and, as if following Howard's script, "men were assigned to different localities, and took care of the demonstration work against the principle pests of staple crops all over the United

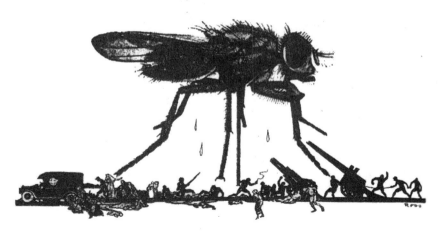

"THE FLY MUST BE EXTERMINATED TO MAKE THE WORLD SAFE FOR HABITATION"
During World War I, Leland Howard brilliantly highlighted the connection between the war in Europe, which was being fought to make the world "safe for democracy," and the war against insects. In this image, printed in the *St. Louis Post-Dispatch,* the killers of the giant fly, not coincidentally, are soldiers.

States." Insect attacks were no longer about agricultural production per se, but about national security. This distinction was critical.

Howard had placed his profession's men on the front lines of this new battle, reminding those who ever doubted the entomologist's place in war that the naysayers

> failed to consider not only how damage by insects to growing crops influences the food supplies of armies, but also how greatly grains and other foods stored for shipment to the front or on the way to the front may be reduced in bulk by the work of the different grain weevils and other insects affecting stored foods. In addition, they did not think of the damage done by insects to the timber which enters into the building of ships, into the manufacture of wings for the airplanes, and that which is used for oars, the handles of picks and spades, and which even occurs in wooden structures and implements after they have been made.

Before it even began, Howard understood that the war was about more than conventional weaponry.[28]

The work of insect control during wartime, as well as the public expectation for success, was daunting. "It is hoped," wrote Royal N. Chapman, "that our country may be aided in its effort to conserve the food needed to win the war." The *New York Times* upped the rhetoric, noting with respect to the food supply that "winning the war is the absorbing and imperative business of the American people. Twenty-four hours of every day must be devoted to it." An unlikely crop, the castor bean, offered a case in point of the sacrifices and flexibility involved. With the onset of hostilities, the Department of War encouraged Americans to show their loyalty to the cause by planting castor beans because castor oil was crucial to lubricating airplane cylinders. Small planters throughout Florida and Texas, in particular, read up on planting castor beans (which had traditionally been cultivated in Mexico) and transformed thousands of acres of land into fields choked with stringy legumes. It was not long, however, before the Department of War was calling on the Bureau of Entomology for urgent advice. Patriotic farmers soon found their plots of new crops under attack by an unfamiliar pest. One writer noted that his castor beans were "riddled and cut short by boll worms." Howard called the culprit a "southern army worm." Whatever the insect, entomologists fanned across the nation to ensure that castor oil continued to lubricate not only airplane valves, but also, in the *New York Times*'s phrase, "the feelings of a cross and anxious nation."[29]

The storage of grain and wood posed problems, too, ones that government entomologists were ready to handle. "The damage to stored grain and grain in shipment," wrote Howard, "soon came to the front." Massive quantities of grain were being transported from farms to a single warehouse in Brooklyn, New York, before being packed and shipped to Europe. Fearful that insects would infest these supplies as they sat in the warehouse, the Department of War yet again contacted the Bureau of Entomology for advice, going so far as to establish an on-site laboratory for entomologists to conduct appropriate investigations. "Many of the millers and dealers who handle the cereals which the Food Administration is now requiring as substitutes for wheat flour," wrote one entomologist, "have always recognized them as being subject to insect attack." The collective aim of the entomologists was, through on-site inspections, to minimize those attacks. Entomologists on the west coast and in the South undertook similar investigations in the mills and warehouses that temporarily stored grain destined for New York. No less attention was paid to the nation's supplies of forest products. Entomologists advised the War and Navy Departments after bark-boring beetles caused "serious losses of forest resources"—in this case, 1 million feet of Mississippi ash logs intended to become oars and handles. In the Pacific Northwest, entomologists dispatched by the Spruce Production Board worked with loggers to ensure that the spruce needed for airplane production remained free from infestation. Their work against tree borers, as entomologist Maurice C. Hall put it, was "warfare as real and as systematic as the conflict of militarism and democracy."[30]

Animals required their own share of entomological attention. Demand for cattle reached a new level during the war, as did efforts to protect them from insect-borne diseases. "Texas fever" and scabies (caused by ticks), as well as diseases caused by worms in pigs and sheep, had always been chronic agricultural problems. But now, rather than ranchers, the Bureau of Animal Husbandry, Army Veterinary Corps, and Chemical Warfare Service were entrusted to devise ways to prevent the outbreak of these diseases. Again, tapping Howard's corps of entomologists, the government drew on their expertise to concoct "suitable dips to be readily applied by swimming the cattle through a dipping vat" and thus assure the nation that "the last tick will be dipped or collected as a museum specimen, or whatever is appropriate for a last tick, and the tick and Texas fever will become extinct in this country." While bovine diseases had once "cost the country dearly," it accurately could be said that "the savings in wool, leather, mutton, and beef, all unusually valuable in these days of war, constitutes an indemnity that repays us many times for the outlay which it has cost to prosecute this campaign."[31]

Castor beans, grain, wood, and cattle were not the only assets that wartime entomologists sought to protect. There were soldiers as well. Malcolm Evan MacGregor, a scientist at the Bureau of Scientific Research, best captured the novelty of this responsibility when he wrote, "During the last few years medical entomology has been rapidly establishing itself as an invaluable branch of preventive medicine." With "the numerous biting flies and mosquitoes find[ing] exceptionally favorable opportunities for breeding in war areas," the Bureau of Entomology charged even deeper into the fog of war. The dangers of "trench fever," infectious jaundice, scabies, wound infections, typhus, dysentery, and a number of other ailments that commonly afflicted soldiers led to body lice, mosquitoes, spiders, maggots, flies, ticks, and any other insects capable of transmitting parasites and germs being declared enemy combatants. It was widely agreed, as Hall put it, "that entomologists versed in medical entomology be attached to our army units." By 1918, scores of expert entomologists, commissioned and noncommissioned, had left for Europe, where they worked in military camps to control insects that carried diseases. Howard proudly noted that "every suggestion that came to the War Department in regard to control of the body louse was referred to the entomological committee, or to the Bureau of Entomology."[32]

Entomologists thus embraced Howard's opportunity with enthusiasm and organizational prowess, fulfilling the role of "competent investigators whose advice and help meant everything in the war against insect life."[33] Their method of combat, most importantly, was never in question: they were going to fight the insects with poison. The urgency of war, the requirement that every task be accomplished with the utmost efficiency and accountability, and the implicit expectation that the latest technology be employed were only three factors that ensured that World War I was a boost for government entomologists and chemical insecticides. If early tests on and experiences with them had proved anything conclusively, it was that these compounds could work well in the short term and under controlled circumstances. Cultural and biological methods of insect control—both of which required more time, patience, adaptability to unique situations, and appreciation of the long-term consequences—were simply out of the question under the unique conditions of war. Chemical poisons were the only option seriously entertained. In turning to them as decisively as it did, the Bureau of Entomology helped establish chemical insecticides as the primary means through which any modern scientist intent on eliminating insects undertook his work. It perpetuated this opinion, moreover, with the consolidated support of the federal government, substantial economic backing from nonprofit foundations, the eager endorsement of insecticide manufacturers, and widespread public approval. The war precipitated all these developments and, in so doing, helped further standardize chemical insecticides as the most conventional means of fighting insects, whether abroad or at home, in the trenches or, fatefully, in the fields.

"the best in the world"

The use of poison on the front lines did, indeed, reverberate on the home front during and after World War I. The manipulation of the nation's crops to meet wartime demand, in addition to the onset of unusually mercurial weather in the spring of 1917, created agricultural conditions whereby "certain insects not hitherto notably conspicuous appeared in great abundance and added new problems to the production of certain crops." While there was nothing novel about rapid environmental change leading to larger insect populations, what did differ during the war was the rapidity and thoroughness of the nation's response with chemical insecticides. The deliberation of the past, a "chaos of experimentation" that had brought farmers and editors to the discussion, was quickly receding as a characteristic of insect control.

Consider an incident in Kansas. When, in the autumn of 1917, Kansas state entomologists predicted the worst outbreak of grasshoppers in five years, there was little time to discuss the most judicious solutions, much less experiment with a variety of methods, to combat the infestation, as might have happened before the war. Instead, extension agents, guided by the Bureau of Entomology, alerted farmers throughout the state that their wartime wheat production was under threat by an imminent "great hatching." Then, without discussion, they ordered from the government 36,000 pounds of white arsenic and 336 tons of wheat bran, mixed them into "poison bait," and distributed 900 tons of the insecticide packets to farmers. Extending these chemically based policies even further, the entomologists soon thereafter ordered 25,000 more pounds of the poison bait to protect sugar beets and alfalfa destined for Europe. In 1917, a plant louse called the potato aphid, which before the war had done little crop damage, began to infest not only potatoes, but also eggplant, tomatoes, spinach, and peppers. The response was, again, the "poisoned bait method," executed as a bureau decision, without discussion or debate. When castor beans came under attack by worms or weevils, entomologists responded with arsenical insecticides as well, as they did with "the proper fumigation" when wheat stores were threatened. While there was really no way to measure the effectiveness of these chemical solutions, state officials were not shy about touting them as having saved millions of dollars in crops, not to mention the entire war effort itself.[34]

The insecticide manufacturers, needless to say, were enthralled with the government's embrace of chemical insecticides. The Insecticide and Disinfectant Manufacturers Association, for one, worked diligently to secure government contracts and keep prices at a premium while taking advantage of wartime emergency measures to oppose bills that mandated "the proper labeling of products." In 1916, anticipating the imminent deployment of chemical insecticides to protect the nation's food supply, officers from Hemingway and Company, Worrall Manufacturing Company, and Interstate Chemical Company met with J.K. Haywood, chairman of the Insecticide and Fungicide Board of the Department of Agriculture, to gain his support in helping their companies "slaughter humanity's flying and crawling enemies." Haywood assured them a brisk business, touting American-made insecticides as "the best in the world" and praising the companies for recognizing "the scientific nature of the business." The War Trade Board put money where Haywood's mouth was one year later by placing a comprehensive embargo on arsenate of lead, arsenate of soda, kerosene, and several dozen chemicals used to make insecticides. With foreign competition neutralized, William C. Piver's small chemical company went from supplying about twenty customers a year in 1915 to manufacturing 40 tons of calcium arsenate in 1918. Solidifying the already sturdy base being built under insecticides were charitable foundations. In 1917, in a sign of much largesse to come, the Mellon Foundation for the first time offered an industrial fellowship to a scientist who was studying insecticides.[35]

Reports from Europe on the effectiveness of American insecticides only strengthened the increasingly popular reputation of these products. Hydrocyanic acid gas, as long as its application was "carried out by competent people," supposedly controlled for bedbugs across the western front. Sulfur baths laced with napthol removed mites, but with the drawback that the sulfur "spoils utterly all enameled and metal baths." Sulfate of iron or formaldehyde battled typhoid-spreading flies. In his discussion of lice and soldiers in the trenches, Leland Howard, well aware of A. E. Shipley's adage that "lice are the constant accompaniment of all armies," noted that "at the request of the Army War College and the medical department, as well as the chemical warfare service, tests were made of a new poisonous gas." Perhaps, Howard surmised, "the gases used in warfare" could be used "as fumigants for the control of insects and diseases." Whether experimental gas or the more traditional dosages of benzene, petrol, and turpentine were used, lice, which carry typhus, forced soldiers to visit "de-lousing stations." There they (and their clothes) would take a bath in "washing suds" studded with chemicals "capable of penetrating the chitin of the [louse's] body wall during the period of washing, and toxic enough to produce its death." Medics then sterilized the men with blasts of dry heat. The production of kerosene boomed, as did that of xylol and paraffin oil, as soldiers sprayed these chemicals to control for mosquitoes and lice. "In some instances," wrote Hall, "companies of men have been stripped and thoroughly sprayed with these substances, with excellent results in controlling the trouble and with but little personal discomfort on the part of the men." In other instances, soldiers donned "muslin underwear impregnated with sulphur and napthaline" in an effort to keep lice out of their pubic hair.[36]

At home and abroad, insecticides became associated with the word that everyone wanted to hear: "victory." Thomas A. Dunlap writes, "It is impossible to prove, but difficult to resist, the idea that this constant image of battle and war influenced people to choose, or to accept, the use of the only weapons that promised 'victory.'" The evidence is compelling. Although A. L. Melander, a professor of biology at City College, made the following remark in 1933, the seed of his idea had sprouted during the war. With respect to controlling insects, he wrote, "Chemical warfare on insects has become an accepted part of our yearly life. It will never be outlawed . . . and in time to come it will decide whether man or insects will dominate the world." A few years earlier, Edward O. Essig, an entomologist at the University of California, had confirmed the prevalence of the man-versus-nature aspect of the insect war within the war. Speaking of entomologists, he wrote, "Ours is not the menial chore so often alluded to as 'chasing bugs,' but rather the gigantic task of saving the human race." In the highly charged wartime atmosphere, with stakes raised so high, insecticides proved to be the most potent weapon of choice. The upshot at home was nothing less than an arms race, and in an arms race, the combatants rarely look back or keep their conflict in perspective.[37]

"people were beginning to slap"

Insecticides saturated the war effort. The urgent need for the short-term protection of crops, the involvement of and unified support for chemical methods of insect control by several powerful governmental bureaus, the unique needs of Allied soldiers fighting a trench war, the powerful advocacy of chemical insecticides by companies and foundations, and a prevailing sense that "men of science" were just as critical as soldiers in winning the war—all contributed to the emergence of insecticides as the primary means of battling insects during World War I.

Neither Leland Howard nor the promoters of chemical insecticides were prepared to stop their campaign once the war came to an end. To argue that wartime use of chemical insecticides created an inertia that carried over into peacetime agricultural practices is not to suggest that cultural and biological methods of pest management became completely irrelevant. Entomologists and farmers have continued to pursue these approaches, often with great passion, into the present day. On balance, however, they became peripheral to the scientific and agricultural focus on arsenic-, petroleum-, and lead-based compounds as the most efficient means to eliminate destructive insects. During the war, American entomologists and chemists not only applied familiar insecticides but tested more than 100 new ones. They had no intention of simply forgetting about these products because the ink had dried on the Treaty of Paris.

To understand the postwar implications of insecticide use, it helps to return to the issue of mosquito control. After the war, John Smith's approach was not completely abandoned. Government officials, agricultural scientists, university entomologists, and public-health experts certainly entertained the idea of draining swamps whenever it was feasible. Still, any chance of public-health officials systematically promoting a cultural approach to mosquito control diminished significantly. What decisively moved to the foreground were, in unparalleled amounts, chemicals. The questions had changed. No longer was it: Chemicals or no chemicals? Instead, it was: What kind of chemicals? What to use to spray them? How much to spray?

The most decisive move away from Smith's proposals for the eradication of mosquitoes, ironically, took place in his own backyard. It was in Morris and Middlesex Counties where the New Jersey Agricultural Experiment Station, in 1926, first tested an insecticide composed of pyrethrum and oil. Reporting the same concerns that had dogged experimenters in the nineteenth century, the station (although ultimately approving the mixture) noted that "only on a still day when there is no appreciable wind to blow away the [insecticide] from its course" would "the killing film of oil" effectively work. Another substance that came into vogue as a mosquito killer was Paris green. "For chemical control," wrote an Alabama civil engineer, "Paris green is dusted over the surface of the water and poisons the larvae of the Anopheles which eat it." The dust earned accolades as a viable insecticide, even though it had to be "applied in such large quantities as to become an actual menace of poisoning animals and men." The Alabama agent also promoted oil, preferably kerosene, but reiterated the ongoing challenge of finding a solution "light enough to spread well, yet heavy enough so that it will not evaporate for several days." He did not discount the conventional cultural methods of clearing land, draining swamps, and importing fish, but he revealed his sympathies when he stressed that "for *complete control*—by which is meant maximum possible prevention—of mosquito breeding, dependence is placed on oil and chemicals."[38]

As with the earliest agricultural insecticides, there was just enough seemingly objective evidence to convince most critical observers that these chemical applications were basically effective. One well-publicized case suggests how public-health experts and public opinion came to accept the chemical approach as, at the least, reasonable. In late June 1936, heavy rains saturated the Passaic Valley of New Jersey thoroughly enough for the Essex County Mosquito Commission to predict "a heavy brood in the [Newark Schools Stadium]" during a scheduled concert for which a crowd of 18,000 was expected. As predicted, both people and mosquitoes arrived in droves. But so did agricultural experiment station agents, and they were armed with an unnamed insecticide and 400-gallon sprayers. At 6:00 P.M. they doused the stadium. The chemicals kept the mosquitoes at bay until 8:50 P.M., when an agent noted that "people were beginning to slap." It turns out that a brood had "flown in with the wind from the south side." Without hesitation, the agents leaped into action. "The area in the south section of the stadium was immediately fogged with [insecticide] from a shade tree gun which threw a stream thirty or forty feet high against the wind," recalled one of the agents, Thomas Headlee. "This produced a blanket mist of [insecticide] drifting with the wind, high above the audience and slowly settling downward," he continued. How these concertgoers felt about being coated with chemicals went unrecorded, but the agents were not at all shy about publicizing the fact that "protection from mosquito bites" was "practically complete for the entire period of the concert."[39]

While World War I and its emphasis on insecticides thoroughly altered the nature of the mosquito battles, the impact of the transition from biological and cultural to chemical methods of pest control did not affect one crop, one region, or even one group of people. It touched all humanity. As Howard had said, the problem "was of the greatest importance to all mankind." And, therefore, so was its solution. When mosquitoes became the targets of insecticides, and when the war validated that approach, any remaining doubts about the place and prominence of chemical methods were put to rest at the Bureau of Entomology. The bureau's emphasis was now systematized. In 1922, W. R. Walton, praising the discipline that Howard had imposed on the bureau, wrote, "Every executive who possesses even a rudimentary knowledge of his calling knows that the first important requirement is harmony, because this means teamwork, which is a long stride toward success in any organized effort. . . . The entomological service is not so very different from that of the army; we have our battles to fight, and organize and mobilize our forces accordingly."[40] In other words, uniformity was now paramount. As mosquito control went chemical, it became much easier for American farmers—whose crops had been thoroughly covered in insecticides during the war—to abandon variable plans for biological, chemical, or cultural control of insect pests and, with the utmost federal assurance, accept insecticides as a necessary aspect of agribusiness, government, and public health. Whatever concerns about insecticides may have existed before the war, in short, were put to rest afterward. The tools in the box had changed, replaced by the chemicals that were said to have won the war, the airplanes that doused the swamps, and the sprayers that left a group of concertgoers sitting in a haze of insecticides.

The advances in the development, manufacture, and use of insecticides made in the decade after World War I comprise an extensive catalog of evidence showing that Americans were soon spraying with a vengeance. A year-by-year summary reveals a steadfast adherence to many entomologists' insistence that one should not rest easy as the war on insects raged:

1920 Farmers dropped 10 million pounds of calcium arsenate dust to control the boll weevil, doing so with recently designed ground equipment.
1921 Orchard keepers began to spray lead arsenate to control for maggots in fruit, while scientists recommended arsenic as a mosquito larvicide.
1922 Farmers began to use rotenone-bearing insecticides used to control the cattle grub, calcium arsenate to manage the Mexican bean beetle and boll weevil, and calcium cyanide as an insecticidal fumigant, while air-

planes first complemented ground sprayers in applying calcium arsenate to cotton.

1923 Millers began to use carbon tetrachloride and ethylene dichloride to fumigate grain mills.

1924 Officials in Louisiana used airplanes to drop Paris green on state swamps; others employed blimps to dust crops with insecticides; and scientists tested colloidal sulfur, fluorine compounds, and cryolite as potential insecticides.

1925 Scientists synthesized an ethylene dibromide fumigant, and airplanes began to dust orchards with insecticides.

1926 Fish oil began to be used as a binder to help insecticide dust stick to plants; lead arsenate, against the clover leaf weevil; barium fluosilicate and thallium sulfate, as insecticides; and chloropicrin, as a fumigant in flour mills.

1927 Sodium arsenate sprays hit the market; farmers used calcium arsenate on blueberries; and warehousemen saturated stored grain with an ethylene dichloride–carbon tetrachloride mixture.

1928 Farmers and householders began to use ethylene oxide and alkyl formates as fumigants on crops and in homes, and cryolite was doused on coddling moths in the Pacific Northwest.

1929 Insecticide manufacturers, which now included DuPont and the Hercules Powder Company, produced $23,505,000 worth of chemical insecticides.

1934 Farmers were saturating crops with 90 million pounds of arsenicals, 73 million pounds of sulfur, 10 million gallons of kerosene, 21 million pounds of naphthalene and paradichlorobenzene, 21 million pounds of pyrethrum, and 2 million pounds of rotenone.

As a result of mosquitoes, World War I, and Leland Howard, the United States—in just a generation—had become an insecticide nation.[41]

Leland Howard retired from the Bureau of Entomology in 1927. He left behind fifty years of tireless service, a stack of honorary degrees, awards galore, a publication record a mile long, and a sense that, as one colleague put it, "none had so powerful an impact on the world-wide study of harmful insects." Howard always approached the insect wars with creativity, ambition, and the best of intentions. Time after time, however, he ultimately came back to the chemical approach. There were entirely sensible reasons for him to have made this decision. As a progressive scientist, it was imperative that he seek quantifiable solutions to problems of insect management that fit within the tight framework of progressive reform. As the director of the Bureau of Entomology, it was imperative that, when mosquitoes were linked with the transmission of malaria, he pursue approaches to their eradication that were quick, relatively inexpensive, and consistent with the interests of industry and government. As the federal entomologist during World War I, it was imperative that he promote measures that met comparatively urgent short-term needs with solutions that were universally applicable, expedient, and powerful. Ever the pragmatist, ever the career man, ever the believer in the power of science to control the environment in the interests of humanity, Howard placed insecticides on a pedestal. Industry and government ushered them into the 1920s and beyond, fully convinced that the only way to win the insect wars was with increasingly powerful chemical compounds.

As a result of this multifaceted transition, a transition that Howard was critical in orchestrating, the nation's battle against invasive insects changed in several basic ways. What had once been a decentralized attempt to manage the insect paradox became a bureaucratized effort to eradicate the problem. What had once been an approach sensitive to local contingencies devolved into a more simplified, universally applicable solution. And, finally, what had once joined professional entomologists and struggling farmers in a cooperative quest for control now placed agribusiness at the mercy of an increasingly powerful government–corporate–scientific complex that promoted the chemical option as an irreversible path toward the extermination of insect pests. To be sure, a cohort of American farmers and entomologists would always preserve the localized "chaos of experimentation," but by the 1920s a chemical path of dependence had been set for mainstream commercial farmers. All that was left to do was defend it.

"And I must now stop."[42] So Leland O. Howard concluded his professional autobiography, *Fighting the Insects,* in 1933. But stop he hardly did. Whereas the men who fought the hardest for insect control policies based on biological and natural approaches had a tendency to die prematurely—

John B. Smith, after all, was only fifty-four when he succumbed to Bright's disease in 1912—Howard lived for another twenty-three years after his retirement, long enough to ride his bicycle many miles; become a formidable billiards player; spend countless afternoons at the Cosmos Club in Washington; and, should he have been reading the news, realize that even the best intentions can go awry. The paradox he had hoped to solve once and for all turned out to spawn new paradoxes of its own.

CHAPTER 6

"vot iss de effidence?":
RESIDUES, REGULATIONS, AND THE POLITICS OF PROTECTING INSECTICIDES

Through his tireless work as director of the Bureau of Entomology, Leland O. Howard disrupted the continuity that had characterized the field of economic entomology from 1840 to 1900. The mosquito menace and the world war were pivotal events that allowed him to envision a way to end, rather than manage, the paradox of commercial expansion and insect proliferation, thus liberating the field from the commandments that its founders, including Thaddeus William Harris and Charles V. Riley, had articulated. The commandments favored experimentation with a wide variety of means to control predacious insects, communication with farmers and farm journal editors, and decentralized approaches to insect management that respected local conditions. Howard's reorientation of the bureau around the advocacy of chemical insecticides refocused it on a bureaucratic, unified, and universal approach to insect control that required the active participation of chemical manufacturers and political operatives. The transition to chemicals, as a result, replaced cooperation with contention and brought to the fore those who were passionately in favor of chemical insecticides and those who dedicated their lives to opposing them. With the end of war in Europe, new battle lines were redrawn at home. While no world war, it was ruthless in its own way, if for no other reason than that the new weapons had to be protected by any means necessary.

"There is no question in my mind"

Harvey W. Wiley stood against the rising tide of insecticides. He lived for eighty-six years and spent fifty of those years fighting to keep food clean. Born in 1844 on an Indiana farm, he worked his parents' land until the age of nineteen, when he joined the 187th Regiment, Indiana Volunteers. In 1867, he graduated from Hanover College and attended Indiana Medical College, earning an M.D. in 1871. Instead of pursuing a conventional medical practice, however, Wiley specialized in food chemistry. He studied the structure of sugar at the Bismarck Institute in Germany and returned to the United States to teach in the Department of Chemistry at the newly established Purdue University. His passion for pure food led him to abandon academia in 1883 and assume the post of chief chemist in the Bureau of Chemistry in the Department of Agriculture. In this position, he introduced dozens of pure-food bills to Congress and gradually assembled a powerful team of lobbyists intent on protecting the nation's food supply. One of his fans was President Theodore Roosevelt, who in 1906 signed into law the culmination of nearly thirty years of Wiley's work: the Pure Food and Drug Act. A year later, Wiley addressed directly the dangers of insecticides, mentioning arsenic-residue poisoning in his book *Foods and Their Adulteration* and insisting that consumers peel apples before eating them. When Wiley retired from the Bureau of Chemistry in 1912 to test foods at the Good Housekeeping Institute, a newspaper headline blared: WOMEN WEEP AS WATCHDOG OF THE KITCHEN QUITS AFTER 29 YEARS.[1]

Anton J. Carlson was another advocate of food safety who turned his scientific mind to the dangers of arsenical sprays. A Swedish-born professor of physiology at the University of Chicago, Carlson rigorously applied the scientific method to virtually any issue that crossed his nimble mind. He was outspoken, did not shun the spotlight, and routinely punctuated his work with his signature, heavily accented query: "Vott iss de effidence?" In 1917, he wrote to Walter C. O'Kane, an entomologist who encouraged the use of chemical insecticides, and sternly outlined his concerns about traces of arsenic and lead on crops treated with insecticides. "Speaking as a physiologist interested in public health," he explained, "I should say that the question is not how much of the poison may be ingested without producing acute or obvious chronic symptoms, but how completely can man be safeguarded against even traces of poison." This was a bold claim to make as the nation prepared for war, one made even more jarring by the letter's next sentence: "There is no question in my mind that even less than so-called toxic doses of lead and arsenic have deleterious effects on cell protoplasm, effects that are expressed in lower resistance to disease, lessened efficiency, and shortening of life."[2]

Women may not have wept upon Carlson's retirement, but his early warnings about insecticide residues helped turn scientists' attention to a public-health concern that was otherwise a distant blip on the national radar screen. Poison residues, food safety, and public health were, as a result of the work of men like Wiley and Carlson, becoming major topics of discussion. They would play an important role in the ultimately frustrated attempt to limit the use of insecticides throughout the twentieth century.

"go on as you used to go on"

There were also those who rode the wave of insecticides. In the spirit of progressive interest in public welfare, Harvey Wiley, Anton Carlson, and other public-health officials were passionate in their concern about foods adulterated with insecticide residues. Other men, however, were equally passionate in their defense of corporate freedom, bureaucratic standardization, and concern for profit. Two examples suggest the nature of the opposition that Wiley and Carlson would face as they confronted the dangers posed by insecticides. In 1907, a year after the Pure Food and Drug Act went into effect, the Bureau of Chemistry informed the Department of Agriculture that dried fruits preserved in sulfurous acid qualified as adulterated food under the new law. James Wilson, the acting undersecretary of agriculture, charged with enforcing the provisions of the Pure Food and Drug Act, disagreed. As a government official entrusted with guarding the interests of American farmers, Wilson responded that nothing be confiscated, no one be punished, and business proceed as usual. Addressing the fruit growers directly, he instructed them to "go on as you used to go on. . . . I will not take any action to seize your goods or let them be seized, or take any case into court."

It was a promise he kept. If Wilson's tactics seem extreme, however, there was always Llewellyn Banks to put them in perspective. Twenty years after Wilson and the Department of Agriculture made clear their strategy of tacking around regulations, Banks, a prominent fruit grower in Medford,

Oregon, met representatives from the Bureau of Chemistry with a threat they had yet to encounter. He promised to shoot to kill the agent who tried to confiscate his insecticide-laden fruit. The bureau wisely opted not to call Banks's bluff on his Oregon turf. Instead, it confiscated his lead- and arsenic-tainted apples and pears as they traveled through Illinois, met him in court when he appealed, and relied on the testimony of Carlson to uphold the confiscation. Banks and his company, Suncrest Orchards, lost more than $75,000 in revenue.[3]

It was a rare victory for reformers. Knowing as we do the health hazards posed by arsenic- and lead-based insecticides, it would be easy to portray men like Wilson and Banks as uncaring and rapacious demons who tried to stymie the efforts of high-minded crusaders. But the conflict over insecticide residue and food regulation did not play out in such simplistic terms. Given the minimal regulation that the federal government had practiced throughout American history, the nation's laissez-faire approach to land acquisition and use, and the radical environmental transformations that American farmers had undertaken across the contiguous United States for more than three centuries, Wilson and Banks—the threat of gunplay notwithstanding—were behaving in a way that was entirely consistent with the country's defining value system. The insult of having the federal government use the Pure Food and Drug Act to restrict growers' customary freedoms seemed especially unfair, considering that since the early twentieth century, the Department of Agriculture, the Bureau of Entomology, and dozens of experiment station researchers and university professors had praised chemical methods of pest control, taught farmers how to use the appropriate insecticides, and extolled the power of modern science. In a nation where farmers in several states were actually *required* by law to spray their crops, it must have struck average growers as grossly paradoxical to suddenly find federal chemists and public-health officials upholding the regulations on adulterated food and questioning the use of insecticides. As farmers justifiably saw the matter, such actions contradicted custom and threatened livelihoods.

The emphasis on poisonous residues and public safety, moreover, came at an especially difficult time for farmers—orchard keepers, in particular. After the Civil War, the fruit-growing industry gradually moved from New York to the west coast. Supported by extensive and environmentally devastating irrigation systems, the industry expanded exponentially to meet a constant demand. By 1911, Washington State alone was producing more than 7.5 million bushels of apples a year, a figure that rose to 38 million bushels by 1930. The war had been especially good to fruit growers, and production, as well as land values, soared. But there were underlying problems that began to manifest themselves in the 1920s, at the very time when the rhetoric on residues was beginning to heat up. Labor costs skyrocketed, doubling and even tripling over a few years. In addition, consumers were beginning to prefer processed rather than fresh fruit, leading unscrupulous marketers to mix inferior east coast fruit with highly valued and carefully cultivated west coast produce, a ploy that depressed prices overall. Finally, there was the simple matter of oversupply. The high returns of the 1910s drew so many growers into the industry that, by 1930, they were competing for an increasingly limited market. Under such circumstances, orchardists had little incentive to consider the long-term consequences of frequent spraying. The last thing they needed to be added to their growing list of concerns was an unusually bad outbreak of codling or tussock moths. Thus the growers piled on the arsenical insecticides in unprecedented amounts.[4]

It is against this larger backdrop of American agrarian history that the effort to establish acceptable levels of insecticide residues—arsenic and lead, in particular—became more complicated. Skeptics had been expressing sporadic and poorly publicized concerns about residues ever since Paris green had hit the market in the 1870s. It was not until the 1910s, however, that these disparate grumblings started to cohere into a sustained argument that caught the attention of the public-health lobby, government, and, to a much lesser degree, general public. Improved and cheaper spraying equipment and better insecticide binders fabricated with fish oil, combined with farmers' increased addiction to spraying, led to residues lingering on crops more tenaciously than they ever had. Concerned public officials, government scientists, chemists, physicians, and academics dutifully hoisted red flags to warn about threats to the safety of the nation's food supply, citing case after case as evidence of their worries. As necessary advocates of agribusiness, the federal and state governments were not eager to chase down farmers who violated potential regulations on residue levels. Under pressure, though, government officials would reluctantly and temporarily join the growing chorus of concern about the effects of insecticide residues on public health. On the surface, the efforts seem to have yielded results.

Several state legislatures passed laws that addressed the dangers of insecticide residues, as did, eventually, Congress, which enacted the Food, Drug, and Cosmetic Act of 1938. The Food and Drug Administration, formed in 1927, incorporated the Bureau of Chemistry, took a steely-eyed view of residues, and lowered the acceptable level of arsenic on crops every year from 1927 to 1932. Hundreds of confiscations of tainted fruits and vegetables, including the apples and pears of Llewellyn Banks, reiterated the government's apparent willingness to protect public health. From a distance, the legislative response seemed commensurate with the problem, leaving the public to enjoy a worry-free diet.[5]

On closer investigation, however, these regulatory measures obscured a thick stew of politics, power, and profit. Earnest efforts to protect American crops and consumers from dangerous insecticides repeatedly ran into special interests, a divided federal bureaucracy, public indifference, and the impossibility of finding a direct connection between insecticide residues and health risks. Perhaps most important, there was history, specifically the history of staying one step ahead of the insect paradox. American farmers, led in large part by the Bureau of Entomology, had taken insecticides mainstream. Several factors encouraged and rewarded the adoption of chemical solutions to insect problems: results were just successful enough to maintain a modicum of faith in insecticides, which fit squarely within the progressive political framework; early insecticide-purity laws promoted the legitimacy of the compounds; government agencies actively advocated their use; and World War I made them central to national security and public health. The larger history of environmental control further broadened and legitimated the chemical path. Americans had been transforming their landscape since colonial times to grow crops for distant markets; this manipulation invariably involved extensive farming and monocultural specialization; and it was a way of farming that, when it ran into trouble, looked to improved chemical insecticides rather than turning to older cultural or biological methods to manage predacious pests. As a result, the paradox was, like a mounting debt, effectively ignored rather than directly faced. Insecticides paid the interest, and everyone went on as they used to go on. Given the cost of the alternatives—agricultural reform, decentralized insect control, and path reversal—it was worth a little residue.

Sometimes the most important events in history are nonevents. Nothing happens; nobody acts; life continues. On the whole, as the concerns of some scientists and officials about the safety of crops coated with lead and arsenic became increasingly evident, Americans in general quietly stuck to the well-worn and comfortable path of chemical dependence. They chose to optimistically embrace Wilson's dictum to stick with an appealing custom rather than ponder Carlson's more pessimistic plea to question the "effidence" behind it. The result was a habit of reacting to a rash of insecticide scares by downplaying their threat and searching for new and better chemicals. "The proper way to deal with the health hazard like lead," wrote P. J. Hanzlik, a pharmacologist at Stanford University, in reference to lead arsenates, was "to develop new insecticides of an entirely different kind."[6] American scientists would succumb to this path dependency and embrace this logic until, years later, it dead-ended into a wall of poison known as DDT.

"an insidious menace to the health of the people"

Chemical insecticides evoked concerns about their safety from the moment they appeared on the market. A doctor in Boston reacted in 1888 to the widespread use of Paris green with the assertion that "we consider this unsafe, as there is no intimation of the fact that Paris green is a compound of arsenic and copper, and a deadly poison." The entomologist C. H. Fernald recalled in 1877 that "reports of cases of poisoning . . . were startling in the extreme," with many consumers wondering if "with each meal it would be necessary to take an antidote for the poison." A writer for *Country Gentlemen* mentioned in 1891 that he was constantly "making calculations of the chance of poisoning." Also in 1891, the magazine *Garden and Forest* calculated "hundreds of tons of a most virulent mineral poison in the hands of hundreds of thousands of people," and declared the situation "a danger worthy of serious thought." Robert Clark Kedzie, a chemist in Michigan, found that arsenic in fruit "might produce slow poisoning." Examples such as these demonstrate that, from a relatively early date, farmers were at least intermittently reminded that they could be poisoned by the insecticides they were applying to their crops.[7]

Early efforts to draw attention to the dangers of arsenic enjoyed the added backing of a vocal minority of doctors. In 1872, a crusading group of toxicologists led by Frank Winthrop Draper condemned "the evil effects of the use of arsenic" and read carefully the work of John Ayrton Paris and his journal *Pharmacologia*. As early as 1822, Paris had written that "the pernicious influence of arsenical fumes" created an atmosphere in which "horses and cows commonly lose their hooves, and the latter are often to be seen in the neighboring pastures crawling on their knees and not infrequently suffering from cancerous infection in their rumps." By 1888, James J. Putnam, a neurologist in Boston, had diagnosed several patients with chronic arsenic poisoning and documented their rapid decline into paralysis and, in some cases, death. The dermatologist Jonathan Hutchinson, in the 1880s, drew a suggestive connection between arsenic and skin cancer. While there was no scientific consensus in the nineteenth century on the question of the hazards of arsenic, the possibility of agreement in the future did not seem entirely out of the question.[8]

The general public, for its part, was not especially attuned to these fragmented scientific discussions. For all the medical concern, writes historian James Whorton, "the potential dangers of an arsenical environment were clearly not a matter of universal alarm." Indeed, as often as arsenic and lead set off alarms, the blaring always managed to become background noise, never continuing loudly or long enough for the people to seek redress. As late as 1920, for example, the *New York Times* could report that "recently four persons were poisoned by eating pickles" that had been served from second-hand barrels containing arsenic—and nothing would be done. A case of likely arsenic poisoning in New Jersey in early 1925 was blithely dismissed by public officials as intestinal flu. In both cases, there was no public outcry. When it came to concerns about insecticides, the reticence of the people allowed the government to remain dormant. Many observers, especially those who opposed arsenicals from the outset, understood the nature of this reality. They knew full well that the only way the federal and state governments would mandate limits on the amount of arsenic and lead that could remain on produce would be if a significant number of consumers gathered around the issue. If consumers expressed fear, then the fruit industry's markets would be endangered and officials would take action.[9]

This finally happened in October 1925, when the Hampstead Board of Health in England linked two instances of acute poisoning to American apples that had been sprayed with arsenic. The Ministry of Health began doling out stiff fines to English merchants who sold American produce. The incident blanketed the newspapers. Summarizing several articles from London newspapers, the *New York Times* reported that the precise problem with American apples was that farmers sprayed not only the blossoms but also the fruit in order to protect trees from the codling moth. The *Times* of London wrote that the apples contained "¹⁄₁₀ of a grain per pound" of arsenic and noted that this amount was "especially large." The press, relishing the hysteria, reported that efforts to wash the produce were evidently nonexistent or, if carried out, ineffective. Trade relations immediately chilled. In January 1926, the British government fired another salvo at American fruit, warning that "quantities of dangerously contaminated apples are still on sale." The same month, health officials in Wales confiscated apples in order to "investigate the possibility of arsenic poisoning owing to the spray used by growers." By April, the press had dubbed the incident the "poison apple scare," with the *New York Times* reporting that "Covent Garden commission merchants have ceased handling apples from the United States for the time being." This was all the federal government needed to hear. By February 1927, with the threat of an international boycott looming, the Department of Agriculture finally established an arsenic-tolerance level of .025 grain per pound, an overly generous level according to critics, but a federal restriction nonetheless.[10]

Although this restriction would prove to be relatively innocuous, it gave reformers a berth. The more critical outcome of this international-trade dustup was intensified media attention to arsenic's true character—attention that nineteenth-century crusaders, for all their passion, were never able to muster. A national press corps that had been reticent to emphasize this possibly controversial topic started to address the health risks associated with arsenic with greater sensationalism after 1925. The *New York Times* alone demonstrated a newfound dedication to this deadly substance. A well-placed report in 1928, for example, informed readers that "analysis of the contents of a cow's stomach . . . found that the animal had died of arsenic poisoning." The cow had eaten arsenic that a railroad company had sprayed near train tracks to kill weeds—hardly a big news item, but there it was on the front page. The same year, the paper reported that a three-year-old boy "playing near some rose bushes" had inhaled arsenic and died at his family's summer house. The following year brought news that a Canadian family "was saved from illness and possibly death" when postal officials stopped an arsenic-laced fruitcake from being delivered to the family's home. The Virginia woman who baked the fruitcake had "accidentally mixed [a bag of flour] with an arsenic insecticide which was stored on the premises." As a media matter at least, arsenic was becoming popular copy in the wake of the "poison apple scare." If only over coffee and doughnuts, the American people were devouring that copy and, in the process, developing a healthy fear of arsenic's toxicity.

Consumer-protection groups rallied around these reports, as well as the widely publicized trade debacle, to further stoke public awareness of arsenical dangers. Few advocates for public health were as blunt about arsenic as the secretary of Consumers' Research, the indefatigable Arthur Kallet. In "a profit economy," he railed, "the jeopardizing of a dozen lives to save a few pennies is entirely normal." Having highlighted corporate indifference to public health, he then launched into a litany of horrors inherent in the fruit and vegetable industry. "In order to protect fruits and vegetables from the ravages of insects," he wrote in 1934, "increasing quantities of lead arsenate and other insecticides, poisonous to man as well as to insects, are being used." Producers could remove residual arsenic with a "chemical wash" made with hydrochloric acid, but they usually relied on rain to wash away the residue, which, of course, it rarely did. Even if the growers had been willing to use industrial washing machines to clean fruit, the machines were prohibitively expensive and difficult to maintain. As a result, "a dozen different fruits and vegetables reach the consumer's table bearing residues, mainly of arsenic and lead, in amounts sufficient to constitute a serious hazard." In case consumers missed the implications of these remarks, Kallet added, "Arsenic has been known to cause cancer, and the possibility that arsenic residues are contributing to the rising cancer death rate certainly cannot be ignored." As for lead: "the effects of lead, even where acute poisoning does not occur, are frequently serious, sometimes leading to obscure nervous ailments difficult of diagnosis." The bottom line on arsenical insecticides, as Kallet saw it, was that "there have been many deaths from the eating of produce carrying heavy residues of poisonous insecticides."[11]

Conventional scientists who, after the "poison apple scare," were inching closer to a consensus on the dangers posed by insecticides, joined Kallet in becoming unusually strident in their denunciations of lead and arsenic. P. J. Hanzlik lambasted insecticides as "an insidious menace to the health of the people." Lead, he continued, is a cumulative poison, "a fact which has been repeatedly corroborated by the most rigid investigations." Hanzlik routinely returned to the theme of scientific authority, explaining, for example, that while "some people, not experts, have claimed these things to be trivial," "most practicing physicians" recognize that "many people residing in localities where insecticidal sprays are used show symptoms which can be ascribed to lead or arsenic or both." These symptoms included weight loss, joint pain, anemia, paralysis, and constipation. Even fruit that had been carefully washed posed health hazards, as "there is a certain unremovable residue of lead arsenate which remains on the skin, around the stem, and in fractures and inaccessible parts." It was more than theoretically possible, Hanzlik claimed, to eat several carelessly cleaned apples in one sitting and suffer "acute poisoning." Average consumers need not be medical specialists to appreciate the risks associated with these toxins. All they had to do, Hanzlik insisted, was observe the world around them and note that "turkeys allowed to run in sprayed apple orchards . . . developed lead poisoning," that "horses nibbling on alfalfa or other vegetation grown between the sprayed trees" died within three or four years, or that "drift carried to pastures during aero-plane-spraying of adjoining ranches has resulted in repeated payments of indemnities for poisoned live stock."[12]

A final notable effort to raise public awareness about arsenic and lead in food came from Ruth deForest Lamb, the chief education officer at the Department of Agriculture and the author of a book appropriately titled *American Chamber of Horrors*. The matter of "spotty coating" was to Lamb "a most deadly serious problem." Her book effectively described how growers dodged and downplayed state regulations on residue levels, doing so with implicit disdain for the common good. Shippers, she wrote, "took little stock in the government's contention that the residue was dangerous." Industry routinely created smokescreens around studies that posited arsenic and lead poisoning, taking advantage of "how very difficult it is to diagnose poisoning from lead and arsenic." Lamb provided her readers with a compendium of local headlines from all over the country about the hazards of insecticide residue on produce, reminding them that the problem was national in scope. She also portrayed the farmers themselves as normal human beings—which is to say, flawed. "Growers are much like other people: some of them can be shown [how to apply spray], while others will never learn; some are painstaking in anything they attempt, others sloppy or reckless." With the International Apple Association practicing "unadulterated, if misbranded, obstructionism," any legislation "proposed in the interests of consumers" was destined for defeat. "The statutory authority of everyone," she wrote, "is hamstrung in some way."[13]

Given this onslaught of criticism between 1925 and 1940, it would have required a strong set of blinders not to have been at least somewhat aware of the threats that insecticides, especially those that contained arsenic and lead, posed to public health. Newspaper reports, books, scientific studies, consumer-interest articles, and everyday conversations combined to initiate an inchoate public-health campaign not unlike contemporary efforts to reduce smoking, halt global warming, and fight obesity. Informed American consumers would have known by the 1920s that arsenic can be absorbed through the skin, is transferable through breast milk, was embossed in wallpaper ("walls of death," one crusader dubbed papered walls), is an old-fashioned poison used to murder people, and is linked to cancer.

As is often the case, however, the impact of these scientific claims was systematically, and often brilliantly, thwarted by advocates of insecticides. Although the lines of contention are not always easy to disentangle, it was generally the case that fruit producers, industry lobbyists, and even entomologists defended insecticides as the victims of misinformation, hysterical overreaction, shady science, and naïve idealism. Through well-honed rhetorical strategies, they ultimately succeeded in preventing the passage of effective anti-insecticide legislation because they clung tenaciously to a narrow, profit-driven notion of progress while aiming to protect their professional reputations, co-opting government agencies that were responsible for agricultural productivity, and working diligently to keep the general public as indifferent as possible to these potential harms that were increasingly hidden in plain sight.

Path dependency thrived on short-term economic success within an agricultural model of intense environmental transformation. It demanded that farmers "go on as you used to go on," never reevaluating the past for missed opportunities, never contemplating reversal, and always believing that the future would have the answers to present problems. And, ultimately, it depended on a populace that, even when moderately informed, felt little need to actively question the axioms of modern agrarian capitalism. Despite a full decade of constant chatter about the dangers of arsenic and lead, Hanzlik still felt justified in remarking that "the public is not awake to the real situation." And for this, he believed, there would be consequences.

"good will is a sensitive plant"

As objections to the use of insecticides became more strident, a well-oiled publicity machine lurched into motion to counteract the post–"poison apple scare" denunciations of chemical dependence. Advocates of chemical defense had been sharpening their public-relations strategies since the nineteenth century, when they had learned that the most effective approach to defending their product was to exploit the smallest hint of uncertainty in order to undermine the entirety of the opposition's case. In 1877, a writer in the *American Agriculturalist* admitted that "it is true a few individuals have claimed that they were made sick by eating sprayed fruit," and then in the same sentence insisted that these claims "were absolutely without foundation."[14] This simple response masked a powerful tactic. Supporters of insecticide use well understood that consistently finding a direct link between arsenic- and lead-based insecticides and human disease was nearly impossible and that tangential evidence, however compelling, would not hold up in the court of public opinion. Thus when such early critics as Frank Winthrop Draper, Robert Clark Kedzie, James Putnam, and Jonathan Hutchinson—not to mention Benjamin Walsh and Charles Riley—questioned the safety and effectiveness of insecticides, advocates stridently demanded a burden of proof they knew to be unattainable. Early defenders of insecticides ingeniously realized that they could get away with this line of defense because history was on their side. Insecticides, after all, had arrived with minimal opposition, had achieved a strong foothold in the agricultural community, had won scientific and government backing, and had become standardized before serious opposition developed. If the early advocates of chemical solutions to insect infestations had learned anything about protecting their turf in these early years, it was that the naysayers could raise all the objections they wanted, but, barring replicable and incontrovertible scientific evidence, they would be unable to marshal the will to initiate reform.

Advocates also recognized the strategic value of fear. Medical professionals in the late nineteenth century did not necessarily enjoy the high status they do today. They were often as intent as entomologists to establish their well-deserved place in the more prestigious ranks of "hard science." Well aware that many men of science were especially sensitive to charges of quackery, the pro-insecticide camp was quick to use mockery as a means of intimidation. Charges of hysteria must have gone a long way toward keeping many medical professionals quiet for fear of being labeled as irrational cranks. When, for example, doctors in Boston began to question the safety of arsenicals, defenders of insecticides began to speak of the "Boston fad," insisting that the medical experts at Harvard answer the facetious question of whether "arsenical poisoning . . . occurs only in Boston." Beyond highlighting the supposed flakiness of this "fad" in the halls of an elite institution, supporters of insecticides were quick to label their opponents as fickle scientists prone to bursts of irrationality, not a reputation that medical men wanted to cultivate. As the editors of the *Medical Record* wrote, "Having passed safely through disquieting exacerbations of homeopathy, mind-cure, Ibsenism, and psychical research, the profession has been brought up sharply with an attack of arseniophobia." "In fact," the piece continued, "there seems to be an appalling possibility that Massachusetts is being systematically poisoned by an inoculable, malignantly infective, and extremely prevalent form of arsenical poisoning." James Whorton, who has thoroughly explored this rhetoric, argues that their "uncertainty about the degree of danger from sprayed arsenic" prevented doctors from acting in a unified fashion. It seems likely that this uncertainty was fostered as much by genuine fear as by evidentiary doubt.[15]

Having both established an impossible burden of proof and branded their opponents as loony cranks, the members of the insecticide lobby judiciously assumed a contrasting pose of common sense. They were the hard realists, the rock-ribbed pragmatists, and the reality that they confronted was undeniable: the United States was an agricultural powerhouse under siege by insect enemies that, if not met with the strongest possible weapons, would cause incalculable economic hardship. However one felt about insecticides, the fact remained that "no other poison than arsenic in some form is effective, or applicable on the large scale." Rather than deny the challenges that farmers faced from invasive insects, rather than attempt to craft defenses that differed radically by region or farm, proponents of insecticide use insisted that chemicals did not kill people, but people killed people. All farmers had to do was learn to properly use these poisonous compounds.

It was not an altogether specious argument. Indeed, a writer in the *American Agriculturalist* tapped this logic when he described arsenic as a "sharp axe" in that it "needs to be handled with care but should no more be tabooed on this account than the axe." As long as farmers were cautious about not swallowing dust when mixing or spraying, they would be safe. As long as farmers washed the fruit they sold, consumers would be safe. As long as farmers sprayed only twice a year and according to extension-agent calendars, the environment would be safe. This line of thought—which stressed individual responsibility over government intervention—held considerable sway. Furthermore, the advocates cited the opponents' own studies demonstrating that arsenic did not infiltrate the fruit, but only coated it. Even when washing was inadequate or incomplete, residual amounts of arsenic were, according to entomological reports they cited, generally harmless. Advocates well understood that anyone who was undecided about the benefits and risks of arsenic would presumably choose the logical side on hearing that, based on the residue found on one shipment of apples, "74 apples would convey a poisonous dose." Such extraordinary figures eased public fear, and it did so while further castigating the opponents of arsenic-based insecticides as extremists who were out of touch with the realities of agricultural life and the American economy.[16]

These lessons were not lost on the generation of insecticide advocates forced to carry the arsenical torch into the arena of public opinion. Perhaps the most conspicuous advantage that proponents of insecticide use enjoyed was a corporate environment that lived by the dictum that the business of America was business. Corporate pragmatism meant, on the one hand, that producers had no choice but to grudgingly accept the inevitable regulatory legislation. Corporate pragmatism also meant, on the other hand, that industry would work behind the scenes to minimize the legislation's negative impact on profits. The trick was to allow—and, even better, praise—the passage of federal, state, and municipal laws that ostensibly vowed to protect consumers from dangerous insecticide residues, and to do so while protecting corporate interests. Arthur Kallet fumed at the fruit industry's success in striking this balance: "To be sure, there are . . . food and drug laws to protect both the consumer's pocketbook and his health, [b]ut these laws are written and enforced so as to prevent 'unnecessary' business losses to the larger and more influential producers." His anger peaked when he claimed that the fruit industry and its pro-insecticide allies were waging "a battle to the death against laws requiring grading and informative labeling." That was not quite the case. These post–World War I pragmatists were not radical libertarians cut in the mold of the gun-toting Llewellyn Banks. Rather than greet federal regulation with the threat of gunplay, they did something more effective: they embraced a legacy of "common sense," resigned themselves to the eventual enactment of regulatory laws, and put their considerable resources into ensuring that the legislation was as toothless as possible.[17]

Agribusiness enacted this strategy with cold efficiency. On the state level, corporations allowed, and even encouraged, laws to pass while quietly gutting them of enforcement and punitive powers. They correctly predicted that the public would applaud the law without reading the fine print. In Florida, the state legislature passed a signal law in 1927 that prohibited the artificial maturation of citrus fruit with chemical sprays and fertilizers. Counter to what one might expect, the citrus industry secretly championed the legislation as a means to "impress the foreign consumer with the fact that Florida law guarantees the consumer's health against the dangers of arsenic residues in fruit." The growers, in other words, capitalized on the public-relations benefits that the law automatically conferred while downplaying the underlying reality that, as the *Yale Law Journal* put it, "the larger purpose of the law is the resolution of an economic conflict." Although few industry lobbyists were actually going to come out and say it, they knew that the resolution was taking place "behind the veil of consumer protection." Using an arsenal of carefully crafted legal weapons, the citrus lobby hobbled the law with caveats (such as "reasonable use" exceptions) that led the *Yale Law Journal* to lament that "the teeth of the law have been extracted." The Florida law further typified the producers' machinations in its implicit recognition that acceptable legislation had to place the interests of the industry ahead those of the consumer. "The arsenical spray law problem," the article concluded, "is but one evidence of the inadequacies of the legislative protection of the consumer." This inadequacy was ultimately due to a single fact: "the legislation is dictated by producer interests."[18]

The fruit industry successfully applied these state-level techniques to the federal context. The passage of the Food, Drug, and Cosmetic Act of 1938 officially granted to the Food and Drug Administration the power to determine residue-tolerance levels. While the bill was an ostensible victory for the anti-insecticide crusaders, it underwent the same vivisection that so many state laws had endured at the hands of corporate lawyers and lobbyists. David F. Cavers, a law professor at Duke University and an adviser to the Department of Agriculture who had worked to shape a legitimate bill, described his five-year effort to get the bill passed as "a campaign of attrition." The fruit industry's goal in Washington was much the same as it had been in Florida. "The tactic of the opposition," Cavers wrote, was predictable: "Some bill would have to be enacted, and the problem was to restrict the measure narrowly enough to avoid the risk of embarrassing changes in merchandising and industrial practices while at the same time establishing in the public mind the belief that an acceptable law had been adopted." Once again, industry understood that "consumer good will" was necessary to business success while also realizing that "good will is a sensitive plant." Thus growers and their allies pragmatically accepted the reality that they were going to have to swallow a major federal law. After making this concession, however, "that form of cooperative effort known as logrolling" went into effect. When the bill's architects invited industry representatives to offer input in April 1933, lobbyists came to the table without concrete provisions, dragged their feet at every turn, and behaved in a way tersely described by Cavers as "unsatisfactory." After the bill graduated to the Senate Commerce Committee in 1934, Royal Copeland of New York, the chairman of the committee and a physician wedded to the fruit lobby, admitted to never having read it before he shunted it into a subcommittee that opened it to "a swelling tide of protest" from industry lobbyists. After the bill finally limped into law, Cavers conceded, "a more ambitious undertaking would not have been politically practicable." His ultimate feeling about the law was one of frustration more than accomplishment.[19]

The Food, Drug, and Cosmetic Act of 1938 was not a completely toothless piece of legislation—it could, after all, mandate acceptable levels of residual insecticides on produce for the nation as a whole. But it was denied at least one critical power. In 1937, Representative Clarence Cannon of Missouri became chairman of the Subcommittee on Agriculture of the Appropriations Committee. Cannon was also a toady to the large producers and had gone on record as saying that "lead arsenate on apples never harmed a man, woman or child," despite overwhelming evidence to the contrary. When the bill landed in his committee, Cannon proceeded to deny its supporters one of their most important tools: the power to conduct proper scientific investigations. "No part of the funds appropriated by this act," his committee concluded, "shall be used for laboratory investigations to determine the possible harmful effects on human beings of spray insecticides on fruits and vegetables." That job would go to the Public Health Service.

Once responsibility for testing insecticides was transferred from the Food and Drug Administration to the Public Health Service—a stipulation that the fruit industry had lobbied hard for—the growers had their victory. One example indicates why. By 1937, almost all medical experts agreed that chemical residues did indeed have the potential to cause a major public-health hazard. Every significant medical group, including the American Medical Association, endorsed this assessment. Only the federal government, however, had the resources necessary to undertake studies comprehensive enough to provide evidence to support the lowering of arsenic and lead levels in insecticides. The Food and Drug Administration, for its part, had already made clear what it would do given the chance—undertake an experimental study to evaluate the impact of the *accumulation* of insecticide residues in the body. Its concern was quite properly with the effects of *chronic,* not acute, poisoning. The distinction was critical. Scientists had theorized since the nineteenth century that lead and arsenic were corrosives that did their work slowly. "The consumer's concern," wrote Kallet, "is not that he will be keeled over by the arsenic residue on a single apple or head of lettuce, but that he will store up in his body sufficient amounts of arsenic and lead from the residues remaining on apples, pears, lettuce, cabbage, celery, broccoli, and many other fruits and vegetables." In an address to the Pennsylvania Horticultural Society, W. R. M. Wharton, a scientist in the Food and Drug Administration, reiterated the scientific assessment that "lead and arsenic are cumulative poisons, that is, they store up in the system, and untoward results may not be noticed until there has been an accumulation of a sufficient quantity to cause some serious manifestation." Another official in the agency pointed out that "traces of poison continuously consumed may manifest results only after a period of years." In order to determine the nature of this slow rate of poisoning, any serious study would have to be done with animals, over a long period of time, and would have to test for chronic poisoning. This is exactly what the Food and Drug Administration was intending to do when Cannon's committee pulled the plug.[20]

The Public Health Service had no intention of undertaking such a study. To the contrary, it quickly designed and executed a three-year evaluation of the residents of the heavily sprayed river valley near Wenatchee, Washington. It was never clear precisely how the fruit lobby held sway within the Public Health Service. What is clear, however, is that its fingerprints were all over this investigation. The Wenatchee plan was, yet again, a finely calibrated strategy that worked strongly in industry's favor. On the surface, the study would effectively ease public concerns about the safety of insecticide residues, lending to the government an image of proper public vigilance. The Public Health Service, after all, was investigating the toxic impact of lead arsenate on 1,231 people living in an area that annually absorbed 4.5 million pounds of the chemical. But the evaluation purposefully overlooked what scientists were most concerned about—the long-term accumulation of poison. Through clinical and laboratory (not experimental) examinations, the Public Health Service hoped to discover the patterns, if any, of poor health that resulted from the consumption of lead arsenate spray. The precise tests the investigators ran were geared to find evidence for acute rather than chronic poisoning, even though acute poisoning was, according to historian Thomas Dunlap, "the very thing that no one claimed was a serious problem." Not surprisingly, a mere seven people tested positive for having suffered from an intake of lead arsenate. The scientists, however, could not find among them a single pattern that suggested a connection between poor health and lead arsenate. In the end, the fruit industry, in the words of Ruth deForest Lamb, could now taunt the naysayers with a challenge: "Let the Government point to even one death or even one illness!"[21] For the casually concerned consumer, this rejoinder seemed convincing enough.

This brand of machination—Cannon's transferal of testing powers from the Food and Drug Administration to the Public Health Service—never made the headlines. It was the sort of insider bureaucratic maneuver, in fact, that only the most dedicated political newshound would dig up and read about. Which is exactly how corporations wanted it. In addition to gutting state and federal laws of their most potent powers, the fruit industry benefited from public indifference about the deleterious effects of insecticide residues. Granted, much of this indifference was due to the growers' successful whitewashing of the issue, and, granted, the public battle over lead and arsenic residues had reached a crescendo at the onset of the Depression. But still, as Cavers noted, the matter "was of consequence to the health and pocketbook of every citizen of the country" and, therefore, deserving a genuine national discussion. Even so, he continued, "the public at large, including persons ordinarily well-informed on national affairs, knew little or nothing of what was transpiring in Congress." The Food, Drug, and Cosmetic Act, as well as the internecine battles swarming around it, "never became the object of widespread public attention, much less of informed public interest." After the law passed, Cavers conceded, "I suspect that today only a small fraction of the public knows that a new law has been enacted."

The press was also partially to blame for the public's nonchalance. While American newspapers had vigorously promoted consumer rights before the passage of the Pure Food and Drug Act in 1906, they appear to have lost their muckraking nerve by the 1930s. Major newspapers were happy to pepper their copy with arsenic stories ranging from the scintillating to the bizarre, as they had begun to do after 1925. But an in-depth exposé of the procedural tactics that the fruit industry employed to protect profits at the expense of public health, in the midst of the Depression, seemed irrelevant. One imagines that such stories might not have made for the most engaging copy. Cavers, however, had a different notion: "[T]his policy [of not covering the politics behind the food laws] was due in no small degree to the fact that the measure was widely represented as menacing to advertising revenues."[22] With the approach to insect control as consolidated as it was, such a charge was quite plausible.

"wonderful!"

"It is doubtful," wrote Raymond D. Tousley, a business professor at the College of Washington, in 1941, "if any piece of recent legislation has been of greater significance to the general public than the Food, Drug, and Cosmetic Act."[23] Tousley's sanguine sentiment was widely shared. But by the time Congress passed the legislation, the terms of the insecticide debate had fundamentally changed in a way that overwhelmingly favored the advocates of chemical methods of pest control. Whereas a wider range of concerned parties had once argued over chemical versus biological and cultural approaches to the management of destructive insects, now the government and a handful of corporations argued over the percentage of these chemicals that consumers should be allowed to ingest. Conflict replaced cooperation, but the conflict revolved around a cohesive national approach to insect control. Biological and cultural options were by no means relegated to history's dustbin, but they were no longer driving the public debate. Although voiced less than a generation earlier, the objections to spraying that Harvey Wiley and Anton Carlson so passionately conveyed would have seemed quaint, if not ludicrous, to the politicians, scientists, and consumer groups framing the insecticide debate of the 1930s in a context that assumed that chemicals were the only viable weapons to use against insects.

The objections would have seemed quaint because not only had the terms of the debate shifted, but the advocates of insecticides again had taken the initiative to define the new terms. Figures such as James Wilson, Llewellyn Banks, Royal Copeland, and Clarence Cannon and the leaders of the International Apple Association, however shameful their lack of sympathy for the consumer, were highly ambitious, pragmatic, and intelligent men. They recognized that Banks had become a local hero after the Bureau of Chemistry had seized his insecticide-laden fruit. They recognized that William Frederick Boos, a prominent scientist, had grudgingly admitted how "very difficult it is to diagnose poisoning from lead and arsenic." They recognized David Cavers's lament that the general public's attention span was minimal. They recognized that few politicians were willing to buck the immediate economic interests of America's heartland in the name of something as amorphous as "public health." And they knew—in fact, they excelled at—how to exploit these recognitions to their great advantage. The gutting of the funding for the Food and Drug Administration to evaluate the risks posed by residual insecticides, the pockmarking of state laws with industry-friendly loopholes, the perpetuation of dangerous chemicals through improper testing, and the portrayal of advocates of insecticides as level-headed realists protecting American values were only a few of the strategies employed to convince the general public that all was safe in the state of American agriculture.

But there was something even more subtle behind the success of the champions of chemicals. While the terms of the debate may have changed, the overall strategy of simplification remained the same. The opponents of insecticides, dispersed as they were, never appreciated the power of keeping things simple. They worked within a complex matrix of considerations, much as their forebears had done when juggling the pros and cons of biological, cultural, and chemical control of insects. Their calculations incorporated numerous variables, such as the nature of poisoning, public health, consumer rights, ecological impact, corporate profitability, environmental stewardship, local contexts, and nuanced scientific and agricultural disparities. Their approach was generally one of asking questions, lots of questions, and never resting easy with the answers. They raised doubts, assumed that chemicals were guilty until proved innocent, and called for additional studies to explore even the slightest safety concerns. Like the skeptics of the nineteenth century, who frequently settled on different strategies of pest control for different times and places, the pessimists of the early twentieth century never sought a single solution to the nation's myriad insect problems. They pined for a centralized authority that would oversee and enforce hundreds, if not thousands, of site-specific remedies administered by field agents with local roots and knowledge of local conditions. They aimed to balance consumer safety with producer freedom, tipping the balance to the former but never neglecting the latter. Their tone tended toward that of Cassandra and Jeremiah, routinely warning hardworking Americans that their actions, however nobly conceived or historically justified, would ultimately be their undoing. It was not the prophesy that hardworking Americans wanted to hear. History does not reward messages of complexity and defeat coming from the bottom up.

The advocates of insecticides, backed by a consolidated bureaucracy, embraced optimism and simplicity. Aware that this was a battle that would eventually be determined in the court of public opinion, insecticide proponents studiously avoided nuance and kept the matter deceptively basic. They did so by carefully linking the right to use insecticides to the practice of successful agriculture, and thus to the nation's economic health as a whole. Their logic came from the top down and was streamlined and upbeat:

- Some chemical compounds, at least under certain circumstances, worked well to keep pests out of crops.
- Crops were essential to the nation's corporate and financial well-being.

- Science was murky at best on the question of insecticide dangers.
- Farmers had been using chemicals on their crops for generations.
- So use them; use them carefully, but use them, and the nation would benefit.

This equation—an equation born of the bureaucratic and corporate push for homogeneity—knowingly omitted the indeterminate variables. The fact that the United States had become a nation with hundreds of types of farms that required thousands of specific solutions to the threats they faced from insects was never considered. Instead, advocates of chemicals promoted the fiction that, when it came to insecticides, one size fit all. They touted the inherent goodness of science and technology, at least in the hands of American scientists, and were content to permit the use of questionable substances before submitting them for proper investigation, thus assuming that chemicals were innocent until proved guilty. When real concerns did arise, such as suggestions of risks to public health from the widespread use of insecticides, proponents stonewalled under the guise of pragmatism, insisting that every bushel of produce be tested individually (a logistical impossibility) or—after the passage of the Food, Drug, and Cosmetic Act—ensuring that the proper tests never be conducted. With a general public otherwise preoccupied, promoters of insecticides were able to complete their goals with such bold assurance that most Americans failed to recognize that options other than chemicals were available. In a nation that was now more urban than rural, and in which farmers had traded on-the-ground ecological knowledge for chemical dependence, few people were even aware that pests could be managed without resorting to insecticides. The path of dependency, in short, had triumphed, clearing the road of debris.

Several telling hints that this road would once again be strewn with obstacles, however, came on June 25, 1938, the very day on which the Food, Drug, and Cosmetic Act passed into law. As legislators celebrated this supposed victory for public health and safety, and fruit growers breathed a collective sigh of relief because they knew that things could have been much worse, a number of state entomologists were reporting to the Department of Agriculture bits of information that would prove to become routine news. C.J. Drake, the state entomologist of Iowa, had this to say about the woolly apple aphid: "Heavy infestations were reported at Oskaloosa . . . just east of south-central Iowa." Writing from New York, S. W. Harmen noted that the "emergence [of the codling moth] in western New York was hastened by hot weather, resulting in first-brood worms being injurious in heavily infested orchards." Reporting from Kansas, H. R. Bryson said of the grasshopper that "heavy infestations are spotted." L. A. Stearns, in Delaware, wrote that the Japanese beetle was "noted in abundance on heads of mature wheat near Smyrna." From Indiana, G. E. Gould told of "a severe outbreak [of chinch bugs] in southwestern Indiana." In Kansas, R. H. Painter wrote of the Hessian fly that "eggs were laid under ideal conditions which resulted in almost 100-percent hatch." A full overview of the reports from entomologists across the country on this single day would make it clear that every part of the United States, despite habitual and heavy spraying, remained under attack by insects. For all the arsenical insecticides that farmers were employing, it was like the nineteenth century all over again.

A generation earlier, entomologists might have joined farmers and editors in going back to the drawing board. But this was a post–Leland Howard world, and times had changed. In light of the reports of infestations, hundreds of which poured into the Department of Agriculture every month, and in light of the approval of chemical insecticides by entomologists, it would not take long for the insecticide industry to conclude on its own that arsenic and lead—as well as the scores of other dangerous chemicals that farmers were applying to their crops—had to be supplemented. Fortunately for all these interests, when a chemist praised a new insecticide called DDT as "wonderful!" they would know just how to usher it past the American people, into the hearts and minds of entomologists, and—as many of the initial backers of DDT would live to learn—directly into the nation's soil, air, and water.

CHAPTER 7

"complaints are coming in":
A YEAR IN THE LIFE OF AN INSECTICIDE NATION, 1938

June 25, 1938, was not the only day that year on which entomologists filed frustrating accounts of insect infestations with the Bureau of Entomology and Plant Quarantine. No fewer than 126 agents representing all 48 states sent an estimated 6,000 such reports from March to November. The agents toured their territories on a monthly basis to survey the status of insect infestations and send their findings to Washington, D.C., where the bureau concentrated these dispatches into the monthly *Insect Pest Survey Bulletin*.[1] The eighteenth volume, published by the Department of Agriculture in 1938, was issued at the end of a year that was not especially unusual in terms of its reports of infestations. What is unusual, however, is how close the bulletins bring us into the insect battle as it was being waged not in corporate board rooms and legislative chambers, but in the fields, where, as we shall see, insects were evading a barrage of insecticide solutions with disturbing regularity and pushing scientists to devise even more powerful and homogenized weapons.

"the gigantic task of saving the human race"

Farmers in the United States endured $1,601,527,000 worth of insect damage in 1938. Summarizing the devastation from that year, S. W. Frost, an entomologist at Pennsylvania State University, wrote, "It has been said that it costs the farmer more to feed his insects than it does to educate his children." Indeed, he continued, "the yearly loss by insects is equal to two times the annual loss by fire, or about two times the capital invested in manufactured farm machinery, or about three times the value of all fruit orchards, vineyards, and small fruit farms in the country." That farmers were spraying more than 8 million pounds of arsenicals and 4 million pounds of Paris green on their crops, as well as millions of pounds of other insecticides, only intensified the impact and irony of Frost's remarks.[2]

The broader context of this practice was critical. By 1938, agriculture in the United States had undergone a major transformation that proved highly supportive of insecticide use. Beginning in the 1920s, American farming began to embrace the spirit of industrialization that was coming to dominate American manufacturing. Paved roads, tractors, bank credit, immigrant labor, and formal commodity markets were just a few of the factors that encouraged agricultural industrialization. Business leaders and agricultural experts overwhelmingly pushed farmers to embrace "the efficacy of rational management" inherent in assembly-line production. The concerted move toward mechanization and standardization certainly paid off for farmers who were able to industrialize without going broke. But it also further divorced farmers from the ecosystems they once knew intimately while overextending many small farmers to the brink of bankruptcy. As the Depression gripped the nation, it was generally the factory farmers who could afford to remain standing, and—given the imperative that they meet yearly quotas—they were in no position to experiment with alternative ways to fight pests. Much as in the 1870s, when Paris green and London purple had appeared on the market just as farmers were mechanizing production, "the twin forces of science and technology" in the 1920s and early 1930s, as well as new economic pressure, predisposed farmers who were streamlining production along factory lines to embrace the promised benefits of insecticides. The insecticides, in turn, were coming to be regarded as routine "inputs" into an industrialized venture, one step in the process of Fordist production.[3]

It was in this context that field agents paid their visits to farms and wrote their reports. On reaching the Bureau of Entomology and Plant Quarantine, the dispatches entered a maze of divisions and subdivisions. Between 1920 and 1935, the bureau had fragmented into tiny shards of expertise. No fewer than nineteen divisions—including the Insecticide Division and the Division of Insecticide Investigations—broke down into fifty-three subdivisions that employed several hundred entomologists, chemists, horticulturalists, arborists, botanists, and engineers. An analysis of these offices reveals a bureau that had organized itself around insecticidal solutions to insect problems. The bureau lacked a single division, subdivision, or scientist dedicated to biological management, including parasitism and predation, and cultural control. By contrast, it dedicated fully staffed offices to cover such concerns as organic insecticides, inorganic insecticides, oil emulsions, fumigants, spray residues, toxicological impacts, and equipment for applying insecticides. When young entomologists, toxicologists, and plant pathologists contemplated their futures, they realized that they were more likely to find employment in this well-funded agency as government bureaucrats than at universities as research professors. If there was a polar opposite to the "chaos of experimentation" that had characterized farmers' fights against insect pests, the Bureau of Entomology and Plant Quarantine was it. Path dependency—in this case, the idea that agribusiness, the federal government, and the insecticide industry had chosen to fight insect infestations exclusively with chemicals—limited the way in which scientists and farmers framed the pest situation and contemplated their options. By 1938, the structure of the Department of Agriculture alone strongly influenced the national response to the problems that the current batch of insecticides was solving with problematic results at best.[4]

Further narrowing the insect-relief options was the chemical industry's emphasis on insecticide research, development, and production. More than thirty-five chemical-manufacturing companies had been established between 1913 and 1935, absorbing even more of the nation's young scientific talent while compensating for the decline in production in Germany after World War I. These new firms joined dozens of older manufacturers to produce millions of pounds of calcium arsenate, lead arsenate, sulfur, rotenone, pyrethrum, cryolite, and various copper compounds. Sales reached over $100 million annually, a figure simply unimaginable had other forms of control been popularly pursued. The companies now included Dow Chemical, Merck, Black Flag, Niagara Sprayer and Chemical Company, Graselli Chemicals, DuPont, Chadeloid Chemical Company, Eastman Kodak, Verona Chemical, and General Chemical Company. They diverted resources from other product divisions to develop and patent hundreds of insecticides in the hope of finding the next Paris green. The insecticide boom was such that a company like Atlas Powder, which since 1812 had manufactured gunpowder, could switch production to insecticides (and cosmetics). These corporations worked from the safe premise that when farmers became consumers of their products, profits grew exponentially. They could count on an upward trend because the first application of an insecticide tended to be the beginning of a gradually intensifying process. For example, apple growers who started with an annual application of lead arsenate were, in a few years, spraying their trees six or seven times a year. Simple as the point might be, there were no profits to be made if farmers systematically adopted mixed strategies of insect control that selectively incorporated chemical insecticides. The companies, in short, profited when farmers eschewed alternative approaches and the chemicals they relied on either failed or required increasingly frequent applications. This business model persisted in part because the manufacturers enjoyed the support of the federal government, state extension agencies, agribusiness, and the entomological profession as a whole.[5]

The emergence of a federal bureaucracy specifically dedicated to the needs of insecticide interests, in conjunction with an unprecedented corporate emphasis on insecticide research and development, highlight the increasingly passive position of growers. Having spent the previous fifty years gradually adopting the chemical insecticides that scientists, corporations, and the government encouraged them to use, farmers found themselves dependent on agribusiness methods that were by no means universally effective. One consequence of path dependency was that it severely undermined the farmers' traditional methods of responding to the ineffectiveness of chemical methods of pest control. In the nineteenth century, before their systematic use of the first chemical insecticides, farmers relied on their inherited flexibility and their "untutored ingenuity" to engage in the "chaos of experimentation" and grapple with infestations of insects in creative and cooperative ways. Influenced by an ecological sensibility that derived from their own experience in the field, they forged solutions informed by intimacy with the environment and tempered by humility. Both forms of power—acute ecological knowledge and unfettered freedom to experiment—had severely eroded over the years. By the late nineteenth century, the expansion and industrialization of agriculture, increasing addiction to monoculture, massive invasions of insects, demand for an all-inclusive solution, and willingness to see the bright side of insecticide experiments gradually drove American agriculture into chemical dependence. Farmers and ranchers may not have been kicking and screaming along the way, but they were carried along by the powerful wave of corporate power, government fiat, and desire for short-term capital maximization.

The expansion and industrialization of American agriculture, accompanied by the rejection of local experimentation, thus rendered American farmers—once proudly independent—dependent on forces over which they had little control. By 1938, farmers no longer spoke of *managing* insects. They were now running factories in the fields, and this responsibility dictated that they eliminate, rather than control, insects. A small sampling of their contributions to this effort can be glimpsed in the insecticides that hit the market in 1938: methyl bromide for fumigation, "bark penetrating sprays" for beetles, DuPont's "IN-930" spray for flies, and oil–insecticide treatment for corn. It was, in short, business as usual in the fields of America, and, unfortunately for farmers, profits that once belonged to them were padding the pockets of men and women whose fingernails remained clean as the farmers watched their crops fall to pests old and new alike. A close look at the *Insect Pest Survey Bulletin* brings the frustration of their experience, as well as the experience of national insect control on the eve of the introduction of DDT, into sharp focus.[6]

"*damage occurred despite extensive control measures*"

In 1938, farmers learned many little lessons about the insecticides they were spraying and dusting on their crops. The big lesson they were reminded of, however, was the most familiar: no matter how bent on extermination they may have been, a natural force as uncontrollable as the weather could be the ultimate arbiter of pest control. It was, on average, an unusually warm year in the United States. The nation's interior was especially hot, and winter everywhere remained mild, with October being particularly warm and dry. Rain poured in abundance, especially in the spring and summer, but periods of drought plagued the Dakotas and Minnesota at different times of the year. During the summer, the Rocky Mountains and the Great Plains, despite unusually high temperatures nationally, briefly cooled to record lows. This shifting regional patchwork of extreme weather conditions coincided very well with the precise needs of many insect species.

The case of grasshoppers offers the most direct example of the weather fostering insect infestation patterns that rendered insecticides effectively useless. The cool and wet month of May prevented hatching while permitting the growth of a profusion of range grasses. Grasshoppers hatched in June and, benefiting from the relatively large food supply, rapidly reached maturity as a warm front parched the land. Seeking relief, the grasshoppers promptly flew in hordes to Canada, the Red River Valley, and eastern Montana. Farmers who were not planning to spray until October were deluged with grasshoppers throughout these regions. R. A. Sheals, the field agent in Montana, confirmed the impact of this regrettable turn of events when, on July 30, he wrote, "Considerable crop loss occasioned during the last few weeks in eastern Montana by hordes of grasshoppers, which had migrated into agricultural areas from nearby breeding grounds in range lands and waste areas." Farmers rushed to respond, but, as Shears lamented, "flights were so heavy that damage occurred despite extensive control measures." Nature, in this case, remained well beyond the reach of the farmers' best weapons.[7]

The grasshoppers were not alone. The very insects that had inspired the earliest research into chemical insecticides—the Colorado potato beetle, chinch bug, San Jose scale, and Hessian fly—had also eluded the poisons intended for them. With the exception of the Rocky Mountain locust, these pests—all of which had destroyed crops on a large scale in the nineteenth century—devastated crops throughout 1938. In April, the Colorado potato beetle showed no signs of relenting. It blanketed the Norfolk, Virginia, area and thoroughly infested the Eastern Shore of the state. Reporting from North Carolina, agent W. A. Thomas wrote that "adults have been very abundant on potatoes around Chadbourn," while "heavy oviposition began the second week in April and the larvae are becoming numerous." In Mississippi, potato beetles were "plentiful in tomato plant beds," whereas in Louisiana they were "seriously injuring Irish potatoes" and in Washington were "abundant in the soil." By May, the insect pest was causing turmoil in New York and New Jersey, where "they are emerging in large numbers." In Kentucky, the beetles were "more abundant than usual"; in Tennessee, they were feeding on Irish potatoes; and in Missouri, they "began to appear with the development of early potatoes."

The West, the pest's original home, was not spared. From Idaho, B. F. Coon reported "one adult beetle found on a small volunteer potato plant 5 miles west of Buhl." While the specimen seemingly was innocuous, Coon noted that "this is believed to be the first report of this insect in south-central Idaho." By June, complaints were pouring in from the already infested states, but also from Connecticut, Delaware, Florida, Kansas, and Utah. Especially discouraging was that in Connecticut, Utah, and southwestern Idaho, the beetles had been scarce or absent for several years. By July, laments were heard from not only the usual agents, but also those reporting from Maine ("more numerous than they have been for several years"), Wisconsin ("after several years of comparative scarcity, reported in quite serious numbers throughout the state"), North Dakota ("moderately abundant at Fargo"), and Oregon ("defoliating potato plants at Alicel"). South Carolina and Nebraska were hit in August, and in September, Oklahoma joined the ranks of the infested states.[8]

The year 1938 was also excellent for the chinch bug. Sixty-four years earlier, a Milwaukee newspaper had slandered the chinch bug as a "national calamity." The description was still apt. In the winter, C. Benton, an agent in Indiana, filed an ominous report suggesting that the short chinch bug hiatus that Midwesterners had enjoyed was about to end: "Chinch bugs were observed sluggishly moving around in clumps . . . in grass and plant litter near Lafayette." In March, the pest could no longer be called sluggish, causing considerable damage in Illinois, Iowa, and Oklahoma. Again, most notable was the pest's resurgence after an encouraging spell of inactivity. "[T]he winter mortality," wrote C. J. Drake from Iowa, "has been considerably lower than during the last 4 or 5 years." As a result, "threatening populations of overwintering bugs" were prepared to menace the eastern half of the state. In April, an agent in Indiana observed that "these fields situated near bunch-grass areas are known to be rather heavily infested." Missouri reported trouble in the central portion of the state. May saw a vast proliferation. The outbreak remained particularly acute in Indiana, where "considerable mating and other activity has been observed on warm days." The concentrations of chinch bugs found would, according to Benton, "produce moderately severe infestations if weather conditions are favorable." Illinois was thankful for recent rain, which would "certainly lessen the damage," but Iowa reported that "chinch bugs occur in threatening numbers," comparing the situation with the "infestation in the spring of 1934." Agents in Kansas and Oklahoma complained of chinch bugs attacking spring barley. The congregation of the bugs persisted into June and July, despite the rain, with the heaviest concentrations in the Midwest. Iowa suffered "heavy local infestations"; the bug "continued to attract some attention" throughout Missouri; and Ohio, Michigan, and Texas were new members of the growing group of states that suffered invasions of chinch bugs in 1938.[9]

The San Jose scale, a pest that specializes in attacking fruit trees and had made its journey east as a stowaway on a shipment of pear trees to New Jersey in 1868, had by 1938 proceeded to infest the entire nation. It had even moved into the Deep South, a region once thought to be too hot for the scale. The year began on a somewhat dour note for farmers concerned about the scale population. As early as February, five states—Vermont, Illinois, South Carolina, Georgia, and Missouri—were reporting potential trouble. Especially discouraging was Georgia's claim that in one peach district, 85.2 percent of the overwintering scale larvae were alive. By March, matters had somewhat improved. T.L. Bissell was able to note that "a correspondent reports that he finds almost no San Jose scale" in the Hawkinsville region. The same month in Mississippi, the scale, while abundant, had caused "severe damage to [only] unsprayed fruit trees." The agent in Kentucky found that "the San Jose scale is very abundant on peach trees." Likewise, in South Carolina, "injury is being complained of by peach growers." Mississippi reported "heavy infestations," while Washington lamented that 95 percent of the wintering scales were alive on apple and pear trees. Arizona, an unusual state to host scales, filed a report of "a heavy infestation . . . in a large rose garden." By July, the situation had grown dire in Ohio, where "the outbreak was rather severe." Wisconsin and Georgia reported welcome reprieves, but Texas noted with some alarm that scales were in pecan trees. In October, Arizona was concerned about a "light infestation"—and that was the good news. Kentucky found the scales "unusually abundant this year"; Ohio had "large numbers of apples marked by scale"; Mississippi saw "considerable damage to peach trees"; Illinois worried that "late broods will aid greatly in bringing up the infestation"; Maryland reported "an unusual buildup of this scale"; and Georgia, after a short break, noted with horror that its peach orchards were "now encrusted with scale."[10]

The Hessian fly, an imported insect, had begun to torment wheat farmers in the 1770s. In 1938, it, too, was still a fact of farming life. The flies made their seasonal debut in March, when they infested 33 percent of the fields in the territory of an agent in Missouri. J.R. Horton, an agent in Kansas, reported heavy outbreaks in several counties, with up to 66 percent of larvae having pupated. The next month brought complaints from Pennsylvania, where light concentrations prevailed throughout most of the state, except in the east, where "heavily infested fields" were common. Horton lamented the fly's "very good ability to withstand unseasonably low temperature in the pupal stage." By May, the Hessian fly had made it onto the short list of major concerns of the Bureau of Entomology and Plant Quarantine. "The hessian fly," began this month's *Insect Pest Survey Bulletin*, "has been favored by the spring weather and populations have built up considerably." Indiana's crops seem to have borne the brunt of the bad news. Two agents collaborated on a report that noted "the unusual early development of the flies," adding that "in a number of fields the infestation is severe enough to reduce the yield materially." The Hessian fly was still on the bureau's short list in June, as it had "multiplied greatly," spreading to such a point that "it may be sufficient to menace the crop to be planted this fall." Six somewhat frantic reports, all from the Midwest, reiterated the fear of a difficult seeding season in the autumn. Dispatches were more encouraging from July through September, but any farmer lulled into a sense of complacency would have been alarmed by the reports sent to the bureau in October. Indiana, for example, admitted "moderate to severe infestation," while an agent in Kansas noted infestations in every field in his territory except three.[11] The Hessian fly, as was true for the other major insect pests that had tormented nineteenth-century farmers, was even more of a fact of agricultural life than it had been before the proliferation of chemical insecticides.

"The pest is definitely on the increase"

In 1938, not only did the nation's more established insect pests continue to exact significant agricultural damage, but farmers witnessed continued and often intensified depredations from recent imports. The United States had suffered crop losses wreaked by more than 100 imported insect species before 1900. But even with quarantine efforts in place, foreign insects continued to infest American crops. Several unwelcome immigrants from Asia—including the satin moth (introduced into Massachusetts in 1920), the Asiatic beetle (introduced into Connecticut in 1920), and, most notably, the Japanese beetle (introduced into New Jersey in 1916)—persisted in plaguing farmers throughout 1938.

Their ravages could be maddeningly indiscriminant. The Japanese beetle—a plump, 0.5-inch metallic-green insect, with copper-colored wing covers and twelve tufts of white hairs projecting from under the wing covers—was especially troublesome. While gradually adapting to cold winters, the pest—which got its start in the United States around Riverton, New Jersey—had spread to more than seventeen states by 1938. It emerged from the pupal stage slowly, but then attacked en masse, doing about $10 million a year in agricultural damage. In April, the Japanese beetle could only be found mildly infesting wild strawberries in Maryland, but by July it was causing alarm up and down the east coast. An agent in New England reported that "this pest has been much more abundant and destructive . . . than in previous years." Connecticut was "heavily infested in good-sized local areas," while Rhode Island noted that the "Japanese beetle increase [was] tremendous . . . in quite a number of new places." Delaware had a "marked increase in infestation," and agents in Virginia were aghast that "several hundred beetles were collected on smartweeds in a field of potatoes in less than 30 minutes." Especially frightful was the situation in Maryland, where the beetles were no longer simply feeding on a few strawberry patches. Instead of filing their own report, agents simply sent in a choice excerpt from the *Baltimore Sun*: "Corn, apples, soybeans, and other crops ruined by the beetle in Cecil County. Over 20,000,000 beetles were captured on two farms alone."[12]

Pests recently arrived from Europe were also very much at large in 1938. They included the cabbage worm, codling moth, and European corn borer. The corn borer, introduced into the United States around 1910, began to emerge in April, with New Jersey and Virginia filing especially concerned reports. A large number, ranging from 40 to 80 percent, of the pinkish larvae were surviving to the pupal stage into May. Tan moths about 0.5 inch long emerged in June, leading A. M. Vance to say of Ohio, Indiana, and Michigan that "pupation and emergence in these States is more advanced in the spring of 1938 than in any previous year on record." Throughout the rest of the summer, no fewer than six states filed reports claiming such conditions as "very serious damage to early sweet corn," "severe infestations," and "the pest is definitely on the increase." By the autumn, "late corn was heavily infested" in Connecticut; every corn plant was overrun with second-brood borers on Long Island; and in Adams County, Indiana, "every stalk was infested." The cabbage worm, which in larval form is 1 inch long, with a sharp yellow stripe down its back, supplemented the borer's efforts. It caused "severe injury" in Mississippi and "considerable injury" in Louisiana in April. By midsummer, adults were "flying and laying eggs in great abundance" in Maryland, while there was "an increase in cabbage worms at Columbia [Missouri]." The codling moth—perhaps the most conspicuous target of arsenical insecticides—remained an aggressive presence alongside its companions from the Old World. It debuted on the *Insect Pest Survey Bulletin*'s short list in April, with the ominous remark that "this insect seems to have passed the winter successfully over most of the country." By May, the short list had the moths "pupating generally over the country." The next month, they were again highlighted on the short list. The agent wrote that the pests were "emerging earlier than usual," only to be held in check by a couple of well-timed cold fronts in July.[13]

Imports from Mexico completed what amounted to a three-pronged attack from nonindigenous insect invaders. The harlequin bug, sugarcane borer, potato tuber worm, and, of course, boll weevil fed relentlessly on American crops throughout 1938. In the third week of March, a heavy infestation of harlequin bugs—sap-sucking, red-and-black stink bugs that infest crucifers throughout the United States—were observed in North Carolina. "By the last of the week," wrote W. A. Thomas, "most of the [mustard and cabbage] seed stalks were dead or dying." By April, the bugs had emerged in fields ranging from Virginia to Oklahoma, having overwintered under forest debris. An agent in Mississippi mentioned "noticeable damage to turnips," while mustard greens fell prey to the bug throughout Tennessee. By late spring, harlequin bugs were destroying gardens and cabbage crops in the Deep South, continuing their work through the summer and even into the early autumn, when warmer weather brought them out in Ohio, Indiana, and Kentucky. In Louisiana, the sugarcane borer began to pupate in early March. On March 2, 13 percent of the larvae in Bayou Teche had pupated; on March 15, that figure had exploded to 50 percent. By May, infestations were scattered across the state, with the borers even partaking in "February-planted corn in some fields"; by midsummer, "spotted locations" were "very heavily infested." The potato tuber worm—a slender moth with yellow-brown wings—showed its ability to adapt and migrate as well, moving from Mexico to Maryland, where, in July, it "was found attacking potatoes . . . on the Eastern Shore," while in Florida it was consuming tobacco. By October, it had risen to the ranks of the short list, cited as infesting North Carolina potatoes to California tomatoes. The boll weevil, for its part, renewed its attacks after a "downward trend" that had begun in 1933. Not only did the usual Cotton Belt areas suffer from "the larger number of weevils entering hibernation," but farms on the eastern seaboard experienced weevil damage that was "greater than it had been for a number of years."[14]

"a real tussle with spring cankerworms"

In addition to the persistence of established insect pests and the resurgence of recently imported species, the Bureau of Entomology and Plant Quarantine kept close tabs on insects never before reported in the United States. It chose to highlight, for example, "a single male moth" that had emerged from a larva "boring in the bark of white pine at Hartsdale, N.Y." Another report of what appeared to be a moth of the same species had come from Cross River, New York, in 1934, and the bureau now suspected the worst—that it was "well-established in Westchester County." A lone leafhopper elicited concern because it was "the first individual of this species recorded from the United States." When specimens of a strange conifer sawfly were discovered under pine straw in Lamington, New Jersey, the bureau reported that "this apparently represents the first identification of the species from the United States." Farmers on the west coast could not have been pleased with the news that "a European leafhopper . . . not previously reported from America was collected in abundance in the Pacific Northwest" or when the curator of the Boston Society of Natural History reported his discovery of a "relatively uncommon European species" of blowfly.

Equally worrisome for farmers was the appearance of a particular insect in and its adaptation to a radically new geographical location. "A weevil previously recorded in only two or three localities in the eastern states" created a stir when agents found it "in abundance on the porches of houses" in Portland, Oregon. An infestation of Colorado potato beetles in Utah caused consternation because "these beetles have been scarce on potatoes [in the infected area] to date." Likewise, when the Oriental rat flea, which typically confined itself to relatively warm urban centers, showed up in Ames, Iowa, and St. Paul, Minnesota—and then Urbana, Illinois, and Youngstown, Ohio—the Bureau of Entomology and Plant Quarantine contemplated the public-health consequences. The flea had to be taken seriously because it spreads bubonic plague. Concerns could become so serious that an innocuous report—say, of a longhorn beetle "having emerged from the woodwork of a living room in Camden. N.J."—would cause considerable fear of an impending infestation.[15]

Another revealing aspect of these reports is their frequent references to a particular pest having reached an extreme level of infestation after a long period of quiescence. In Arizona, C. C. Deonier interviewed a rancher from Gadsden who observed that "this is the worst year for screwworms" he had ever experienced. In Texas, the stable fly became "quite abundant . . . more so than during the same season in other years." Savannah, Georgia, suffered the worst outbreak of sand flies in five years. The cloverleaf weevil was "appearing in much more than normal numbers" and "ragging clover and alfalfa in many fields." The European elm scale, wrote an agent in Indiana, "has been gradually spreading and increasing its destructiveness for several years." Considering nymphs of the cotton fleahopper in Texas, two agents noted, "This is the earliest we have ever found adults in eastern Texas." After "5 or more years of comparative scarcity," wrote an agent in Kansas, "boxelder bugs are more abundant than usual." With the vegetable weevil being "more destructive in Alabama, Mississippi, and Louisiana than usual," with Mississippi experiencing "the heaviest infestation of variegated cutworm in several years," and with cutworms in Tennessee being "much more severe than the heavy infestation of 1937," it was clear to entomologists that they were not fighting a war of attrition. Instead, they were engaged in a battle to which the enemy came unpredictably and, due to factors beyond human control, in waves. A field agent in Missouri must have understood as much when he wrote of the cankerworm: "At Columbia this week I have seen more male spring cankerworm moths than I have ever seen in any similar length of time in the 30 years I have been connected to the department." Undaunted, he added that "apparently we are to have a real tussle with spring cankerworms in central Missouri this year."[16]

So much about the relationship between farmers and insects remained the same. Not unlike the earliest settlers, Americans in 1938 were often left holding up their hands in frustration while insect pests destroyed not only their crops, but also their animals, homes, and material possessions. In February, a rancher in Arizona whose sheep were infested with the sheep-biting louse watched in desperation as his "band of sheep . . . rubbed off most of their wool along the fence." Another Arizona rancher noted that the sheep ticks numbered "1 per square inch" on his flock. Many reports tended toward the seemingly irrelevant. Saw-toothed grain beetles invaded packages of cupcakes in Utah. Pine flooring throughout Peterboro, New Hampshire, succumbed to the furniture beetle. "Boxelder bug annoyance in homes and schoolhouses" was common in Missouri. An agent in Logan, Utah, claimed that "ants were causing annoyance to workers, and are invading food-products storage rooms." In Nebraska, reports of "damage or annoyance in houses by ants" came in from several counties, along with the observation that they preferred to "mine extensively in structural timbers." The clover mite was proving to be "a real household nuisance in food, dishes, and bedding in the Denver area." One poor agent, while researching an infestation of dog ticks, ended the day picking the bloodsuckers off his own body. While minor, these inconveniences highlight the point that any hope of extermination on the farm could not possibly apply to other areas of life, including the workplace and the home. Eradication, at least with the current batch of chemicals, was a long shot.[17]

"owing to the reduction by natural enemies"

Given the popularity of insecticides and the accompanying rhetoric of extermination, one would have expected them to have made more of a dent in the insect population. Specific references to insecticides in reports collected in the *Insect Pest Survey Bulletin* are extremely rare; when agents did mention them, it appears that the insecticides, as they had in tests conducted in agricultural experiment stations in the late nineteenth century, were meeting with mixed results. On the one hand, a number of examples suggest that spraying helped very little. In March, a massive and unexpected outbreak of plum curculios in Georgia led an agent to mention that, in several counties, "growers have been spraying for more than a week," obviously with few positive results. In April, the agent could write only that "it is probable that Elberta peaches in Georgia will be subjected to a second brood of the curculio this year," thus reiterating the bad news.

And by May, he was reporting an infestation rate of an astounding 42 percent. None of this news suggests the effectiveness of insecticides. When purple mites started to emerge in Florida "considerably earlier than usual," spraying commenced. But the agent had to report nonetheless that "the infestation is heavy over much of the state." Although "some poisoning" was being done around Candelaria, Texas, the cotton leafworm remained unaffected, "causing some damage to young cotton." Such juxtapositions between instances of intense spraying and agricultural decline were commonplace throughout the reports.

There were, however, a few mentions of success with insecticides. In Missouri, "a heavy carry over of larvae" was expected, but only in "unsprayed orchards." In Texas, goat lice were abundant, but only on "undipped goats." With regard to the San Jose scale, an agent in Mississippi again made an important distinction when he reported "severe damage to *unsprayed* fruit trees." Of course, these bits of tangential evidence for the effectiveness of insecticides hardly hold up against the larger reality that thousands of reports of insect depredations were coming into the Bureau of Entomology and Plant Quarantine at a time when farmers were spraying and pouring more insecticides on their crops than they ever had. Nevertheless, the chemicals do seem to have performed well enough in some circumstances for many farmers to accept the traditional glass-half-full interpretation of their effectiveness.[18]

Interestingly, these lukewarm endorsements of insecticides paled in comparison with the overwhelmingly positive evaluations that field agents had been making fifteen years earlier. From the perspective of 1938, the earlier assessments suggest that these chemicals were in fact becoming less effective over time. In April 1924, the bureau, summarizing the reports of several agents, wrote that "the application of 500,000 pounds of paradichlorobenzene in the Georgia peach belt last fall appears to have been highly successful in controlling the peach borer." Agents called the results "uniformly good" and noted that "growers are greatly pleased with the control." Such an overtly optimistic portrayal is conspicuously absent in the records from 1938. Concerning the grape mealybug, an agent in Michigan wrote in 1924 that "preliminary experiments using nicotine, strong lime-sulphur, and Sunoco spraying oil were made" in April. The next month, he reported the good news that the lime-sulfur had "killed practically all that were hit." In Ohio, T. H. Parks reported that "calcium-arsenate and gypsum dust mixture is being used successfully in controlling [striped cucumber beetles]." Control of the grasshopper in Texas was directly linked to the extension entomologist R. R. Reppert, who "has done a large amount of work in demonstrating to groups of county agents the mixing and distribution of poisoned bran mash." As a result of his efforts, "white arsenic has been bought by the carload." Potato beetles in Texas were subjected to heavy spraying in early April, and, as the agent reported, "where spraying had not been done the first brood of potato beetle larvae is doing serious damage"—an unequivocal endorsement. Agents sent in hundreds of similarly optimistic assessments of sprays and dips in 1924, suggesting that by 1938 the insecticides that once worked effectively had, over the intervening years, diminished in power.[19]

Not only did agents in 1938 fail to routinely report positively on the efficacy of insecticides, but they expressed optimism about the very methods that the Department of Agriculture was not systematically exploring, methods that had fallen out of favor—biological control, including reliance on parasites and predators of insect pests, and cultural control. When agents recovered a parasite of the alfalfa weevil in every field examined in Jackson County, Oregon, they were hopeful that "it may develop into an important factor for alfalfa weevil control." Louisiana may have been "very heavily infested" with the sugarcane borer, but agents were thrilled to find that "68 percent of the eggs on cane and corn are parasitized [by tricogramma parasites]." The hemlock looper in Idaho underwent a "marked decline in the severity of this season's infestation" because a "large percentage of overwintering eggs are parasitized." A "fungus disease" killed "from 50 to over 75 percent of the aphids" in parts of Virginia, a development that agents wanted to pursue. Predatory control received similarly praiseworthy feedback. "Around Orlando," wrote the Florida agent H. Spencer, "the Chinese ladybeetle (*Leis dimidiatus* F.) is appearing in larger numbers than in previous years and is the most important controlling factor [of the green citrus aphid]." Agents in Louisiana were pleased that the turnip aphid was becoming less abundant around Baton Rouge, "owing to the reduction by natural enemies." Thus while the chemical industry and the Bureau of Entomology and Plant Quarantine primarily promoted the effectiveness of chemical insecticides, agents on the ground, as well as farmers themselves, remained aware of methods of biological control that had been overshadowed by the grand promise of insecticides.

Agents demonstrated an equally keen attention to clean farming and cultural control of insect pests. Despite regular pleas by extension agents for farmers to avoid sloppiness, an agent in Vermont reported larvae of the European corn borer to be "moderately abundant in stubble and waste stalks." The implication was that clean farming, as agriculturalists had long argued, reduces pest populations. Dwight Isely, author of the popular textbook *Methods of Insect Control*, ranked clean farming as the ideal method of controlling damaging insects. "Prevention of insect damage," he wrote, "involves making the environment for a given species as unfavorable to it as possible." This condition could be achieved by "crop management and the use of various cultural methods." Due to unkempt farms, according to the agent, "cutworms were abundant in rubbish in and around fields." In a thorough report on the Hessian fly, the Bureau of Entomology and Plant Quarantine scolded growers: "In general, where the safe-seeding dates were adhered to, light or no infestation were recorded."[20] All of this was information that farmers had known for centuries, and thus it was also telling testimony that knowledge on the ground could be strikingly timeless, as well as at odds with the federal and corporate response to insect infestations, a response that reinforced the role of farmers as homogeneous consumers of chemicals rather than flexible stewards of their fields.

Another important theme running throughout these reports, as indicated, was a sober reverence for the power of weather to determine the level and timing of an insect infestation. In its comprehensive weevil report, the bureau had nothing to say about insecticides but everything to say about "the effects of climate on the weevil." In essence, it drew a direct correlation between weather and insect destruction, noting that "the amount of damage in any locality depends on whether spring weather favors or hinders weevil development." More than the weevil was directly susceptible to changes in the weather. The chinch bug population took a temporary dive in June because "the cold wet weather last month has been unfavorable for chinch bug development." One reason that intense spraying for the plum curculio in Georgia failed was that heavy rains made "conditions in peach orchards . . . favorable for the matured larvae to enter the soil." Spraying proved ineffective when wet soil "facilitated the entrance of larvae into the soil and the construction of their soil cells." Spray schedules in New Jersey had to be scrapped when "very warm weather" in April caused the European corn borer to pupate two weeks earlier than it did under normal conditions. A "mass emergence" of strawberry weevils in North Carolina was deemed a direct consequence of "the sudden advent of spring," a climatic quirk that the agent predicted would lead to "a sudden infestation . . . much earlier than usual."[21] Here, again, we can see reminders that farmers who sprayed crops could not possibly have felt as confident in their chemicals as did the bureaucrats, scientists, and manufacturers whose ecological sensibility, such as it was, had been formed in offices, laboratories, and factories.

In 1938, finally, a few scientists were beginning to make empirical headway on the vague question of why insecticides were not performing as well as they perhaps once did. Henry Joseph Quayle, an entomologist working at a citrus experiment station in California, published a study that year documenting the resistance of certain scale insects to hydrocyanic acid. Although the precise genetic mechanisms of resistance remained in doubt, the article went so far as to label the California red scales that were not responding to insecticides as "the resistant strain." Quayle was not the first scientist to propose and test this potentially devastating hypothesis. In 1908, A. L. Melander noticed that lime-sulfur washes that had been moderately successful in controlling the San Jose scale on certain kinds of apples had eventually lost effectiveness. "Can insects become resistant to sprays?" he asked in 1914. Honest as the question was, it was not an especially popular conjecture in the midst of World War I. For the next two decades, entomological researchers dedicated their work to insecticidal efficacy rather than insecticidal failure. When scientists like Quayle switched research gears in the 1930s, however, it quickly became evident that Melander's question was more than just a provocative one. In 1941, Harry S. Smith placed the theme of resistance at the center of his presidential address to the American Association of Economic Entomologists. "Will continued, intense pest control operations result in the development of insect races which become harder and harder to control?" he asked. The question was purely rhetorical. Entomologists knew the answer. "The fact of the existence of [a] resistant race is," Smith claimed, "no longer in doubt."[22]

"like a cheetah moving at full speed"

Placed in the larger context of the insect wars, this view from the agricultural trenches in 1938 takes us to the busy intersection where factory farmers, insecticides, and insects converged. Farmers and field agents, for their part, demonstrated minimal overt opposition to the use of chemical insecticides, preferring as much as anyone a one-shot solution to the problem of infestations. Since the 1870s, these substances seem to have worked just well enough under the circumstances for American farmers to continue their spraying regimes without considerable concern or protest. The expansion and commercialization of agriculture during and after World War I, accompanied by a "farm crisis" in the 1920s that weeded out smaller and potentially more flexible operations, rendered noninsecticidal techniques especially inefficient and unprofitable for the nation's larger farms. The equation that agribusiness adhered to was based on the premise that it was cheaper to cover crops with manufactured insecticides that would accommodate expansion than to employ alternative techniques that, while more ecologically responsible, would ultimately prove less profitable by inhibiting expansion. Growers thus sprayed. Every year, in fact, they sprayed more aggressively than they had the previous year, assuming it was either a necessary evil that was worthy of the risk or a positive good that could only get better. Path dependency, described by one writer as "like a cheetah moving at full speed," thus determined the fate of insect control as powerfully as it ever had in the year 1938, and it did so as crops succumbed as they always had to invasive insects.

As the *Insect Pest Survey Bulletin*s also attest, however, the familiar kinks in the insecticidal armor remained. Farmers, despite their support of chemical solutions to insect infestations, remained acutely aware of their inadequacies. Alternative approaches to pest control—while more impractical, less understood, and minimally supported—continued to show signs of promise. And, finally, the inability of chemical insecticides to overcome the unpredictable size and timing of insect outbreaks in response to weather conditions tempered popular optimism for them. Over the years, habitual spraying eroded the subtle ecological sensibility that earlier farmers had tapped during their decades-long "chaos of experimentation." But, as the agents' repeated references to parasitical, fungal, and cultural control strongly suggest, it did not eliminate it altogether. No matter how distant farmers removed themselves from the ecological subtleties of the farm, they still observed and sometimes even swayed to their rhythms. Considered from the sole perspective of the *Insect Pest Survey Bulletin*s, the building blocks, if not the underlying motivation, for a flexible system of integrated pest management were manifestly available. These blocks may have been scattered, worn, and hidden, but they were intact, waiting to be assembled.

The structural opposition to doing so, however, remained daunting. A specific example of a potential noninsecticidal approach to pest control that showed considerable promise but was not seriously pursued involved the Japanese beetle. The search for insecticides that would effectively control the beetle was turning out to be a difficult one. The Department of Agriculture tested a lime-aluminum sulfate spray, pyrethrum, and lead arsenate, all of which proved to be unreliable. However, the discovery in 1921 of a bacterium that attacks the Japanese beetle and, in 1929, of a nematode that parasitizes it, seemed to provide possible answers. The bacterium, called milky spore, attacks the beetle in the larval stage, causing it to assume a milky appearance and die before ovipositing. Farmers who were finding milky-white grubs were also finding that the Japanese beetles were doing much less damage in the fields where the grubs appeared. It was, in many respects, a typical observation that farmers traditionally would have explored in the pages of their farm journals. But now they were beholden to the government and, by extension, the chemical industry. Although it was established right away that the beetles were experiencing a "classical example of microbial control," it was not until 1939 that the Bureau of Entomology and Plant Quarantine, the agency on which farmers so directly depended, produced 109 tons of spore powder and sent it to nine agricultural experiment stations for testing. The results were encouraging. A Maryland golf course treated with the manufactured spores saw a dramatic decline in Japanese beetle grubs: from between twenty and sixty a square foot to one to three a square foot. Matters further improved when scientists discovered a nematode that was killing Japanese beetles on a New Jersey golf course. Laboratory tests demonstrated a 94 percent success rate of these parasitic worms killing beetles, a performance that led a scientist at the Rockefeller Institute for Medical Research to conclude that "*Neoplactana* appears to be ideally adapted as a parasite for the Japanese beetle." Neither the government nor agribusiness, focused on chemicals, took notice of this alternative method to control Japanese beetles. In fact, Dow Chemical and Shell, in 1945, introduced soil fumigants designed to wipe out nematodes, failing to distinguish between harmful and beneficial species.[23]

In such ways, a more hopeful vision of a more ecologically grounded system of pest control was dashed by the force of path dependency and the power of bureaucracy. Indeed, considered from the context of path dependency, any emerging awareness that past and present methods of biological and cultural control deserved a closer look quickly became irrelevant. In 1939, sales of pesticides in the United States exceeded $40 million as research scientists, capitalizing on the weaknesses being documented by the *Insect Pest Survey Bulletins*, sought better and more effective chemical fixes. Since the late 1910s, the considerable research and loyal faith in chemical insecticides had come with the unstinting support of several federal agencies—including the Chemical Warfare Service, Department of Agriculture, and United States Army—as well as increasingly powerful and generally unregulated corporations. The nature of this support was such that alternative methods of pest control never received a fair trial. As a result, popular agricultural perceptions came to idolize insecticides as the best and most practical killer of the thousands of insect species that damaged hundreds of crops. For example, when Dwight Isely mentioned the many virtues of cultural control in his textbook, he followed his compliments with the remark that "although often very beneficial, [cultural methods] are seldom striking and do not appeal to the popular imagination." Insecticides, by contrast, sometimes produced results that "may be striking" and "hence appeal to the popular imagination."[24] It was a point that held considerable sway with the general public.

The prospects for alternative methods reached rock bottom the following year. In 1939, Paul H. Müller made an unexpected discovery that would soon refocus the scramble for better insecticides. Müller was a Swiss chemist who worked for J. R. Geigy Aktien Gesellschaft, a chemical manufacturer based in Basel. His particular area of focus was the synthesis of dyes. One of his research projects in 1939 was to find a moth-proofing solution that could be worked into woolens as they were being dyed, an assignment that led him to experiment widely with carbon, chlorine, and hydrogen. One afternoon, he synthesized a talc-like substance and placed it in a container with flies. The brood showed no response, and a frustrated Müller went home, resigned to starting from scratch in the morning.

In the morning, however, he found the flies dead. The compound evidently had worked at its own pace. He repeated the experiment—more dead flies. For days, Müller continued to add flies without even adding more dichloro-diphenyl-trichloroethane, as the talc was called, and he continued to sterilize the box with soap and water. The flies kept dying. Müller suddenly forgot about woolen dyes and reported to his bosses that he may have inadvertently produced a new insecticide—a potentially effective one. Nine years later, in 1948, Paul Müller—a one-time laboratory foot soldier making chemical dyes—was awarded the Nobel Prize in Physiology or Medicine. Every American farmer who had endured what he had endured in 1938 surely would have applauded this well-deserved accolade. After all, by the late 1940s, they had become addicted to the "wonderful" derivatives of this chemical agent known affectionately around the world as DDT.

"never . . . had a chemical been discovered that offers such promise"

What happened to Paul Müller's discovery over the next two decades is, in one sense, a tale of triumph. DDT saved millions of lives worldwide. It protected much of the world's population from malaria, especially in Indonesia and Sri Lanka. It saved thousands of more lives from typhus during World War II. It protected otherwise vulnerable troops against malaria and yellow fever. DDT was mixed with kerosene and sprayed across the breeding grounds of *Anopheles* throughout the Western world, with wildly successful results. According to a report published by the American Chemical Society, DDT "reduced the number of malaria cases in India from an estimated 75 million to 100,000 in only twelve years." DDT was more than an apparent public-health panacea. It also proved to be an agricultural one. It successfully exterminated potato beetles throughout Switzerland and, in the United States, soon replaced arsenates as the main insecticide used by farmers, with tremendous results on everything from grapes to vegetable crops. Needless to say, the federal government, chemical manufacturers, and the world's wealthiest foundations and health organizations put their economic and moral weight behind this chemical. These forces set in motion the same apparatus that had pushed arsenicals to the forefront of agricultural practice. "The world's insect problem," writes Frank Graham, "seemed to be solved." Sievert A. Rohwer of the Department of Agriculture agreed, adding, "We feel that never in the history of entomology had a chemical been discovered that offers such promise to mankind for relief from his insect problems as DDT."[25]

Whereas the government and the insecticide industry had dithered for decades over nonchemical approaches to insect control, they investigated and produced DDT at warp-like speed. The government, well aware that it may have found the ultimate one-shot solution, wasted no time in contacting Geigy for access to this supposed miracle insecticide, and on October 16, 1942, the Division of Insecticide Investigations received its first shipment of DDT. The Department of Agriculture initially intended to evaluate it for military applications. Scientists quickly affirmed its effectiveness as a louse deterrent and mosquito larvicide before going on to promote it as a control for flies, bedbug exterminator, and treatment for scabies. By 1943, less than a year after testing began, the division officially recommended DDT to the armed services as a safe chemical to use on soldiers to keep them pest-free. At a laboratory in Orlando, Florida, the Division of Insects Affecting Man and Animals immediately began to synthesize the first batches of DDT to be channeled to the armed forces. At this point, the precedent of Leland O. Howard's insinuation of the Bureau of Entomology into the effort to win World War I paid off. In 1943, the Bureau of Entomology and Plant Quarantine worked seamlessly with the Army Air Forces to adapt chemical-warfare tanks to disperse DDT. By placing such new wine in old bottles, the Air Forces were able to cover "thousands of acres" of the malaria-infested Pacific theater with liquid DDT dropped from combat airplanes. Just as the atomic bomb promised "total victory" over the Axis powers, so DDT fed the fantasy that Americans could likewise annihilate the insect empire. It was certainly with this martial connection in mind that Edward O. Essig, in his presidential address to the American Association of Economic Entomologists in 1945, remarked, "There should be no doubt in our minds as to the great value of entomology as a science essential to the winning of the war."[26]

Indeed, Essig and the entomological community were by no means prepared to stop exploring and exploiting the potential of DDT. As had happened after World War I, just because World War II ended did not mean that the war on insects came to a halt as well. Indeed, it intensified. "In view of the great opportunities that appear to await us," Essig continued in his presidential address, "we as individuals and as organizations should adopt an all out insect control program!" The exclamation point was as well placed as it was well timed. Essig knew very well, in fact, that R. C. Roark and N. E. McAdoo of the Division of Insecticide Investigations had spent the previous two years exploring the potential of DDT for use in American agriculture. Somewhat unexpectedly, however, Roark and McAdoo urged caution, hoping to put the brakes on the headlong rush to embrace DDT as an all-inclusive insecticide. "At present," they wrote in December 1944, "the Bureau of Entomology and Plant Quarantine does not recommend DDT insecticides for the control of any species of insects injurious to agricultural crops." Despite its wartime heroics, "the insecticidal action of DDT is not universal," the authors duly warned. It should have been a red flag, but neither Essig nor the entomological community was going to be cowed by such caution. "For the first time in history," explains John Perkins, "entomologists envisioned the possibility of controlling or eradicating malaria, typhus, yellow fever, filariasis, denge, and other diseases transmitted by insects." Plus, Roark and McAdoo were making mere recommendations, which the market could easily ignore.[27]

And the market did just that. In August 1945, the War Production Board authorized private manufacturers to market DDT. "The war against winged pests," it declared in making this grant, "was underway." As the American Society of Chemists recalled, "The first samples of DDT reached the U.S. in 1942, and U.S. industry quickly made it the cheapest insecticide available." Between 1939 and 1954, the number of companies that produced insecticides and fungicides grew from 83 to 273. They manufactured not only DDT, but also a variety of organic insecticides, including benzene hexachloride (BHC), chlordane, dieldrin, aldrin, parathion, and toxaphene. Magazines such as *Business Week*, *Fortune*, and *Barron's* anointed chemicals as "the premier industry of the U.S." The war and the chemical industry's role in it created a new "big five"—DuPont, Union Carbide, Dow, Allied, and Monsanto—and these corporations relied heavily on the production of insecticides for their booming profits. With industrialists like George W. Merck, of Merck and Company, serving on the Biological Warfare Committee, and George Perkins, the president of Merck, serving as a colonel in the Chemical Warfare Service, there was little doubt what the big five would be producing. Thus it mattered very little what Roark and McAdoo thought about DDT and its related insecticides.[28]

The Special Committee on DDT of the Department of Agriculture ensured that the federal government and the entomological profession supported the chemical industry in bringing the wonders of DDT and its counterparts into the heart of American agricultural production. The "promise" of DDT, it explained, "covers three chief fields: public health, household comfort, and agriculture." As it elaborated, "Agriculture includes not only farms, gardens and orchards, but forests, livestock and poultry." DDT and BHC, in particular, had virtually replaced arsenic- and lead-based insecticides by the late 1940s. In a sense, the entire effort of controlling pests, which had come to center on insecticides in general, now came to orient itself specifically around DDT. Not only did the production of arsenic- and lead-based insecticides diminish, but so did the already minimal research into cultural and biological methods of control. According to an overview of research articles published in the *Journal of Economic Entomology*, studies of biological control appeared at half the rate they had before 1939. What emerged was the attitude, expressed by one entomologist, that "it is better to live without than to learn to live with destructive insects." And what legitimated that attitude was a behavior—fostered by the fact that less than 10 percent of Americans were farmers, that they were out of touch with nature, and that they were getting crucial agricultural advice from pesticide salesmen. That behavior was, irrespective of the particular conditions on a particular farm, to spray.[29]

The hysteria surrounding the promise of DDT, as well as the corporate response that capitalized on it, might easily lead one to forget that the DDT panacea was also something of a mirage. Problems arose—serious and familiar problems. Insects began to show an alarming resistance to DDT. Concerned scientists began to draw links between DDT and birth defects. Periodic reports of DDT killing wildlife other than insects began to pepper the popular press. In 1945, DDT was found capable of directly poisoning humans. Tentative reports of chronic toxicity as well as claims that DDT had been found in breast milk began to circulate. As in the 1920s, the public remained generally unmoved throughout the 1940s and 1950s, and the conventional forces of obfuscation worked to keep it that way. As with earlier concerns about arsenic- and lead-based insecticides, the power of path dependency pushed emerging concerns about DDT and its related organophosphates aside and focused on agribusiness benefits and corporate profits.

This time, however, the insecticide lobby encountered a woman who wrote a book. It was a book driven by a deep understanding of nature. "I am," Rachel Carson wrote, "so much interested in the things I write about that there is no real dividing line between my work and my recreation."[30] It was a book, moreover, that inspired a movement that gave the cheetah pause. With the publication of *Silent Spring*, the growth of modern environmentalism, and the example of Carson, a powerful minority of previously slumbering Americans came to live by a new creed: "Ecology has become the thing." The insect wars would be cast in a new, environmentalist light, and, if for only a brief historical moment, Americans seemed poised to confront directly the insect paradox, which they had been avoiding for almost two centuries.

CHAPTER 8

"Let's put our heads together and
start a new country up":
SILENT SPRINGS AND LOUD PROTESTS

Few books published in the United States have enjoyed the influence of *Silent Spring*. Rachel Carson's attack on DDT and related insecticidal compounds had an impact that has been compared with that of Thomas Paine's *Common Sense* and Harriet Beecher Stowe's *Uncle Tom's Cabin*. First serialized in the *New Yorker* in 1962, Carson's book has sold millions of copies, has been released in numerous editions (the most current one with an introduction by Al Gore), and sparked the modern environmental movement in the United States. No matter how one feels about the viewpoint of *Silent Spring*, it ranks as one of the most important books ever published by an American writer.[1]

When it came to the insect wars, Carson's work had a particularly transformative impact. It removed the topic of insect control from the track it had been on since 1870, while creating the genuine possibility that a diversified and ecologically responsible system of insect management might someday come to fruition. Her book, and the developments surrounding it, allows us to enter the twenty-first century cautiously optimistic that the path dependency that strapped blinders on insect control and sent it galloping past potential methods of biological and cultural control might eventually hit a dead end.

"little tranquilizing pills of half truth"

On the surface, there was nothing especially novel about what Rachel Carson was doing. Research scientists who questioned the legitimacy of insecticides were common figures by 1962. Critics ranging from Benjamin D. Walsh and Charles V. Riley to Arthur Kallet and Ruth deForest Lamb had been howling in the wilderness since the nineteenth century about the dangers posed by chemical insecticides. Nor was there anything particularly unusual in the opposition's attempt to undermine *Silent Spring*. Corporate and government scientists who jumped to their products' defense and castigated naysayers as dreamy idealists were also familiar figures with a long heritage. Advocates ranging from Leland O. Howard and Walter C. O'Kane to researchers for the International Apple Association and DuPont backed insecticides as unsafe for pests but safe for humans. Finally, there was nothing new about the political protection afforded the chemicals that *Silent Spring* condemned. When it came to the federal government promoting a product that appeared to do wonders for the nation's agricultural and economic health, several political entities—including the Bureau of Entomology and Plant Quarantine, the Public Health Service, and members of Congress with chemical-manufacturing plants in their jurisdictions—were seasoned veterans. In short, the dialectic of points and counterpoints that had characterized the insect wars since the advent of chemical insecticides followed a familiar beat between Paul Müller's discovery and Rachel Carson's exposé of DDT. All of which bring us to the pertinent question: How did *Silent Spring* break the barrier of public consciousness, awake incipient environmentalists, and permanently change the nature of the insect wars?[2]

There is no simple answer to this question. Nevertheless, it seems safe to assume that *Silent Spring*, much like *Common Sense* and *Uncle Tom's Cabin*, tapped an emotion deeply embedded in the American psyche, a belief ineradicable and genuine, but also repressed and neglected by the culture's most conspicuous representatives. For Thomas Paine, it was the underlying and long-violated sense that freedom and tyranny could no longer coexist. For Harriet Beecher Stowe, it was the underlying and long-violated "self-evident truth" that—as the Declaration of Independence insisted—"all men are created equal." And for Rachel Carson?

RACHEL CARSON

Again, it is hard to say. Carson's eloquent and well-researched prose reveals a number of obvious qualities—but none of them altogether new. With muckraking vigor, for example, she removed the veil from the "elixirs of death" by peppering her text with real stories: "A tank truck driver was preparing a cotton defoliant by mixing diesel oil with pentachlorophenol. As he was drawing the concentrated chemical out of a drum, the spigot accidentally toppled back. He reached in with his bare hand to regain the spigot. Although he washed immediately, he became acutely ill and died the next day." She also employed the common rhetorical strategy of shocking her readers with figures that reached stratospheric multiples. She wrote, for example, that "some 7,000,000 pounds of parathion are now applied to the fields and orchards of the United States," contextualizing her figure with a choice reference: "The amount used on California farms alone could, according to one medical journal, 'provide a lethal dose for 5 to 10 times the whole world's population.'" There was also the occasional indulgence, the well-crafted quotable gem designed to remind readers of the topic's profundity. In perhaps the most notable case, Carson wrote, "The question is whether any civilization can wage relentless war on life without destroying itself, and without losing the right to become civilized." All these techniques opened many eyes to the myriad problems caused by DDT and chlorinated hydrocarbons. Even collectively, however, they did not empower *Silent Spring* to achieve the seemingly impossible task of rousing Americans from their "mesmerized state."[3]

What ultimately did that was Carson's appeal to a concept that goes even deeper to the core of human existence than political freedom or racial equality—ecological balance. Carson's signal accomplishment, an accomplishment that had eluded her predecessors, was to contextualize her alarming facts, figures, and rhetoric in a more fundamental reality, one that insists that humans are not masters of the universe but integral elements of an ecosystem. It was a profound and humbling corrective, and if readers were willing to accept it, their worldview was bound to change radically. Environmentalists of preceding generations, such as John Muir, had written brilliantly about the American wilderness, but they had done so as though humans were distant observers of a beautiful but vanishing landscape that they were compelled to conserve, honor, and nurture back to health. Ecologists of the preceding generation, such as Barrington Moore, had conceptualized intricate systems of plant and animal life, but they too had done so with humans as a distinct entity, outside participants in disembodied processes.[4]

Carson denied to humanity the luxury of such distance. No matter how "civilized" we had become, no matter how ensconced in the shell of technology, we are, she insisted, fundamentally inseparable from the natural world around us. By thrusting human beings to the vital center of an ecosystem, she imposed on us more than a responsibility to treat the ecological community with respect. Instead, she saddled us with "an inconvenient truth"—that is, when we meddle with the environment, when we saturate the soil and douse the water with "elixirs of death," we are doing nothing short of committing suicide. To deny the consequences of this environmental interconnectivity was, as Carson turned the phrase, to be hooked on "little tranquilizing pills of half truth."[5]

No scientist had ever framed the matter in such starkly emotional and ecological terms. What remains most surprising about Carson's success, however, is that her bold reorientation did not so much assume the special sanctity of human life, as her predecessors had implicitly done, as place humans in the realm of the earthworm, sage grouse, springtail, and bacterial and fungal subcultures. To reduce humanity to such an animalistic level, to bring us down from an evolutionary pedestal as a necessary precondition for us to directly see the potential harm we were causing ourselves, was an enormous risk. Not only did it violate a deep tradition of environmental control dating back to the first setters, but it threatened to alienate readers who were, perhaps understandably, none too thrilled to have their lives so abruptly resituated in the same context as the amoeba. It was a risk, in short, that could be minimized with only the most compelling and persuasive scientific evidence and literary style.

It was not only the content and style of *Silent Spring*, however, that empowered critics of insecticides to find a voice strong enough to productively confront their opponents. It was also the way that the book became a pivot on which several broader intellectual developments transitioned from relative obscurity to popular consciousness. Between the discovery of DDT and the publication of *Silent Spring*, several ideas simmering below the surface of mainstream culture were intensifying to the point of breaking through. In universities, the field of ecology was growing into a discipline that was gradually placing the human being at its interpretive core; from the Sierra Club to the National Audubon Society to the Wilderness

Society, environmentalists were blending the once mutually exclusive concepts of "wilderness" and "civilization" into a more holistic vision of human interaction with the natural world; and, finally, ethnographers, historians, and anthropologists were coming to praise rather than condemn the Native Americans for their traditional integration of society into nature's cycles. *Silent Spring* helped to both consolidate these intellectual trends under the rubric of "ecology" and launch them into the activist-minded 1960s where, due to a series of environmental disasters that made *Silent Spring* look like a prophesy, they became the basic tools in a new, politically relevant, and culturally powerful movement that we continue to call environmentalism.

"the whip of hysteria"

On August 31, 1945, DDT hit the American market. It came with the tacit endorsement of the Food and Drug Administration and an arbitrary and generous legal limit of 7 parts per million in foods. DDT enjoyed unparalleled advance publicity. Any opposition to DDT in 1945 would have cut sharply against the grain of a conventional wisdom that praised the white dust as a panacea for a historically intractable problem. Due largely to its stellar performance during World War II, the insecticide—which was distributed under licenses granted by J.R. Geigy Aktien Gesellschaft—had become known as a veritable hero: a "killer of killers," "the atomic bomb of the insect world," and a product that "had every householder pawing the ground in eagerness." Consumers were said to be "raiding the stores for every can that shows its top above the counter." According to one newspaper clipping service, more than 21,000 articles on DDT appeared in 1944 and 1945. Only a handful of them expressed doubts about the insecticide that was already driving arsenicals off the market. The bulk of these articles accurately attributed to DDT the end of a massive typhus epidemic in Naples in 1943—an epidemic that had felled up to 60 people a day and had promised to kill more than 250,000 if it was not stopped. Americans anticipating access to this supposed wonder drug read about soldiers remaining free from lice, ticks, fleas, and mosquitoes thanks to the pocket aerosol bombs that they exploded in their barracks and mess halls as a matter of military course. The potato beetle and gypsy moth no longer bothered European farmers because, as the news reported, DDT had driven them into at least temporary extinction, a prospect that thrilled American farmers. "The wonder insecticide of World War II" was even used as a substitute for rice at Italian weddings, a gesture to its success in having combated typhus and, as an added benefit, a way to keep the mosquitoes down while the festivities progressed.[6]

Scientists, especially entomologists, were equally enthralled with DDT's seamless transition to civilian life. It took two entomologists from Arizona to adapt the "fog generator"—an oil fogger that had been a secret weapon during the war—to "peace time use as a [DDT] applicator," an innovation that enabled "aerosol production on a field scale." In 1951, Paul D. Harwood, an entomologist at the Ohio Agricultural Experiment Station, lambasted science writers who had the audacity to impugn DDT as potentially unsafe. As one of the condemned writers put it, DDT had unleashed "one of the most devastating biological weapons ever loosed by a people upon themselves," a claim to which Harwood, writing in *Science,* mockingly responded, "Obviously we are dealing here with a disaster inflicted on mankind by criminally careless . . . scientists." He dismissed the writers' investigations of DDT poisoning as marred by "absolute, nonscientific, emotion-packed phraseology, and startling but meaningless metaphors." Addressing the American Association of Economic Entomologists in 1953, Edward F. Knipling suggested that if American farmers did not continue down the path of chemical insecticides—in this case, by embracing a product that, he claimed, had already saved 100 million lives—then the consequences were nothing short of mass starvation, widespread disease, and, of all tragedies, "organic farming." An agent at the New Jersey Agricultural Experiment Station, praising DDT as "this outstanding new insecticide," published a "rapid method for preparing it in the laboratory"; a professor of anatomy explained in the pages of *Science* that laboratory scientists could apply "an ordinary nebulizer to the inner surface of the ear" of laboratory rabbits to protect them from "scab mites"; a scientist at a natural history museum found space in the "Letters" section of *Science* to publish a recipe—10 percent solution of DDT in kerosene—for killing black widow spiders. The good news seemed to be endless.[7]

Government entomologists and chemists whose job it was to keep America's diverse fields, orchards, and nurseries free from damaging insect pests were quick to assume the stance of restrained pragmatism in the face of what they often saw as unleashed hysteria. In his presidential address to the Florida Entomological Society in 1962, W. C. Rhoades honored the "entomologists and chemists" who "placed in the hands of the people who needed to use them . . . new pesticides such as DDT and hundreds of other new chemicals." With that said, he categorically labeled those who were skeptical about DDT's unalloyed benefits as "publicity seekers and misguided individuals" who had "seized upon the idea that the public was being poisoned." The scientists, by contrast, deemed themselves models of composure, men who had "settled down to a detailed analysis and factual study of the problem." Rhoades carefully listed the highlights of their findings, including the claims of the National Academy of Science that without pesticides "the American people cannot be fed adequately"; of the Manufacturing Chemists Association that "misuse and abuse [of DDT] must not be allowed to obscure the fact that these tools are vital to the health and the survival of humanity"; and of a government toxicologist that "DDT could be used safely." These scientific verifications of the safety of DDT thus stood firm against the "whip of hysteria" that, at end of an article that pleaded for calm analysis, riled Rhoades to bellow at the "misguided" skeptics: "STOP!"[8]

The Department of Agriculture bureaucrats and industry representatives, presenting their position as one that responsibly balanced the interests of public safety and economic growth, worked to avoid legal restraints on residue levels allowed on crops. By this point, their methods were well honed and predictable. The president of the National Agricultural Chemicals Association exploded at the mere mention of DDT regulation, complaining, "I know of no other industry which has to comply with more laws and regulations in order to sell its products." These interest groups understood quite well that the image of hardworking American companies being hindered by bureaucratic rules evoked popular disdain, and so they loudly reiterated their innocence and altruism in the face of overzealous and out-of-touch regulators. W. W. Sutherland, a scientist at Dow Chemical, despaired to the American Association of Economic Entomologists that the insecticide industry "was one of the most, if not the most, highly regulated industries in the country," without even remotely acknowledging that the alleged scrutiny may have had something to do with the unique nature of his product. In reference to the Miller Amendment, an innocuous alteration to the Food, Drug, and Cosmetic Act added in 1948 that required companies to register pesticides before selling them, Sutherland ratcheted up his rhetoric. He now insisted that the government had nothing less than "arbitrary powers over the marketing of products." As it did whenever insecticides came under the microscope, industry assumed the easily defensible stance that, as Sutherland explained, "research can go on forever without establishing all the factors which practical use of a product may eventually show up." If the nation's economic progress was to continue, "a reasonable approach must be taken," and it had to be one that allowed American companies to keep pace with "the rapidity of future progress." In the Cold War climate, which brooked only the most cautious critiques of capitalism, this tact went a long way toward keeping the regulation of acceptable levels of DDT residues at bay.[9]

The spectacular accomplishments of DDT during the World War II, in addition to the equally spectacular devastation that insects continued to exact on arsenic-saturated American farms, allowed the defenders of DDT to make an even more compelling argument. Writing in 1931, Leland Howard had noted, "I am afraid that the bug-fight cannot be given the pride, pomp, and circumstance of human war." Edward Knipling aimed to prove him wrong. The title of his address to the American Association of Economic Entomologists suggests the power of this relatively novel rhetorical approach: "The Greater Hazard—Insects or Insecticides." One imagines that there was no question mark following the title because the answer was, in Knipling's mind at least, perfectly obvious. Framed in such "us versus them" terminology, the standoff blended imperceptibly with the wartime dichotomy of democracy versus totalitarianism and the postwar dichotomy of capitalism versus Communism. Nuclear weapons and DDT, as Edmund P. Russell has argued, became reinforcing phenomena after World War II, with the concept of total annihilation uniting them in a common goal of nationalistic supremacy, be it over a foreign country or foreign insects. As these technologies converged in the cultural arena, the environmental imperative that humans radically control the environment by exterminating evil insects became one and the same with the nationalistic imperative that they defeat the evil empire. "The ability of human beings to kill both natural and national enemies on an unprecedented level," writes Russell, "developed in the twentieth century partly because of links between war and pest control." As in the earlier fights over laws that mandated acceptable levels of arsenic and lead residues, industry was infinitely more aware of and quicker to exploit the emotional power of international war to enhance the insecticide agenda. And once again the skeptics, ceding the rhetorical battle to their opponents, never bothered to counter with, say, a comparison of the insect wars to George F. Kennan's containment policy. Perhaps they knew that such a confrontation was a losing battle. Rachel Carson, after all, was frequently called a Communist. Plus, DDT skeptics could hardly retort with the charge that insecticide advocates were, of all things, *capitalists*.[10]

For all these reasons, the wall of protection surrounding DDT and related insecticides was higher than it had ever been. The most effective factor rendering DDT a seemingly untouchable product, however, was perhaps the most mundane: sustained corporate interest. What had been a handful of relatively small and poorly coordinated chemical companies in the 1920s had emerged by the 1950s—another era quite friendly to business interests—as a seamless conglomerate solidified with well-heeled and well-funded lobby groups and industry associations. What had been an industry producing $40 million worth of insecticides in 1939 was, by 1954, manufacturing $260 million worth of such chemicals. DuPont, Olin Mathieson, and Hercules were three of about a dozen corporations that monopolized DDT production both during and after the war. These companies entered political battles with guns blazing, not only financed by an ample war chest, but allied with such organizations as the Manufacturing Chemicals Association and the National Agricultural Chemicals Association. Flawlessly unified would best describe the DDT defense.[11]

Working for the common goal of unregulated insecticides, these corporations and associations left few options unexploited, but they were especially effective, again, in packaging the scientific case for DDT safety in a way that rendered it immune to public scrutiny. Knowing that it would be the pro-DDT scientist Wayland J. Hayes who, as a toxicologist at the Public Health Service, would lead the requisite studies of DDT, the chemical industry worked diligently to make sure that the "right" studies were conducted before helping Hayes take his message of assurance to the public. Accordingly, Hayes performed tests based on the assumption that DDT was an acute poison, which few skeptics were charging, rather than a chronic one, which most critics were insisting. With Hayes having deemed DDT "absolutely safe" after extensive investigations, the chemical industry relied on him to make widely publicized declarations about the safety of DDT, as he did before the House Select Committee to Investigate the Use of Chemicals in Food Products, when he testified that "the use of DDT by the general public was permitted only after a large amount of careful scientifically controlled study convinced the government authorities that DDT could be used safely." Through such publicity, Hayes, according to Thomas Dunlap, "became the mainstay of the medical defense of DDT." It was, perhaps, the final detail of a public-relations fait accompli, one achieved in the face of an audience that, to the great delight of DDT advocates, "took little interest in the matter."[12]

"You ought to write about this . . . you must"

Although the public may have kept incipient concerns to itself, scientific opposition to DDT became as overt and intense as it had ever been against the dangers of arsenic and lead. Starting in 1945, studies that questioned the conventional interpretation of the safety of DDT proliferated throughout the scientific community. But uncertainty, rather than consensus around concrete proof, was the invariable conclusion that investigators reached, and this of course was exactly the conclusion that advocates could spin. In a report published in *Science* in 1945, three scientists documented the connection between the ingestion of DDT and the "intoxication of the nervous system" in dogs, but ultimately admitted that the mechanism of poisoning was "imperfectly known." The Committee on Medical Research published a report in 1948 that provided a less than hopeful overview of DDT's safety, noting the accumulation of the chemical in the body as well as its conveyance through breast milk. Because the study was short term, though, its authors were forced to concede that they lacked hard information on the insecticide's chronic toxicity, conceding that "this problem will require careful observation over a period of years for elucidation." A study reported in the *Scientific Monthly* in 1949 raised doubts about the insecticide, including "the high susceptibility of fish and other aquatic forms to DDT," "the possible hazards to mammals," and threats to birds. In the end, however, it concluded that "there are still many unknowns regarding the biological effects of new insecticides and the many new economic poisons." When it came to DDT "and its remarkable stability under some conditions," another study explained, "one of the most critical needs is for a better understanding of the hazards implied by possible cumulative action." This tendency to admit—in the name of intellectual honesty and scientific caution—that uncertainty was endemic to these studies did little to challenge the defenders of DDT as they were downplaying the potential liabilities of their product.[13]

By the 1950s, however, the scientific opposition to DDT had become more assured. Ornithologists led the charge. After the Bureau of Entomology and Plant Quarantine began to spray DDT in Princeton, New Jersey, in 1949 to eliminate Dutch elm disease, the practice spread to a variety of other places, including cities in Michigan, Wisconsin, and Illinois. Concerned citizens in Lansing, Milwaukee, and Urbana vocally expressed their opposition to DDT when routine spraying appeared to kill off the local bird population. University scientists responded with studies documenting that the dead birds had DDT in their tissues and that nestlings were surviving at a rate of 44 percent in sprayed regions as opposed to more than 70 percent in unsprayed regions. Concentrations of DDT in birds' brains reached up to 252 parts per million, embryos showed signs of DDT poisoning, and nests contained contaminated eggs. After several studies showed a direct correlation between DDT levels and songbird deaths, George J. Wallace, an ornithologist at Michigan State University, decided that he was "inclined to question the whole program [of DDT spraying] on ecological grounds," adding that "any program which destroys 80 or more species of birds and unknown numbers of beneficial predatory and parasitic insects needs further study." Other scientists at other universities reached the same conclusion that DDT killed birds. When it came to birds and DDT, therefore, a scientific consensus had clearly formed by the late 1950s. Nevertheless, the scientists' customary appeal to "further study" helped ensure that the issue never evolved beyond a local matter that pitted elm enthusiasts against bird lovers—an unlikely fight to achieve national attention.[14]

Ornithological concerns, however, did not fade away. For one, they sparked the interest of such organizations as the Audubon Society and the National Wildlife Federation, which began to turn their attention to the impact of DDT on wildlife in general. Scientists and wildlife biologists working for these organizations assumed that the poison did more than target specific animals in circumscribed regions, but invaded the entire food web. It was an important conceptual step toward a better understanding of the full impact of DDT on the environment in ecological terms. Crustaceans at the bottom of Lake Michigan turned out to accumulate in their bodies considerable concentrations of DDT from contaminated mud that was carried into the lake in DDT-saturated runoff. The buildup of DDT in earthworms that fed on sprayed elm leaves at Michigan State University was large enough to kill the robins that ate them. Fish off the coast of Connecticut showed DDT concentrations of 99 parts per million, a rate high enough to compromise the reproductive systems of the herring gulls that ate them. Falcons, ospreys, bald eagles, and even penguins living in Antarctica (DDT dust traveled quite efficiently in wind and ocean currents) showed signs of DDT poisoning, including the gradual thinning of their eggshells due to residue levels that reached 6.44 parts per million (DDT reduced calcium production). In late 1949, the Department of Agriculture sponsored the massive spraying of dieldrin dust pellets to exterminate the Argentine red ant, which had invaded Mobile, Alabama, in 1918 and thirty years later was posing a dire threat to cattle throughout the South. Thirty million acres of ant-infested land were soaked with dieldrin. By the mid-1950s, farmers and hunters were reporting, and wildlife biologists were confirming, the chronic poisoning of livestock, as well as the decline or disappearance of poultry, birds, fish, snakes, lizards, and frogs. The Department of Agriculture promised to spray with "the least possible damage to wildlife and other insects" in mind. But Wallace, offering a taste of the rhetoric to come, called it "one of the worst biological blunders that man ever made."[15]

Skeptical scientists were soon buttressing their case against DDT with information about bio-concentration and prospects of insect resistance. They had learned that cows that consumed DDT-saturated foliage concentrated the insecticide in fat cells that passed to their calves in milk. Building on the ominous proposition that such bio-concentration may not be limited to isolated herds of bovines, scientists began to test for the prevalence of DDT in other species and found, to their dismay, that the most humble plants could efficiently accumulate small amounts of DDT absorbed from the soil and water. So could fish swimming in water poisoned by DDT in runoff and drift. What this meant in concrete terms was that, for example, the crustaceans in Lake Michigan that developed a DDT concentration level of 0.41 part per million did so by consuming mud that had a DDT concentration level of 0.014 part per million; the fish that fed on the crustaceans took it to another level, showing a concentration level of 3 to 6 parts per million; the herring gulls that ate the fish had a concentration level of 99 parts per million; and the eggs of the herring gulls showed an astonishing DDT concentration level of 227 parts per million. And so on and on it went, the buildup of DDT increasing as it coursed through the food web.[16]

Casting the phenomenon of biological magnification into an even more dire light was the growing evidence that the intended targets of DDT already were developing a resistance to it. As early as 1946, entomologists at the Department of Agriculture determined that resistance was more than a theoretical possibility when they selectively bred a housefly that proved dauntingly immune to the power of DDT. The entomologists warned that "it seems possible that, in time, a similar increase in resistance may occur under natural conditions." In 1908, as mentioned, entomologists had become aware that resistance to chemical insecticides was a likelihood, given that San Jose scales that had infested areas of Washington State failed to succumb to a lime–sulfur mixture that had worked well over previous years. By the 1930s, scientists had learned that three kinds of California citrus scales that had traditionally responded successfully to hydrocyanic acid fumigation had become completely resistant, as had codling moths in Colorado to lead arsenate. In 1948, mosquitoes in New Jersey showed resistance to DDT–kerosene spray. Head lice that tormented soldiers during the Korean War—vectors of typhus, no less—had developed resistance to DDT by 1951. Reports from across the nation documented resistance to DDT in the cabbage looper, cabbage worm, tomato hornworm, and spider mite. By 1955, the boll weevil had become immune to the chlorinated hydrocarbons being sprayed on them. Three years later, a Canadian entomologist surveying the damage could note without fanfare that "the golden age" of chemical insect control was over. Thus by the late 1950s, questions about resistance to DDT had led to several answers, further proving the Department of Agriculture's prediction of resistance to be true.[17]

The case being made against DDT was thus formidable. But it did not mean much until the general public paid attention, took the matter to heart, and translated its concerns into a coherent political voice. As long as worries about DDT remained locked away inside elite scientific circles, defenders of the insecticide could continue to rest their case on dubious claims, such as "no human has been proven to be poisoned by DDT."

The woman behind Rachel Carson's decision to challenge this logic was Olga Owens Huckins, the owner of a bird sanctuary in Duxbury,

Massachusetts. Huckins and Carson were longtime friends, and in 1958 Huckins invited Carson to stay at her house the night before an airplane was scheduled to spray DDT across the city to control for mosquitoes—a common practice by then. The next morning, the two women boarded a small skiff and toured the pond located in the center of the sanctuary. What Carson found stunned her. The water and the banks surrounding it were littered with death. Dead birds, fish, and crustaceans were everywhere. Carson actually watched these animals, their nervous systems shot through with poison, die before her eyes. Huckins, equally moved by the carnage, turned to her friend, a marine biologist and already prominent as the author of three books about the ocean—*Under the Sea-Wind, The Sea Around Us,* and *The Edge of the Sea*—and said to her, "You ought to write about this . . . you must."[18]

"We stand now where two roads diverge"

Like many other culturally transformative books, *Silent Spring* did not so much make a profound discovery as place the obvious in a new light. Hidden in plain sight as the nation's farmers and public-health officials saturated the earth with organophosphorus insecticides were humans. Rachel Carson understood that studies confirming the undeniable link between the use of DDT and the death of wildlife had to have an implication for humans beyond being a necessary evil of getting rid of malaria or boosting the agricultural economy. The prospect of bio-concentration directly affecting humans, as well as the likelihood that resistance to DDT could lead to even more powerful insecticides, would be fully appreciated by the general public only once it grasped the human place *within* the ecosystem that DDT had invaded. This reorientation of humanity's place within the environment was the crux of Carson's argument. However, in the face of an entrenched antienvironmentalist cohort that asked, as Robert Wernick did in the *Saturday Evening Post,* "Why shouldn't we spoil the wilderness?" she had her work cut out for her. Indeed, when the opposition believed, again in Wernick's phrasing, that those who regarded humans as one with nature were people who "spit and sweat and boast of their friendship with aborigines,"[19] the cultural hurdles were going to be high. Fortunately for Carson, her book was not ahead of its time, but very much of it. To fully understand how *Silent Spring* altered the insect wars by pushing ecology into the mainstream, one must first appreciate how Carson's work both tapped and helped transform three larger cultural and intellectual preconditions for change.

First, Carson published *Silent Spring* at a time when the scientific conception of "ecology" within the academy was finally including humans in its conceptual framework. From the birth of the discipline in the 1880s through the early twentieth century, humans generally stood apart from the ecological systems they were defining. Ecologists understood their task, as one put it in 1904, as "being nothing less than to explain why each plant is what it is, where it is, and in the company it is." The true ecologist was able to "see for himself and to transmit to others a faithful picture of the vegetation of the earth," perhaps with an eye toward "the application of ecological principles in a great practical industry." By the 1920s, these professional objectives had expanded to include "the biotic balance and interdependence of animals," working under the relatively new assumption that "the classification that ecologists must eventually settle upon will be one in which both plants and animals are included in the same communities." Here too, though, the ultimate end was industrial service. Barrington Moore noted as much when he sung the praises of "animal industrial ecology," a field that had "devised a method of restoring depleted grazing lands without the expense of artificial reseeding or the hardship of closing the range against grazing." Occasional hints that ecologists were attempting to incorporate human welfare into the realm of ecological science do emerge, such as when Stephen A. Forbes published "The Humanizing of Ecology" in 1922 or when H. G. Duncan explored the idea of a "personal ecology." On the whole, though, the profession throughout the 1930s and 1940s continued to find "questionable" the proposition that "ecology should need to be brought into any special or intimate relation to considerations of human welfare."[20]

By the 1950s, however, the intellectual basis for a more inclusive ecology was starting to take shape. In 1956, Francis Evans defined "ecosystem" as a web that "involves the circulation, and accumulation of energy and matter through the medium of living things and their activities." Ecologists once mired in classification and floral distribution patterns were now, as Evans wrote, hoping "to take a world view of life and look upon the biosphere with its total environment as a gigantic ecosystem." As an example "of such a pathway," Evans named a telling phenomenon: "the food-chain."

He anticipated that his ecological vision "will eventually be adopted universally and that its application will be expanded beyond its original use to include other levels of biological organization." Eugene Odum, a professor of zoology at the University of Georgia, started to explore a "philosophy of the ecosystem approach" with his students in order to "provide a means of stressing principles which are basic to good management." His expansive understanding of the ecosystem, he argued, enabled him to study not the ways in which ecologists could serve agribusiness, but such matters as "the role of oxygen in stream pollution, the place of the predator in natural systems, and the problem of the use of insecticides in non-cultivated as compared with cultivated systems." At Western Michigan University, W. C. Van Deventer was offering courses like "Human Ecology, a Study of Man in Relation to His Environment," while his contemporaries were publishing books with such titles as *Man's Nature and Nature's Man*; *Man, Culture, and Animals*; and *The Ecological Perspective on Human Affairs*. Over the course of sixty years, "ecology" as a concept had gone from a static emphasis on isolated taxonomies to a dynamic emphasis on biological interconnectivity.[21]

Second, the publication of *Silent Spring* dovetailed with an emerging view of environmentalism that focused less on wilderness preservation and more on humanity's effort to strike a pragmatic balance within nature. Environmentalists in the early twentieth century tended to embrace a neo-Thoreauvian version of the wild that stressed what Roderick Nash has called "a reversion to the primitive." In many ways, this older environmentalist mentality provided a corrective to the founding ideology of conquest, which assumed that the wilderness was there to be beaten into domestication. The idea was that, as Nash explains, "civilization had largely subdued the continent," and thus humans had to "react against the previous condemnation of wilderness" by indulging the "widespread appeal of the uncivilized." Wilderness became a place necessarily distinct from common experience, an exotic venue ripe for voyeurism, a land where the Teddy Roosevelts of the world occasionally retired in order to replenish the vigor and virility of manhood, the sublimity of quiet contemplation, and the innocent appeal of the primitive. It was the place where in 1913, Joe Knowles, a magazine illustrator, entered the Maine woods, naked and alone, to live for two months, making fire with sticks, killing a bear with a club, filling his stomach with berries and fish, and recording his experiences in charcoal on bark—wood chips of wisdom he sent to newspapers that published his "reversion to the primitive" to a delighted audience of wilderness voyeurs.[22]

But by the time Carson wrote *Under the Sea-Wind, The Sea Around Us,* and *The Edge of the Sea,* environmentalists had little use for such impractical publicity stunts. By the 1950s, environmentalists—still very much an elite group of naturalists associated with a handful of organizations like the Sierra Club and the Audubon Society—sought to strike a balance between the raw and the cooked. Wilderness and civilization, in this reformulated version of environmentalism, had to coexist. In a ringing endorsement of this more pragmatic transformation, the famous outdoorsman Colin Fletcher confessed, "The last thing I want to do is knock champagne and sidewalks and Boeing 707s. Especially champagne. These things distinguish us from other animals. But they can also limit our perspective." Howard Zahniser, the executive director of the Wilderness Society and the future author of the Wilderness Act of 1964, struck a similar chord of compromise when he wrote, "We like the beef from cattle grazed on the public domain. We relish the vegetables from the land irrigated by virtue of the Bureau of Reclamation. . . . We nourish our minds with books manufactured out of the pulp of our forests." John P. Saylor, an environmentalist and a member of Congress from Pennsylvania, further reiterated this emerging sentiment when he said of Americans, "We are a great people because we have been so successful in developing and using our marvelous natural resources, but also, we Americans are the people we are largely because we have had the influence of wilderness on our lives." If the comment needed clarification, there was always Justice William O. Douglas's maxim: "We can have the road and a stretch of Wilderness both."[23]

Third, and finally, *Silent Spring* made its mark at a time when cultural constructions of Native Americans were undergoing a sea change toward appreciation for their supposed environmental sensibility. Throughout the eighteenth century, Anglo-American conceptions of Native Americans favored images of the noble savage—that is, a people living innocently and simply in a wilderness Eden. This myth served the European colonizers quite well in that it lent itself nicely to a "conjectural" view of history that assumed the balm of civilization would soften the noble savages into pliable subjects. Writers such as James Fenimore Cooper and Herman Melville sustained the image into the nineteenth century, but it did not last long. By the 1870s, the myth of the noble savage had fallen out of favor, a development that chronic territorial wars did much to foster as Americans extended their aggressive ceremonies of possession into the West. Few writers captured this cultural transition better than the historian Francis Parkman, whose sweeping hagiographic narratives of that chosen people known as white Americans reduced the "red men" to little more than dangerous wild animals who lived to wield the scalping knife and bellow war cries. "To make an Indian a hero," sneered Parkman in 1901, "is mere nonsense."[24]

By the mid-twentieth century, however, many scholars thought this assessment was itself nonsense. The once despairing version of Native Americans began to yield to a more culturally sensitive and optimistic perspective, one that emphasized their unique balance with the natural world. In the privileged seminar rooms of universities, anthropologists and historians began to analyze what the anthropologist Evon Z. Vogt identified in 1957 as "Indian systems of social structure and culture persisting with variable vigor within conservative nuclei of Indian population." Despite the undeniable historical fact that "European culture came to them in their native habitat and proceeded by force to overrun the continent," Native Americans managed to preserve traditional ways "with variable vigor."[25] Social scientists who worked for the program in Indian Personality and Administration Research investigated contemporary Native American communities and came to admire "the Indians' confidence in their own traditions . . . in the face of strong, disintegrating pressures."

In 1957, the American Philosophical Society published what may have been the most comprehensive study of Native Americans to date. Written by Harold E. Driver and William C. Massey, the report, synthesizing studies from the 1930s to the 1950s, delved into nearly every aspect of Native American culture but showed special interest in the minutest details of the Indians' interaction with their environment. What emerged was a portrait of responsible ingenuity. The authors repeatedly showed that Native Americans carefully crafted provisioning habits to adhere to specific geographical conditions, marveling at their "almost endless variation in details from locality to locality." They also commented on the Europeans' disruptive impact on the Native Americans' traditional practices, writing, for example, that "before European contact the hunting methods known to the Indian . . . were not generally efficient enough to be a menace to the survival of the species hunted." When the Europeans introduced guns into the environmental equation, however, "the result was nearly the same in every area, the differences being only a matter of which species was seriously depleted or exterminated and how rapid was the process." The implication running throughout this study, as in many others from the 1950s, was that precontact Native Americans had struck a rational balance with their environment that future generations of immigrants had gradually failed to maintain. These writers were not environmentalists per se, but the tenor of their message seemed to suggest that there was no reason why modern-day Americans could not learn a thing or two from the Native American example.[26]

Rachel Carson effectively tapped these overlapping wellsprings of intellectual and cultural change. Her debt to recent developments in the field of ecology was perhaps the most obvious. "Man," she wrote, "however much he might like to pretend to the contrary, is part of nature." It was such a basic point, so obvious, but what was remarkable was how badly it had to be said. "For each of us," she continued, "as for the robin in Michigan or the salmon in the Miramichi, this is a problem of ecology, of interrelationships, of interdependence." The "chain[s] of poisoning and death" that insecticides were strapping around the environment "are matters of record, observable, part of the visible world around us." They were chains that "reflect the web of life—or death—that scientists know as ecology." Her debt to modern environmental pragmatism was just as strong. As moderate environmentalism moved toward a middle ground between development and preservation, Carson met it halfway, noting of ecological connections that "sometimes we have no choice but to disturb these relationships, but we should do so thoughtfully, with full awareness that what we do may have consequences remote in time and place." Through her extended plea in *Silent Spring* not to beat back the forces of civilization but instead to "preserve our roadside vegetation," to maintain an environment conducive to "the good will of vacationing tourists," and to ensure that spraying was carefully targeted on those specimens that "must be eliminated," we can see how yet another quiet and well-timed intellectual shift occurring beneath the surface of American life informed a work that would soon break into the mainstream.[27]

Having drawn on these larger but less visible intellectual trends, Carson was able not only to foreground ecology as the guiding principle for understanding chlorinated hydrocarbons, but also to challenge the path dependency that the nation had endured for almost a century. The critics of *Silent Spring* would quickly dismiss Carson's analysis as marred by emotion. For an emotional book, though, it harbored a chilling and irrefutable logic that went something like this: chemical insecticides—DDT and related chlorinated hydrocarbons—could be understood only in the broadest ecological context, this context of interconnectivity included humans and their behavior, and thus humans were subject to the negative impact of insecticides.

It was a logic that authorized Carson, relying heavily on studies published in British journals, to unleash a torrent of bad news on the American public. "For the first time in the history of the world," she explained, "every human being is now subjected to contact with dangerous chemicals." Synthetic pesticides were "so thoroughly distributed throughout the animate and inanimate world that they occur virtually everywhere." Reiterating that "there is an ecology of the world within our bodies," she explained the damage that chlorinated hydrocarbons do to the human liver: "a liver damaged by pesticides is not only incapable of protecting us from poisons, the whole wide range of its activities may be interfered with." She laid out the evidence about the harm that organophosphates and organochlorines do to the nervous system; regarding DDT, she wrote that "its action is primarily on the central nervous system of man; the cerebellum and the higher motor cortex are thought to be the areas chiefly affected." As one scientist who subjected himself to DDT absorption through the skin described the impact: "The tiredness, the heaviness, and aching of limbs were very real things, and the mental state was also most distressing. . . . [There was] extreme irritability . . . great distaste for work of any sort . . . a feeling of mental incompetence in tackling the simplest mental task. The joint pains were quite violent at times." This example of acute poisoning was hardly the end of the discomfort, as an even more insidious danger was the "delayed effects from exposure." Dieldrin had little impact initially, but after it accumulated in the human body it had "long delayed effects ranging from 'loss of memory, insomnia, and nightmares to mania.'" Parathion buildup was documented to have caused "muscular weakness in the legs" and, in one case, "paralysis in both legs and some involvement of the hands and arms." Case after case enabled Carson to conclude her chapter "The Human Price" with this admonition: "Confusion, delusions, loss of memory, mania—a heavy price to pay for the temporary destruction of a few insects, but a price that will continue to be exacted as long as we insist upon using chemicals that strike directly at the nervous system."[28]

These reports were not suggestive, nor were they about how chemical insecticides affect animals. They were definitive, and they were about how chemical insecticides harm humans. In case readers missed the import of her message, however, Carson effectively sealed her case with prose replete with the kind of rhetorical thunderbolts that would elicit mockery were they not backed by facts. *Silent Spring* catapulted into public consciousness due in large part to a literary sensibility that linked the mundane act of spraying chemicals with nothing short of the fate of the earth that sustains us. Chemicals were not simply possible dangerous substances, but "the sinister and little recognized partners of radiation in changing the very nature of the world—the very nature of its life." Those who believed that insecticides were part of some kind of natural teleological arc of technology were warned, "No witchcraft, no enemy action had silenced the rebirth of new life in this stricken world. The people had done it themselves." What had to be avoided was "the sterile and hideous world we are letting our technicians make," and, as Carson assured us, "there is no dearth of men who understood these things." It was, in fact, the work of "men who understood these things," translated with such eloquent authority in *Silent Spring,* that placed humans in a position that they had not experienced since the days of Thaddeus William Harris and Benjamin Walsh. In Carson's words: "We stand now where two roads diverge."[29]

"*Ecology has become an* in *word*"

There were, of course, those who refused to consider another path. Advocates of insecticides recharged their batteries, dusted off the familiar tactics, and sputtered, yet again, into well-funded action as the *New Yorker* serialized Rachel Carson's work. Neither the arguments nor the players had changed appreciably. Agribusiness, chemical companies, and many scientists began the standard counterattack with the old canard that chemical insecticides were critical for the economic and physical health of the nation. Corporate managers and industry scientists, in particular, insisted, yet again, that they were sober pragmatists whose difficult but rational choice to push insecticides was being undermined by a cohort of hysterical hippies composed of "organic gardeners, the anti-fluoride leaguers, the worshipers of natural foods . . . and other pseudo-scientists." They reiterated, as they had been doing for years, that direct evidence of acute DDT poisoning was almost nonexistent, that the insecticide had saved the lives of millions of people who showed no appreciable side effects, and that the toxicity of DDT was, in the grand scheme of poisons, very low—which, in acute dosages, it was.[30]

Economic entomologists, men and women who had staked not just their careers but their entire profession on the promise of chemical insecticides, were similarly defensive when it came to *Silent Spring*. They were hardly going to back off because of an incendiary book written by a supposedly overemotional woman. Thus they tilted at growing windmills of opposition with claims to exclusive authority over insecticides and their application. "The type of responsibility which pesticide workers must assume," a group of entomologists at the University of Wisconsin wrote in an open letter, "is something that some scientists in other areas on the campus do not realize, do not appreciate, and possibly cannot even comprehend." These men were watching the invasion of their turf by supposedly ignorant outsiders armed with trendy ideas that were dangerously contrary to their own. Their fear became transparent in such remarks as "these people possibly don't realize the enormous amount of damage and confusion that can result from their uninformed intrusion into the pesticide field, or realize the dangers of making broad sweeping statements or loosely philosophizing about a subject in which their background information was so inadequate." Thomas Dunlap writes of these economic entomologists that "if they mistook the attack on a method for an attack on the profession itself, the error is understandable." With every entomologist at the University of Wisconsin having signed the letter, however, we can be assured that the error was not an isolated one.[31]

There were, finally, those who opposed Carson on the grounds that any claim that humans should dance to nature's rhythm was patently absurd. Like Robert Wernick, who wondered "Why shouldn't we spoil the wilderness?" many Americans indeed believed that we should "look after our own interests as best we can, and no more consider the feelings of the eagle and the rhinoceros than they consider ours." With considerable flourish and fanfare, Carson condemned the idea that "nature exists for the convenience of man" as one that was "born of the Neanderthal age of biology and philosophy." But one woman's Neanderthal is another person's prophet. One opponent of Carson's laid out the dispute in vivid terms when he wrote that Carson's ecological assumptions spelled "the end of all human progress, reversion to a passive social state devoid of technology, scientific medicine, agriculture, sanitation." He continued, perhaps taking a stylistic note from Carson herself, "It means disease, epidemics, starvation, misery, and suffering."[32]

If derivative, the defense of DDT in light of *Silent Spring* was still powerful, well coordinated, and richly funded. Clarence Cottam, director of the Weldon Wildlife Federation, commiserated with Carson about "the sinister efforts to counteract the effect of *Silent Spring*."[33] This time, however, the counterattack did not stick. That it did not—that popular notions of ecology, environmentalism, and responsible land use translated into concrete political and cultural power—marked a watershed in the history of the insect wars. But the impact must be kept in perspective. It is, of course, tempting to attribute the emergence of a new road toward integrated pest control to the exclusive impact of *Silent Spring*. Perhaps it is possible that Carson's book single-handedly transformed the debate over how to handle insect pests by mining current intellectual lodes; channeling the repressed wisdom of men like Thaddeus William Harris, Benjamin Walsh, and Charles Riley; and evoking the "chaos of experimentation" that once animated American farmers and entomologists, all the while speaking in perfect pitch to the general public. Perhaps this is true, but supporting events that followed the publication of *Silent Spring* certainly helped propel the book into the main currents of public discourse.

Most important, there was water. As a dire reminder that, no matter how hard they tried, people could not completely control the earth, the worst drought since the 1930s gripped the nation in 1962, the year Carson's book was published. It lasted for four years, crippled agricultural production, and threatened the water level of the Great Lakes. The attention generated by concerns about the quantity of the nation's water supply inspired questions about its quality, and on this point the Great Lakes proved to be in more trouble than anyone had imagined. A resource that Americans completely took for granted suddenly turned out to be what *Time* would call "a North American Dead Sea." Raw sewage, garbage, and phosphates from laundry detergent choked Lake Erie, with the phosphates causing massive algae blooms that sapped the oxygen, and thus the life, out of the lake. "In a society which is so affluent," wrote one outraged scientist, "to have to paddle in its own sewage is just disgusting." By 1969, concerns about the condition of Lake Erie reached a climax when the Cuyahoga River, so polluted as to be a fire hazard, went aflame on June 22, leading *Time*, once again, to weigh in with the observation that the river "oozes rather than flows." Just as Carson was saying about DDT, the blame for these environmental disasters could be located in only one place: human behavior. "Cuyahoga" became a cultural touchstone for an environmental sensibility that values the ecological tenets that Carson so admired. It continues to have resonance as musicians such as Michael Stipe of REM have memorialized the river in a song that pleads, "Let's put our heads together and start a new country up."[34]

A large number of activists in the 1960s and early 1970s hoped to do just that. Americans—younger ones, in particular—began to examine the world through new, greener, lenses. Neil Armstrong's photographs of the earth from the moon stressed that the planet that sustains us is indeed finite, a realization that led Americans to ask why we were saturating it with napalm in a war on the other side of the world, coal residue to serve consumerism at home, and DDT everywhere that crops grew and mosquitoes buzzed. Grassroots, green activism fueled itself on the mantra that people should "think globally, act locally," a bumper sticker–size motto powerful enough to help motivate more than 20 million people across the United States to speak out on a single day—April 22, 1970—in favor of ecological responsibility. It was a day when ecology went public, when Kurt Vonnegut blared to a rally in Bryant Park that the "planet may soon die," and when a critical mass of Americans came to consider the once unthinkable proposition that "every day is Earth Day." From this grassroots activism flowed concrete political change. With the birth of the environmental movement came the death of blatant political logrolling over environmental matters, a transition confirmed by the Clean Air Act of 1970, the Water Pollution Control Act of 1972, the Endangered Species Act of 1973, the Coastal Zone Management Act of 1972, and the Toxic Control Substances Act of 1976—to name only a few pieces of Nixon and Ford–era legislation armed with genuine regulatory teeth. Coming from the top down and the bottom up, a powerful and quite mainstream environmental ethos rooted in basic ecological ideas thus convinced the United States that, on its bicentennial, it was time to join the Declaration of Independence with what Roderick

Nash has called "a declaration of interdependence." "Ecology," according to Paul Smith, had "become an *in* word."[35]

It was virtually impossible for DDT and its counterparts—including dieldrin, aldrin, chlordane, benzene hexachloride, dichloro-diphenyl-dichloroethylene (DDE), mirex, and heptachlor—not to mention the old arsenates, to survive this "greening of America." Beginning in the mid-1960s, the insecticide juggernaut, once so firmly solidified by industrial, corporate, and government support, began to show signs of stress. Over the next decade, it would be greatly diminished. The Environmental Defense Fund, formed in 1967, brought together concerned scientists who worked with the Environmental Protection Agency, established in 1970, to hold public hearings on DDT. These hearings, which amassed and publicized evidence that activists, environmentalists, and scientists had been accumulating and presenting to Congress for more than a decade, resulted in the ban on the commercial use of DDT in the United States. That was in 1972. The same year, the Environmental Protection Agency outlawed the use of dieldrin. By 1974, it had prohibited the manufacture of dieldrin and aldrin, and when the Shell Chemical Company appealed, it lost. Also in 1974, the use of heptachlor and chlordane was banned by the Environmental Protection Agency, and in 1979 mirex suffered the same fate—all decisions that withstood a barrage of appeals and counterappeals. Binding these decisions into something of a unified whole was the passage of the Insecticide, Fungicide, and Rodenticide Act of 1972. With this law, the federal government gained extensive power over recall, enforcement, punishment, and intrastate issues regarding manufactured chemicals. To be sure, as Thomas Dunlap (who rightly calls these measures "a qualified victory") argues, loopholes remained. Nevertheless, as he also points out, "the first casualty of the battle over DDT was the old system of federal regulation of pesticides, and the interest-group control over that policy." The fight against pesticides had now changed, "becoming a large, carefully organized, and scientifically backed crusade against environmental damage and environmental hazards." Any doubts that these developments radically altered the nature of the insect wars were put to rest by the creation by "deep ecologists" of an entirely new legal language around the notion that nature—like the African Americans whom Harriet Beecher Stowe wrote about, and the early Americans whom Thomas Paine cajoled to action—has inherent rights that must be treated with the utmost sanctity.[36]

Rachel Carson saw none of this happen. One theme that she extensively explored in *Silent Spring* is the connection between chemicals and cancer. Insecticides, she quoted Dr. W. C. Hueper as saying, lead to "the danger of cancer hazards from the consumption of contaminated drinking water." Humans, Carson explained, can "create cancer producing substances" when they replace a "natural environment" with "an artificial one composed of new chemical and physical agents." As a result of our decisions to saturate the earth with supposedly beneficial chemicals designed to improve our quality of life, "human exposure to cancer producing chemicals (including pesticides) [is] uncontrolled and they are multiple." She insisted that "the history of cancer is long, but our recognition of the agents that produce it has been slow to mature." For Carson, the threat of cancer had risen dramatically ever since Americans started to use DDT and other chlorinated hydrocarbons.[37]

"About two weeks ago I noticed a tender area above the collar bone . . . and on exploration found several hard bodies." Carson wrote these words to her doctor on February 17, 1963.[38] *Silent Spring* became a very personal book for many Americans, but for Carson, and the nation, it took on especially profound meaning when, two years after its publication, she died of cancer. Unlike Benjamin D. Walsh, Charles V. Riley, John B. Smith, and so many other advocates of a responsible approach to controlling insects, Rachel Carson was able to make herself meaningfully heard through a single act—a book—that convincingly demonstrated that sometimes the best answers to the world's problems are those left behind on the cutting-room floor of history.

EPILOGUE

"Some very learned men are
the greatest fools in the world":
IN PRAISE OF LOCALISM

The story does not end with Rachel Carson, of course. Neither does it end on a note of triumph. The most frustrating aspects of the insect paradox are its durability and its persistence. *Silent Spring* certainly changed the way Americans, in particular, conceptualized the environmental consequences of insect control in an ever-expanding landscape of factory farms. Carson provided a timely warning, one that sparked a vocal segment of the American public to rethink the impact of dangerous chemicals on both human and nonhuman forms of life. The banning of a long list of toxic insecticides, the growing popularity of integrated pest management among professional entomologists and managers of agribusiness, and the increasing difficulty and expense of winning approval of new insecticides from regulatory agencies (no matter how politically corrupt) are generally welcome developments in the insect wars. Add to these factors the growing popularity of insect-resistant seeds for cotton, soybean, and corn monoculture, and it is not difficult to appreciate how the current skepticism of many Americans about chemical insecticides speaks to the great potential for a profound structural change in approaches to protecting the environment.

But this assessment is, I think, overly optimistic. *Silent Spring* and its ability to inspire a more popular and human-centered understanding of "ecology" was indeed a breakthrough. But in the grander scheme, one that includes the past, the impact of Carson's book remains necessarily limited.

Readers of this book who have carefully followed the main trajectory of insect control tactics over almost 400 years will have been struck by a rash of wrong turns, but if there is one that stands out, it is the consolidation of authority over environmental issues in the hands of a relatively few powerful decision makers. This development, I would argue, was a recipe for environmental disaster.

Recall the main lines of this story. The earliest efforts to control insect pests were initiated and carried out by farmers. Before the emergence of the Division of Entomology in 1878, the men and women who worked the soil pioneered experimental, flexible, vernacular, and reversible responses to insect invasions that were, in many ways, the result of their own nonsustainable agricultural practices. By no means was this a golden age of insect control—these solutions were haphazard, of variable quality, woefully unsystematic, and somewhat undermined by the farmers' addiction to monoculture. But they were, in essence, local responses to local problems—derived from honest, firsthand observations of the natural world at work—that could be promoted if they worked and discarded if they did not. The first economic entomologists—scientists who were not organized in a single bureaucracy—respected these tenets and, alongside farmers, confronted and negotiated the insect paradox in a variety of provisional ways. They did so with a collective attitude that might best be glimpsed in Asa Fitch's commonplace book, in which the tireless economic entomologist wrote, "Some suppose every learned man is an *educated* man. No such thing. The man is educated who knows himself and taken accurate commonsense views of men and things around him. Some very learned men are the greatest fools in the world." Common sense, a reliance on homegrown knowledge, and an appreciation for the "chaos of experimentation" thus kept the quest for insect control from becoming a monolithic environmental hazard.

Bureaucratizing the national effort to control insects from coast to coast, by contrast, was a critical pivot away from responsible insect control. Rather than serving a regulatory role, entomologists who worked for the Department of Agriculture assumed (as they were charged to do) a promotional and authoritative role. For a wide variety of reasons, most of them perfectly sensible, federal entomologists sensitive to the expansive nature of American agriculture were inclined to adopt a simple, applicable, and moderately effective program of insecticidal control of invasive insects. This program—perpetuated through well-funded channels of information and buttressed by increasingly powerful industrial interests—gradually weakened the flexibility and ingenuity of biological and cultural methods of insect management. The bureaucratic approach not only had little motivation to deal with the insect paradox in its innumerable specific manifestations, but became a target of corporate power in the often insidious effort to protect chemicals that were known to be public-health hazards. The scale and scope of this government–industry conglomerate had no tolerance for the genuinely democratic debates that had characterized insect management under more decentralized circumstances. As a result, a number of possibly effective control tactics—many nonchemical—were pushed to the periphery, while a single, simplistic, and widely applicable approach culminating in the use of DDT became standardized as the only viable way to survive as a profit-minded commercial grower in the United States. Rachel Carson could change the way we think about ecology, but her work, for all its impact, had little influence on the perspective and practices of the federal agencies that continue to oversee and promote an influential and largely monolithic approach to insect control.

So what does this analysis suggest about the future of insect control? Given that American agriculture continues to have little reason to systematically reform its monocultural ways and that the federal government has become more solicitous then ever of corporate interests, the insect paradox persists as a woefully ignored environmental problem. What is perhaps most alarming about the failure to even contemplate a widespread embrace of integrated pest management structured around more sustainable farming practices is the forecast that, come 2050, there will be some 9 billion people for the world to feed. The demand for food will be greater, the markets will be bigger, and the companies primed to profit from a formalized global trade in insecticides will be more powerful and monopolized than ever before. In the past ten years alone, a distinct trend has developed: agribusiness has joined forces with agrichemical and biochemical companies, hired market-research firms to identify global markets for new chemical insecticides, lobbied for approval of new insecticides, and taken the fight against insects to the developing world. This behavior, seen from a benign angle, is consistent with the main currents of American history—a history of innovation, entrepreneurship, ambition, greed, and a belief that happiness comes through a better material standard of living (however superficially conceived). But, from a more ominous perspective, it also points to a discouraging and persistent view of the environment.

Indeed, as I finished writing this book, I could not help but be deflated by the possibility that the small story I have told about a nation's quest to kill insects is a reflection of Americans' behavior toward the environment in general. Considering what we know about the looming consequences of global warming, water limitations, coal mining, and so many other massive, human-driven ecological transformations, it is hard not to think that the obstacles to effective and safe insect control—the failure of centralized solutions, the corporate obfuscations, the inability to face the environmental impact head on, and the tendency to think in the short term—are not limited to this single issue. Likewise, it is equally hard not to wonder what would happen if the spirit of innovation, civic involvement, and cooperation that seems to thrive so well in local contexts was encouraged to address the environmental problems that we have done an excellent job of identifying but, paradoxically, an inadequate job of managing. History, at the least, suggests something of an answer.

NOTES

Introduction. The Insect Paradox

1. Arnold Mallis, *American Entomologists* (New Brunswick, N.J.: Rutgers University Press, 1971), 25–28; Joseph Kastner, *A Species of Eternity* (New York: Knopf, 1977), 3, 273.

2. Quoted in *Entomological Correspondence of Thaddeus William Harris, M.D.*, ed. Samuel H. Scudder, Occasional Papers of the Boston Society of Natural History, no. 1 (Boston: Boston Society of Natural History, 1869).

3. Mallis, *American Entomologists*, 30; Thaddeus William Harris to Nicholas Hentz, November 19, 1828, Museum of Science, Boston.

4. Thaddeus William Harris, *A Report of the Insects of Massachusetts Injurious to Vegetation* (Cambridge, Mass.: Folson, Wells, and Thurston, 1841), v.

5. Solon Robinson, "Something About the Western Prairies" (1841), in *Solon Robinson: Pioneer and Agriculturalist*, ed. Herbert Anthony Kellar (New York: Da Capo Press, 1968), 1:284–285. The frontier was, in the 1840s, very much a place of great mystery and hope to dwellers on the east coast. It was also much praised in print. As Clarence H. Danhof has written, "The prairies of Illinois and Iowa were the subject of one of the greatest spontaneous promotional campaigns of history" (*Change in Agriculture: The Northern United States* [Cambridge, Mass.: Harvard University Press, 1969], 121).

6. *Union Agriculturalist* 1 (1844): 60; *Northern Farmer* 1 (1854): 290–291; Theodore R. Marmor, "Anti-Industrialism and the Old South: The Agrarian Perspective of John C. Calhoun," *Comparative Studies in Society and History* 9, no. 4 (1967): 390.

7. Robinson, "Something About the Western Prairies," 285; Dan Flores, *The Natural West: Environmental History in the Great Plains and Rocky Mountains* (Norman: University of Oklahoma Press, 2001), 50–70.

1. Insects and Early Americans

1. Oliver Goldsmith, *An History of the Earth, and Animated Nature* (Philadelphia: Carey, 1795), 136.

2. Samuel Deane, *The New-England Farmer or, Georgical Dictionary. Containing a Compendious Account of the Ways and Methods in Which the Important Art of Husbandry, in All Its Various Branches, Is, or May Be, Practised, to the Greatest Advantage, in This Country*, 3rd ed. (Boston: Wells and Lilly, 1822), 228.

3. Joseph Clarke Robert, *The Story of Tobacco in America* (Chapel Hill: University of North Carolina Press, 1949), 18, and *The Tobacco Kingdom: Plantation, Market, and*

Factory in Virginia and North Carolina, 1800–1860 (Gloucester, Mass.: Smith, 1965), 34, 37; G. Melvin Herndon, *William Tatham and the Culture of Tobacco, Including a Facsimile Reprint of "An Historical and Practical Essay on the Culture and Commerce of Tobacco" by William Tatham* (Coral Gables, Fla.: University of Miami Press, 1969), 21, 119–121; Catesby, quoted in Herndon, *William Tatham and the Culture of Tobacco*, 21; Deane, *New-England Farmer*, 456–457.

4. Landon Carter, "Observations Concerning the Fly-Weevil, that Destroys the Wheat," *Transactions of the American Philosophical Society* 1 (1771): 211.

5. "Remarks on the Destruction of the Hessian Fly," *Connecticut Courant*, September 8, 1788, 1; "Hessian Fly," *City Gazette and Daily Advertiser*, August 14, 1804, 3; *Morning Chronicle*, June 2, 1807, 2; *Hagers-Town Gazette*, August 21, 1810, 2; "Preservation of Wheat from the Ravages of the Hessian Fly," *New Haven Gazette, and the Connecticut Magazine*, July 19, 1787, 22, 169; *Transactions of the Society for the Promotion of Agriculture, Arts, and Manufactures* (Albany, N.Y.: Webster, 1801), 85; J. B. Bordley, *Essays and Notes on Husbandry and Rural Affairs* (Philadelphia: Budd and Bartram, 1801), 276.

6. In 1846, Asa Fitch, who would become the state entomologist of New York in 1853, wrote in the margin of one of his essays that the Hessian fly "might have existed here prior to the time alluded to . . . without having been discovered" ("The Hessian Fly: Its History, Character, Transformation, and Habits," *Transactions of the New York State Agricultural Society* 6 [1846]: 319; Fitch's annotated version is held in the Ernst Mayr Library, Special Collections, Museum of Comparative Zoology, Harvard University, Cambridge, Mass.).

7. Kim Todd, *Tinkering with Eden: A Natural History of Exotics in America* (New York: Norton, 2001), 40; New Jersey farmer, quoted in Philip J. Pauley, "Fighting the Hessian Fly: American and British Responses to Insect Invasion, 1776–1789," *Environmental History* 7, no. 3 (2002); M.L. Wilson, "Thomas Jefferson—Farmer," *Proceedings of the American Philosophical Society* 87, no. 3 (1943): 219.

8. Bordley, *Essays and Notes on Husbandry and Rural Affairs*, 276; *Transactions of the Society for the Promotion of Agriculture, Arts, and Manufactures*, xii; John Drayton, *A View of South Carolina, as Respects Her Natural and Civil Concerns* (Charleston: Young, 1802), 118–119.

9. Goldsmith, *History of the Earth*, 142; *Thomas Jefferson's Farm Book*, ed. Edwin Morris Betts (Charlottesville: University of Virginia Press, 1976), 158; Peter Kalm, *The America of 1750: Peter Kalm's Travels in North America: The English Version of 1770*, ed. Adolph B. Benson (New York: Wilson-Erickson, 1937), 2:424; *John and William Bartram's America: Selections from the Writings of the Philadelphia Naturalists*, ed. Helen Gere Cruickshank (New York: Devin-Adair, 1957), 274; Deane, *New-England Farmer*, 228; Herndon, *William Tatham and the Culture of Tobacco*, 119; Reverend Israel Acrelius, "Account of the Swedish Churches in New Sweden," in *Narratives of Early Pennsylvania, West New Jersey, and Delaware, 1630–1707*, ed. Albert Cook Myers (New York: Scribner, 1912), 68.

10. Goldsmith, *History of the Earth*, 136–137.

11. Ted Steinberg, *Down to Earth: Nature's Role in American History* (Oxford: Oxford University Press, 2002), 18–20; Shepherd Kretch, *The Ecological Indian: Myth and History* (New York: Norton, 1999), 104, 201; William Cronon, *Changes in the Land: Indians, Colonists, and the Ecology of New England* (New York: Hill and Wang, 1983), 39–50; Gordon G. Whitney, *From Coastal Wilderness to Fruited Plain: A History*

of Environmental Change in Temperate North America, 1500 to the Present (Cambridge: Cambridge University Press, 1994), 156–227; A. L. Kroeber, *Cultural and Natural Areas of Native North America* (Berkeley: University of California Press, 1939), 142.

12. On the burning of land and deforestation by Native Americans, see Whitney, *From Coastal Wilderness to Fruited Plain*, 101–102, 107–109, 114; on the impact of deforestation on insects, see Emily W. B. Russell, Ronald B. Davis, R. Scott Anderson, Thomas E. Rhodes, and Dennis S. Anderson, "Recent Centuries of Vegetational Change in the Glaciated Northeastern United States," *Journal of Ecology* 81, no. 4 (1993): 647–664; on the evolutionary relationship between insects and plants, see V. G. Dethier, *Man's Plague? Insects and Agriculture* (Princeton, N.J.: Darwin Press, 1976), 68, and Terrence D. Fitzgerald, *The Tent Caterpillars* (Ithaca, N.Y.: Cornell University Press, 1995), 210; Michael D. Atkins, *Insects in Perspective* (New York: Macmillan, 1978), 319; on soil fertility caused by burning, see Conrad Heidenreich, *Huronia: A History and Geography of the Huron Indians, 1600–1650* (Toronto: McClelland and Stewart, 1971); Alfred Crosby, *Ecological Imperialism: The Biological Expansion of Europe, 900–1900* (Cambridge: Cambridge University Press, 1986), 147.

13. On the pine weevil, see D. M. Smith, "Changes in Eastern Forests Since 1600 and Possible Effects," in *Perspectives in Forest Entomology*, ed. John F. Anderson and Harry K. Kaya (New York: Academic Press, 1976), 3–20; on insects and Native Americans, see William A. Starna, George R. Hammel, and William L. Butts, "Northern Iroquoian Horticulture and Insect Infestation: A Cause for Village Removal," *Ethnohistory* 3, no. 31 (1984): 199, 202. Although Indians did significantly change the landscape, the impact was localized, as discussed in William M. Denevan, "The Pristine Myth: The Landscape of the Americas in 1492," *Annals of the Association of American Geographers* 82, no. 3 (1992): 369–385. See also Cronon, *Changes in the Land*, 154, and Steinberg, *Down to Earth*, 18.

14. Francis Higginson, quoted in Cronon, *Changes in the Land*, 55. On the rate of deforestation, see Whitney, *From Coastal Wilderness to Fruited Plain*, 151–156; William Cronon, *Nature's Metropolis: Chicago and the Great West* (New York: Norton, 1991), 150–154. On inherited agricultural traditions, see David Grayson Allen, *In English Ways: The Movement of Societies and the Transferal of English Local Law and Custom to Massachusetts Bay in the Seventeenth Century* (Chapel Hill: University of North Carolina Press, 1981), and Charles F. Carroll, *Timber Economy of Puritan New England* (Providence, R.I.: Brown University Press, 1974). It is important to point out that this process affected the South as well as the North. According to Steinberg, by 1737 Savannah's pine forests had been depleted to the point that the city was forced to import wood, perhaps because Southerners were far more open to girdling trees than were Northerners, who were more likely to loath sloppy-looking fields (*Down to Earth*, 35–37, 39–41). On colonial indifference to forest clearing, Alexis de Tocqueville observed in *Democracy in America* that Americans "may be said not to perceive the mighty forests that surround them till they fall beneath the hatchet" (quoted in Roderick Nash, *Wilderness and the American Mind*, 3rd ed. [New Haven, Conn.: Yale University Press, 1982], 23).

15. According to Dethier, "[T]he ancient cultivated crops, corn and squash, and the wild crop, hay, suffered mainly from periodic population explosions of native insects. Army worms and various species of grasshoppers, the red legged locust, the lesser migratory locust, and the devastating grasshopper, had their lean years and fat years. Anything green that grew in their path fell to their appetites. Crops

just happened to be there" (*Man's Plague?* 68). Cronon writes, "Indian fields, which in addition to being both small and diverse . . . , had originally been troubled by diseases and pests. The arrival of the colonists changed this" (*Changes in the Land,* 153). E. L. Jones, "Creative Disruptions in American Agriculture, 1620–1820," *Agricultural History* 48, no. 4 (1974): 519; Steinberg, *Down to Earth,* 43.

16. Yves Bassett, E. Charles, D. S. Hammond, and V. K. Brown, "Short-Term Effects of Canopy Openness on Insect Herbivores in a Rain Forest in Guyana," *Journal of Applied Ecology* 38, no. 5 (2001): 1045, 1056; J. Bruce Wallace and Martin E. Gurtz, "Response of Baetis Mayflies (Ephemeroptera) to Catchment Logging," *American Midland Naturalist,* January 1986, 25; Raphael K. Didham, Peter M. Hammond, John H. Lawton, Paul Eggleton, and Nigel E. Stork, "Beetle Species Responses to Tropical Forest Fragmentation," *Ecological Monographs* 68, no. 3 (1998): 295, 313; Wayne P. Souza, "The Role of Disturbance in Natural Communities," *Annual Review of Ecology and Systematics* 15 (1984): 353, 382; Charles S. Elton, *The Ecology of Invasions by Animals and Plants* (London: Methuen, 1958), 54. Atkins writes, "Often there is something about a new environment that allows a species to increase in numbers much more rapidly once it has become established there than it did in its place of origin" (*Insects in Perspective,* 313). The strong possibility that invasive insects benefit more than native ones as a result of deforestation may explain why burning by Native Americans did not have a similar impact on the insect population.

17. Svata M. Louda and James E. Rodman, "Insect Herbivory as a Major Factor in the Shade Distribution of a Native Crucifer (*Cardamine cordifolia* A. Gray, Bittercress)," *Journal of Ecology* 84, no. 2 (1996): 229–237; C. J. Hawley, "Aphid (Homoptera, Aphididae) Behavior in Relation to Host-Plant Selection and Crop Protection" (available at www.colo.state.edu/Depts/Entomology/courses/en507/papers_1995/hawley.html).

18. Nellie Payne, "The Effect of Environmental Temperatures upon Insect Freezing Points," *Ecology* 7, no. 1 (1926): 99–106; Cronon, *Changes in the Land,* 124–125; Steinberg, *Down to Earth,* 36; Williams, quoted in Cronon, *Changes in the Land,* 122.

19. Barton, quoted in Dethier, *Man's Plague?* 68–69 (emphasis added); Cronon, *Changes in the Land,* 153–155; Whitney, *From Coastal Wilderness to Fruited Plain,* 312; Jeffrey A. Lockwood, *Locust: The Devastating Rise and Mysterious Disappearance of the Insect that Shaped the American Frontier* (New York: Basic Books, 2004), 55; for a worldwide perspective on the relationship between staple crops and insect attacks, see B. P. Uvarov, "Problems of Insect Ecology in Developing Countries," *Journal of Applied Ecology* 1, no. 1 (1964): 159–168. Crosby, *Ecological Imperialism,* 147; Steve H. Dreistadt, Donald L. Dahlsten, and Gordon W. Frankie, "Urban Forests and Insect Ecology: Complex Interactions Among Trees, Insects, and People," *BioScience* 40, no. 3 (1990): 193; Jakob Lundberg and Fredrick Moberg, "Mobile Link Organisms and Ecosystem Functioning: Implications for Ecosystem Resilience and Management," *Ecosystems* 6, no. 1 (2003): 88–90.

20. Cronon, *Changes in the Land,* 125, 146–147; P. L. Marks, "The Role of Pin Cherry (*Prunus penslyvanica* L.) in the Maintenance of Stability in Northern Hardwood Ecosystems," *Ecological Monographs* 44 (1974): 73–88; E. A. Johnson, "Effects of Farm Woodland Grazing on Watershed Values in the Southern Appalachian Mountains," *Journal of Forestry* 50 (1952): 109–113; Lockwood, *Locust,* 246; Norman Hudson, *Soil Conservation,* 2nd ed. (Ithaca, N.Y.: Cornell University Press, 1981), 250; Steinberg, *Down to Earth,* 49–50.

21. "Le Jeune's Relation" (1633), in *The Jesuit Relations and Allied Documents*, ed. Reuben Gold Thwaites (New York: Pageant, 1959), 5:29.

22. The "disorganized group of amateurs" is discussed in George H. Daniels, *American Science in the Age of Jackson* (New York: Columbia University Press, 1968), 7. "Prior to the nineteenth century," writes Dethier, "insect control was not on an organized basis" (*Man's Plague?* 105–109). While true, this assertion should not imply that it was not on a rational basis, which, in many ways, it was. The classic contemporary work on biological control was Thaddeus William Harris, *A Report of the Insects of Massachusetts Injurious to Vegetation* (Cambridge, Mass.: Folsom, Wells, and Thurston, 1841).

23. Crosby, *Ecological Imperialism*, 171–194; Cronon, *Changes in the Land*, chap. 7; Dethier, *Man's Plague?* 68–69; Lockwood, *Locust*, 49–55.

24. A. Hunter Dupree, *Science in the Federal Government: A History of Policies and Activities to 1940* (Cambridge, Mass.: Harvard University Press, 1957), 8. On the virtues of experimentation, see Wilson Flagg, "Plan of a Series of Experiments for Investigating the Potato Disease," in *Transactions of the Essex Agricultural Society for 1859* (Newbury, Mass.: Huse, 1859), 129.

25. Genesee Farmer, *Farmer's Cabinet*, August 1, 1836, 25–26; Mary Griffith, *Our Neighborhood, or Letters on Horticulture and Natural Phenomena: Interspersed with Opinions on Domestic and Moral Economy* (New York: Bliss, 1831), 30–31; Deane, *New-England Farmer*, 183; H.C. Giddens, "To Kill Lice on Swine—Destroy Poisonous Mushrooms," *Southern Cultivator*, July 1852, 294; "Bucks County Intelligencer," *Farmer's Cabinet*, January 1, 1837, 185.

26. The Agriculturalist, *Yankee Farmer*, January 6, 1838, 1; Deane, *New-England Farmer*, 181; "Destructive Insects," *Farmer's Cabinet*, July 1, 1836, 5–6; "The Wheat Fly," *Farmer's Cabinet*, December 1, 1836, 145; "Remedy for the Culicue, or Plum Weevil," *Southern Cultivator*, July 1852, 301.

27. Connecticut River Vermont Farmer, "The Grain Worm," *Yankee Farmer*, March 17, 1838, 82; "Wheat Worm," *Yankee Farmer*, March 10, 1838, 4; Herndon, *William Tatham and the Culture of Tobacco*, 119; Maine Farmer, "Bugs—Bugs—'O! the Bugs,'" *Farmer's Cabinet*, June 15, 1837, 366.

28. Ezra Clap to Oliver Smith, April 2, 1794, box 1, folder 6, number 1A; Perez Bradford to Oliver Smith, July 14, 1794, box 1, folder 6, number 2A; Benjamin Bassett to James Winthrop, Esquire, July 21, 1794, box 1, folder 6, number 2A, all in Papers of the Massachusetts Society for the Promotion of Agriculture, Massachusetts Historical Society, Boston.

29. Herndon, *William Tatham and the Culture of Tobacco*, 22; Deane, *New-England Farmer,*, 66; John Abercrombie, *Every Man His Own Gardner: The Complete Gardener* (London, 1832), 205; Carter, "Observations Concerning the Fly-Weevil," 212; Senex, "Turnep Fly," *Farmer's Cabinet*, July 1, 1836, 6; Griffith, *Our Neighborhood*, 289; *Thomas Jefferson's Farm Book*, 158; Nicholas P. Hardeman, *Shucks, Shocks, and Hominy Blocks: Corn as a Way of Life in Colonial America* (Baton Rouge: Louisiana State University Press, 1981), 200–202.

30. On worms and corn remedies, see Hardeman, *Shucks, Shocks, and Hominy Blocks*, 200–202; "Insects," *Southern Cultivator*, July 1852, 109; Deane, *New-England Farmer*, 178. Senex agreed with Deane, writing that "against the striped bug, another destroyer of melons, a brood of young chickens is a sufficient protection for a whole garden" ("Turnep Fly," 6).

31. "Method of Destroying Caterpillars on Trees," *Boston Gazette*, October 24, 1763, 3; Deane, *New-England Farmer*, 178–179; "To Drive Bugs from Vines," *Farmer's Cabinet*, July 1, 1836, 13; "Turnep Fly," *Farmer's Cabinet*, October 1, 1836, 88; Genesee Farmer, *Farmer's Cabinet*, August 1, 1836, 25; *Farmer's Cabinet*, January 1, 1837, 181; Culpepper, quoted in Ann Leighton, *Early American Gardens: For Meate or Medicine* (Amherst: University of Massachusetts Press, 1986), 242.

32. Thomas Jefferson to Thomas Pinckney, April 12, 1793, and Thomas Pinckney to Thomas Jefferson, January 29, 1794, in *Thomas Jefferson's Farm Book*, 70–71; Griffith, *Our Neighborhood*, 289; Deane, *New-England Farmer*, 178; Senex, "Destructive Insects," *Farmer's Cabinet*, July 1, 1836, 36; *Farmer's Almanac*, August 1, 1836, 24.

33. *The Notebook of a Colonial Clergyman: Condensed from the Journals of Henry Melchior Muhlenberg*, ed. and trans. Theodore G. Tappert and John W. Doberstein (Minneapolis: Fortress Press, 1998), 3; Herndon, *William Tatham and the Culture of Tobacco*, 8–9; "Weevils," *Yankee Farmer*, March 24, 1838, 90; Senex, "Turnep Fly," 6.

34. N. E. Farmer, "Cut Worm," *Farmer's Cabinet*, August 1, 1836, 28; "Boil Your Garden Before You Plant It," *Maine Farmer*, February 3, 1838, 34–35.

2. The Rise of the Professionals

1. Asa Fitch, *The Wheat Fly* (Albany, N.Y.: Niles, 1843), 4; "Report of the Committee on Improvement of Lands," November 23, 1826, 41–44, Plymouth County Agricultural Society Trustees Records, 1819–1864, Massachusetts Historical Society, Boston; Samuel H. Hammond and L. W. Mansfield, *Country Margins and Rambles of a Journalist* (New York: Derby, 1855), 353.

2. William H. Foster, "Letter on Nathaniel Hawthorne," 1860, Ms. Acc. 89062, Phillips Library, Peabody Essex Museum, Salem, Mass.

3. "Grasshoppers," *Baltimore Farmer*, reprinted in *Yankee Farmer*, October 20, 1838, 331; The Agriculturalist, "The Grain Worm," *Yankee Farmer*, March 17, 1838, 81.

4. H. C. Giddens, "To Kill Lice on Swine—Destroy Poisonous Mushrooms," *Southern Cultivator*, July 1852, 294; "Miscellaneous," *Yankee Farmer*, September 14, 1838, 99.

5. Agriculturalist, "Grain Worm," 81; G. Melvin Herndon, *William Tatham and the Culture of Tobacco, Including a Facsimile Reprint of "An Historical and Practical Essay on the Culture and Commerce of Tobacco" by William Tatham* (Coral Gables, Fla.: University of Miami Press, 1969), 211; Connecticut River Vermont Farmer, "The Grain Worm," *Yankee Farmer*, March 17, 1838, 82; "Grasshoppers," 331.

6. Quotes in John C. Greene, *American Science in the Age of Jefferson* (Ames: Iowa State University Press, 1984), 5–8; Thaddeus William Harris, *A Report of the Insects of Massachusetts Injurious to Vegetation* (Cambridge, Mass.: Folsom, Wells, and Thurston, 1841), v.

7. Arnold Mallis, *American Entomologists* (New Brunswick, N.J.: Rutgers University Press, 1971), 28–29; Thaddeus William Harris to Mr. E. Doubleday, September 23, 1829, Ernst Mayr Library, Special Collections, Museum of Comparative Zoology, Harvard University, Cambridge, Mass. Harris's work with the Harvard scientific community is documented in his letters held in the Gray Herbarium Library, Harvard University. See Josiah Quincy to Dr. Harris, May 4, 1818 (or June 14, ?), and Lewis G.

Beck to Thaddeus W. Harris, April 13, 1840, both in box AI, Entomological Correspondence folder, Papers of Thaddeus William Harris. Harris's plea for professional companionship echoed those of many other American scientists scattered across the nation in the early nineteenth century. A South Carolina botanist, for example, complained in 1830, "I am here utterly destitute. . . . We have no botanists in the up country & consequently I have always been alone" (quoted in Robert V. Bruce, *The Launching of Modern American Science, 1846–1876* [New York: Knopf, 1987], 32).

8. Harris, *Report of the Insects of Massachusetts,* v, 152–154; W. Conner Sorenson, *Brethren of the Net: American Entomology, 1840–1880* (Tuscaloosa: University of Alabama Press, 1995), 70. Samuel Scudder, speaking in 1881, rhetorically asked the Entomological Club of the American Association for the Advancement of Science, "How many of us had drawn our first admiration from Harris?" The answer was obvious—all of them ("Address of Mr. Samuel H. Scudder," *Proceedings of the American Association for the Advancement of Science* 29 [1880]: 3). On the removal of "arts" from science, see A. Hunter Dupree, "The Measuring Behavior of Americans," in *Nineteenth-Century American Science: A Reappraisal,* ed. George H. Daniels (Evanston, Ill.: Northwestern University Press, 1972), 28.

9. The biographical information on Asa Fitch, Joseph Lintner, and Townend Glover is from Mallis, *American Entomologists,* 37–43, 52–54, 61–69. Asa Fitch, diary entry of February 17, 1829, group 215, box 1, Asa Fitch Papers, Sterling Memorial Library, Yale University, New Haven, Conn.; for more on Glover, see Charles Richards Dodge, *The Life and Entomological Work of Townend Glover,* Department of Agriculture, Bulletin no. 18 (Washington, D.C.: Government Printing Office, 1898); Ebenezer Emmons to Thaddeus William Harris [letter of introduction], August 26, 1845, box AI, Entomological Correspondence folder, Papers of Thaddeus William Harris. A good example of future entomologists failing to appreciate the importance of these early pioneers can be seen in Leland O. Howard, "Progress in Economic Entomology in the United States," in *Yearbook of the United States Department of Agriculture* (Washington, D.C.: Government Printing Office, 1899), 138.

10. Harris, *Report of the Insects of Massachusetts,* v; Thaddeus William Harris to Asa Fitch, August 14, 1852, Ernst Mayr Library, Special Collections; Sorenson, *Brethren of the Net,* 70. For a sense of the persistence of the older, classificatory approach to entomology and, by contrast, the radicalism of Harris's approach, consider this remark by A. S. Packard Jr.: "Unfortunately, also, so backward is the science of entomology in this country, that the attention of its students is at present fully engrossed with classifying and describing adult insects" ("Injurious and Beneficial Insects," *American Naturalist,* September 1873, 525).

11. Mallis, *American Entomologists,* 38–41, 43–49, 69–79, 139–142; Sorenson, *Brethren of the Net,* 67–68; "The Diaries of Asa Fitch, M.D.," ed. Arnold Mallis, *Bulletin of the Entomological Society of America* 9 (1963): 26.

12. Packard, "Injurious and Beneficial Insects," 525, and "Moths Entrapped by an Asclepaid Plant (*Physianthus*) and Killed by Honey Bees," *Botanical Gazette* 5, no. 2 (1880): 19; Sorenson, *Brethren of the Net,* 70–90; Bruce, *Launching of Modern American Science,* 3–6; Howard, "Progress in Economic Entomology in the United States," 142.

13. Richard Hofstadter, *The Age of Reform: From Bryan to F.D.R.,* quoted in Sean Wilentz, "What Was Liberal History?" *New Republic,* July 10 and 17, 2006, 27.

14. Herbert Osborn, *Fragments of Entomological History, Including Some Personal Recollections of Men and Events* (Columbus, Ohio: Osborn, 1937), 11; Asa Fitch, *Third, Fourth, and Fifth Reports on the Noxious, Beneficial, and Other Insects of the State of New York* (Albany: Van Benthuysen, 1859), iv; "A String of Bugs," *Prairie Farmer*, August 1851, 1; Hae-Gyung Geong, "Exerting Control: Biology and Bureaucracy in the Development of American Entomology, 1870–1930" (Ph.D. diss., University of Wisconsin, 1999), 8.

15. Sorenson, *Brethren of the Net*, 104; William LeBaron, *Second Annual Report on the Noxious Insects of the State of Illinois* (Springfield: Springfield Journal Printing, 1872), 7; "Introduction, "*Practical Entomologist*, October 30, 1865, 4.

16. Mary Treat, *Injurious Insects of the Farm and Garden* (New York: Orange Judd, 1887), 3, 7–9; for an excellent example of Treat's clear writing style, see *Chapters on Ants* (New York: Harper, 1879). "Correspondence," *Practical Entomologist*, October 30, 1865, 6; the writer went on to say that "there is hardly one farmer or fruit grower in twenty who is acquainted with [the borers] in their different states."

17. Treat, *Injurious Insects of the Farm and Garden*, 18; Asa Fitch, *Tenth and Eleventh Reports on the Noxious, Beneficial, and Other Insects of the State of New York* (Albany: Van Benthuysen, 1867), 46–47, 55.

18. Benjamin D. Walsh, "The New Potato Bug, and Its Natural History," *Practical Entomologist*, October 30, 1865, 1, and "Self Taught Entomologists," *Practical Entomologist*, May 12, 1867, 91; Stephen A. Forbes, "Notes from Livingston and Adjacent Counties," in *Miscellaneous Essays on Economic Entomology*, ed. Stephen A. Forbes (Springfield, Ill.: Rokker, 1886), 127; "Report of the Committee on Improvement of Lands," November 23, 1826, 43, Plymouth County Agricultural Society Trustees Records, 1819–1864; Walsh, quoted in "Bugs on Apple Buds," *New England Farmer*, July 1867, 346–347.

19. "The Curculio," *Michigan Farmer*, June 1858, 175; "The Wheat Aphis," *New England Farmer*, January 1864, 16; "Dr. Fitch's Report on the Insects of New York," *Southern Planter*, July 1857, 437, summarizing Asa Fitch, *Second Report on the Noxious, Beneficial, and Other Insects of the State of New York* (Albany: Van Benthuysen, 1856).

20. Fitch, *Third, Fourth, and Fifth Reports on the Noxious, Beneficial, and Other Insects of the State of New York,* vi; Johannes Godartius, *Of Insects* (York, Eng.: John White, 1682), 6, 102; V. G. Dethier, *Man's Plague? Insects and Agriculture* (Princeton, N.J.: Darwin Press, 1976), 141; K. S. Hagan and J. M. Franz, "A History of Biological Control," in *History of Entomology*, ed. Ray F. Smith, Thomas E. Mittler, and Carroll N. Smith (Palo Alto, Calif.: Annual Reviews, 1973), 433–434.

21. Walsh, quoted in Mallis, *American Entomologists*, 44; "Introductory," *Practical Entomologist*, October 30, 1865, 6; Leland O. Howard, *Fighting the Insects: The Story of an Entomologist* (New York: Macmillan, 1933), 91, 93; Paul Popenoe, "Biological Control of Destructive Insects," *Science*, August 5, 1921, 113–114. Even Howard, an entomologist who would eventually become deeply skeptical of biological control, looked back at the recent history of American entomology from the distant perch of 1933 and recalled its early promise. Noting in his autobiography the "many opportunities for the promotion of natural [biological] control work," he reiterated that "I have always contended . . . that it would be an excellent thing to do if we could bring into the United States from all the parallel life zones of the world as many parasitic and predacious insects as can possibly acclimatized"

(*Fighting the Insects,* 113–115). Joseph S. Howe, "Progressive Farming," in *Abstract of Returns of the Agricultural Societies of Massachusetts,* ed. Charles L. Flint (Boston: Wright and Potter, 1872), 30.

22. Treat, *Injurious Insects of the Farm and Garden,* 1–15; "A Wheat or Corn Crop Saved by Chickens," *Southern Planter,* July 1857, 400; Fitch, *Tenth and Eleventh Reports on the Noxious, Beneficial, and Other Insects of the State of New York,* 55; Charles L. Flint, *Fifteenth Annual Report of the Secretary of the Massachusetts Board of Agriculture* (Boston: Wright and Potter, 1868), 124.

23. Walter C. O' Kane, *Injurious Insects: How to Recognize and Control Them* (New York: Macmillan, 1920), 45–46; Sorenson, *Brethren of the Net,* 102; John Taylor, "Ducks and the Colorado Potato Beetle," *Insect Life* 4, nos. 1–2 (1891): 76; Robert Allyn Lovely, "Mastering Nature's Harmony: Stephen Forbes and the Roots of American Ecology" (Ph.D. diss., University of Wisconsin, 1955); Wilson Flagg, "Utility of Birds," in *Abstract of Returns of the Agricultural Societies of Massachusetts for 1861* (Boston: White, 1862), 50; A. S. Packard Jr., "Nature's Means of Limiting the Numbers of Insects," *American Naturalist,* May 1874, 273–275; Stephen A. Forbes, "The Food of Birds," *Transactions of the Illinois Horticultural Society* 12 (1879): 140–145, and "The Regulative Action of Birds upon Insect Oscillations," *Bulletin of the Illinois State Laboratory of Natural History* 1, no. 6 (1883): 20–21; *Twenty-first Annual Report of the Secretary of the Massachusetts Board of Agriculture* (Boston: Wright and Potter, 1874), 314.

24. "The Grass Caterpillar," *Southern Planter,* July 1857, 439; "Pissant vs. Chinch Bug," *Southern Planter,* July 1857, 437; A. S. Packard Jr., *First Annual Report of the Injurious and Beneficial Insects of Massachusetts* (Boston: Wright and Potter, 1871), 1.

25. Clarence M. Weed, "On the Injurious Locusts of Central Illinois," in Forbes, *Miscellaneous Essays on Economic Entomology,* 51; Walsh, quoted in Dethier, *God's Plague?* 142, 135–136; Charles V. Riley, *Parasitic and Predaceous Insects in Applied Entomology* (Washington, D.C.: Government Printing Office, 1893), 141; Packard, *First Annual Report of the Injurious and Beneficial Insects of Massachusetts,* 28.

26. "Enemies of the Plant Louse," *Science,* August 9, 1889, 100–101; Clarence M. Weed, *Insects and Insecticides: A Practical Manual Concerning Noxious Insects and the Methods of Preventing Their Injuries* (New York: Orange Judd, 1908), 23; Charles V. Riley, "Parasitism in Insects" [annual address of the president], *Proceedings of the Entomological Society of Washington* 2, no. 4 (1893): 35; A. E. Verrill, "Parasites of Animals," in *Fourth Annual Report of the Secretary of the Connecticut Board of Agriculture, 1869–1870* (Hartford: Case, Lockwood, and Brainard, 1870), 79.

27. "The Curculio," *Prairie Farmer,* February 1851, 60; "Agricultural Quackery," *Southern Planter,* August 1857, 477.

28. E. O. Essig, *A History of Entomology* (New York: Macmillan, 1931), 403–405, 438–450; E. G. Lodeman, *The Spraying of Plants: A Succinct Account of the History, Principles, and Practice of the Application of Liquids and Powders to Plants, for the Purpose of Destroying Insects and Fungi* (New York: Macmillan, 1897), 5, 148; J. E. Dudley, *Nicotine Dust Kills Cucumber Beetles,* Wisconsin Agricultural Experiment Station, Bulletin no. 355 (Madison: Wisconsin Agricultural Experiment Station, 1923), 1–10; "Wire Worms," *Prairie Farmer,* May 1851, 209; Jabez Fisher, "The War with Insects," in *Twenty-sixth Annual Report of the Secretary of the Massachusetts Board of Agriculture for 1878* (Boston: Rand, Avery, 1879), 210.

29. "Chemistry as Applied to Agriculture," *Southern Planter*, October 1856, 298; Howe, "Progressive Farming," 19.

30. *Southern Planter*, February 1856, 60; "The Bark Louse," *Prairie Farmer*, March 1851, 107; William R. Schuyler, "Wheat—The Hessian Fly," *Michigan Farmer*, July 1863, 9; Howe, "Progressive Farming," 28; Fisher, "War with Insects," 210.

31. Howe, "Progressive Farming," 30; "The Borer," *Prairie Farmer*, August 1851, 336; *Prairie Farmer*, May 1851, 212.

32. C. H. Fernald, "The Association of Economic Entomologists—Address by the President—The Evolution of Economic Entomology," *Science*, October 16, 1896, 546.

33. Quoted in Thomas R. Dunlap, *Saving America's Wildlife* (Princeton, N.J.: Princeton University Press, 1988), 20–21.

34. James A. Tober, *Who Owns the Wildlife? The Political Economy of Conservation in Nineteenth-Century America* (Westport, Conn.: Greenwood Press, 1981), 85–86; Patent Office, *Report of the Commissioner of Patents for the Year 1856*, vol. 4, *Agriculture* (Washington, D.C.: Office of Printers to the House of Representatives, 1857), 52–53; Dunlap, *Saving America's Wildlife*, 21–33; Roderick Nash, *Wilderness and the American Mind*, 3rd ed. (New Haven, Conn.: Yale University Press, 1982), 108–121. Tober and Dunlap write persuasively about the rise of the conservationist ethic, but they do not develop its emergence in the context of the nation's insect infestations.

35. Hull, quoted in *Abstract of Returns of the Agricultural Societies of Massachusetts, 1856*, ed. Charles L. Flint (Boston: White, 1857), 441. On the measuring spirit, see A. Hunter Dupree, "The Measuring Behavior of Americans," in *Nineteenth-Century American Science: A Reappraisal*, ed. George H. Daniels (Evanston, Ill.: Northwestern University Press, 1972), 22–37; agricultural writer, quoted in Alan I. Marcus, *Agricultural Science and the Quest for Legitimacy: Farmers, Agricultural Colleges, and Experiment Stations, 1870–1890* (Ames: Iowa State University Press, 1985), 18–19; Fernald, "Evolution of Economic Entomology," 545. The language changed in other areas, too. Black flies were "little friends that have come to the farmers' rescue" only when they played a parasitic role in controlling an infestation of little enemies. Because ladybugs attacked aphids, they were deemed "beautiful rounded beetles, usually dressed in yellow or orange, and often adorned with black dots and markings, . . . known and admired by all." Otherwise, they were simply harmful bugs. In short, an insect's reputation came to hinge not on its inherent qualities and spiritual connotations, but on the relative reach of its destruction.

36. S. A. Oddy, *The Housekeeper's Receipt Book* (London: Oddy, 1813), 222; Clarissa Packard, *Recollections of a Housekeeper* (New York: Harper, 1934), 121; William Cobbett, *Cottage Economy* (New York: Gould, 1824), 7–8.

37. For Weed on cockroaches, clothes moths, and ants, see *Insects and Insecticides*, 322–323, 326–327, 329. The invective, it should be stressed, began much earlier than the late nineteenth century. In 1859, Fitch admitted, "I sometimes think there is no kind of mischief going on in the world of nature around us but that some insect is at the bottom of it" (*Third, Fourth, and Fifth Reports on the Noxious, Beneficial, and Other Insects of the State of New York*, vi).

38. Thaddeus William Harris to Ebenezer Emmons, ca. August 1845, box AI, Entomological Correspondence folder, Papers of Thaddeus William Harris.

3. Breaking the Plains and Fighting the Insects

1. James W. Whitaker, *Feedlot Empire: Beef Cattle Feeding in Illinois and Iowa, 1840–1900* (Ames: Iowa State University Press, 1975), chap. 2; Henry R. Schoolcraft, *Travels in the Central Portion of the Mississippi Valley: Comprising Observations of Its Mineral Geography, Internal Resources, and Aboriginal Population* (New York: Collins and Hannay, 1825), 303–310; John C. Hudson, *Making the Corn Belt: A Geographical History of Middle Western Agriculture* (Bloomington: Indiana University Press, 1994), 15–24; John R. Borchert, "The Climate of North American Grasslands," *Annals of the Association of American Geographers* 40 (1950): 1–30; Jack R. Harlan, *Theory and Dynamics of Grassland Agriculture* (Princeton, N.J.: Van Nostrand, 1956); John A. Jakle, "The American Bison and the Human Occupance of the Ohio Valley," *Proceedings of the American Philosophical Society* 112 (1968): 300–304; Dan Flores, *The Natural West: Environmental History in the Great Plains and Rocky Mountains* (Norman: University of Oklahoma Press, 2001), 57.

2. Allan G. Bogue, *From Prairie to Corn Belt: Farming on the Illinois and Iowa Prairies in the Nineteenth Century* (Chicago: University of Chicago Press, 1963), 2–7; William Cronon, *Nature's Metropolis: Chicago and the Great West* (New York: Norton, 1991), 99.

3. Solon Robinson, "Something About the Western Prairies" (1841), in *Solon Robinson: Pioneer and Agriculturalist*, ed. Herbert Anthony Kellar (New York: Da Capo Press, 1968), 1:285; Snider, quoted in Jeffrey A. Lockwood, *Locust: The Devastating Rise and Mysterious Disappearance of the Insect that Shaped the American Frontier* (New York: Basic Books, 2004), 116.

4. Wilson Flagg, "Plan of a Series of Experiments for Investigating the Potato Disease," in *Transactions of the Essex Agricultural Society for 1859* (Newbury, Mass.: Huse, 1859), 129; Fitch, quoted in Theodore R. Marmor, "Anti-Industrialism and the Old South: The Agrarian Perspective of John C. Calhoun," *Comparative Studies in Society and History* 9, no. 4 (1967): 390; Rusticus, "Diversify Your Products," *Southern Cultivator*, July 1852, 113; "Wheat 'Starved to Death,'" *American Farmer*, May 1870, 69; "Annual Address of the President," *Proceedings of the Entomological Society of Washington* 1, no. 1 (1885): 3.

5. Quoted in Clarence H. Danhof, *Change in Agriculture: The Northern United States* (Cambridge, Mass.: Harvard University Press, 1969), 24–25; Richard Hofstadter, *The Age of Reform: From Bryan to F.D.R.* (New York: Vintage, 1955), 57.

6. On why migrants in the Midwest did not practice more subsistence-oriented agriculture, Whitaker writes, "The productivity of the land, the options before them, and the developing transportation system encouraged farmers to think in terms of commercial rather than subsistence agriculture" (*Feedlot Empire*, 9); Flores, *Natural West*, 164–182.

7. Quoted in Bogue, *From Prairie to Corn Belt*; Herbert Quick, *Vandemark's Folly* (Indianapolis: Bobbs-Merrill, 1922), 228; see also Danhof, *Change in Agriculture*, 121–129.

8. Gordon G. Whitney, *From Coastal Wilderness to Fruited Plain: A History of Environmental Change in Temperate North America, 1500 to the Present* (Cambridge: Cambridge University Press, 1994), 255–260; Cronon, *Nature's Metropolis*, 220–221; Christopher Widga, "Bison, Bogs, and Big Bluestem: The Subsistence Ecology of

Middle Holocene Hunter-Gatherers in the Eastern Great Plains" (Ph.D. diss., University of Kansas, 2007); Charles Clinton Smith, "The Effect of Overgrazing and Erosion upon the Biota of the Mixed-Grass Prairie of Oklahoma," *Ecology* 21, no. 3 (1940): 384–388; *Cultivator* 4 (1837): 141; quoted in Danhof, *Change in Agriculture*, 252n.6.

9. A. Abril and E.H. Bucher, "The Effects of Overgrazing on Soil Microbial Community and Fertility in the Chaco Dry Savannahs of Argentina," *Applied Soil Ecology* 12, no. 2 (1999): 159–160; Chad L. Engels, "The Effect of Grazing Intensity on Soil Bulk Density" (available at www.ag.ndsu.nodak.edu/streeter/99report/soil_bulk.htm).

10. Department of the Interior, *Report on the Productions of Agriculture as Returned at the Tenth Census (June 1, 1880)* (Washington, D.C.: Government Printing Office, 1883), 1059, and *Surveys of Forest Reserves,* serial set 3600, vol. 11 (Washington, D.C.: Government Printing Office, 1898), 135–137; Gifford Pinchot, *The Fight for Conservation* (New York: Doubleday, Page, 1910); "Mr. Pinchot's 'Fight,'" *New York Times,* September 10, 1910; Whitney, *From Coastal Wilderness to Fruited Plain,* 168–169.

11. Whitaker, *Feedlot Empire,* 13; Cronon, *Nature's Metropolis,* 222–223; Patent Office, *Report of the Commissioner of Patents, for the Year 1849,* part 2, *Agriculture* (Washington, D.C.: Office of Printers to the House of Representatives, 1850), 175.

12. Illinois farmer, quoted in Bogue, *From Prairie to Corn Belt,* 129; Norman Hudson, *Soil Conservation,* 2nd ed. (Ithaca, N.Y.: Cornell University Press, 1981), 36–38; Phillip Jon Copeland, "The Effect of Corn and Soybean Cropping Sequence on Soil Moisture and Plant Nutrients" (Ph.D. diss., University of Minnesota, 1990); *Ohio Cultivator,* quoted in Whitney, *From Coastal Wilderness to Fruited Plain,* 227, see also 230–240; Patent Office, *Report of the Commissioner of Patents, for the Year 1849.*

13. Hugh H. Bennett, "Soil Erosion: A National Menace," *Scientific Monthly,* November 1934, 385, 390–391.

14. Whitney, *From Coastal Wilderness to Fruited Plain,* 241–242; Stanley W. Trimble and Steven W. Lund, *Soil Conservation and the Reduction of Erosion and Sedimentation in the Coon Creek Basin, Wisconsin,* U.S. Geological Survey Professional Paper, no. 1234 (Washington, D.C.: Government Printing Office, 1976); J. C. Knox, "Human Impacts on Wisconsin Stream Channels," *Annals of the Association of American Geographers* 67 (1977): 323–342; Bennett, "Soil Erosion," 385–389.

15. Robinson, "Something About the Western Prairies," 285; Sellers G. Archer, *Soil Conservation* (Norman: University of Oklahoma Press, 1956), 124.

16. Charles V. Riley, *Potato Pests: Being an Illustrated Account of the Colorado Potato-Beetle and the Other Insect Foes of the Potato in North America. With Suggestions for Their Repression and Methods for Their Destruction* (New York: Orange Judd, 1876), 9–12; W.L. Tower, "The Colorado Potato Beetle," *Science,* September 21, 1900, 438–440; Henry Shimer, "Insects Injurious to the Potato," *American Naturalist,* April 1869, 91–93; Arnold Mallis, *American Entomologists* (New Brunswick, N.J.: Rutgers University Press, 1971), 17–24; E. Porter Felt, "Why Do Insects Become Pests?" *Scientific Monthly,* May 1938, 437–440; Mary Treat, *Injurious Insects of the Farm and Garden* (New York: Orange Judd, 1887), 94–99.

17. What the nineteenth-century farmers and entomologists believed about the beetle was not always altogether accurate. Recent studies suggest, for example, that the Colorado potato beetle may have been in Iowa as early as 1811 and moved both west and east. See R.A. Casagrande, "The Iowa Potato Beetle, Its Discovery

and Spread to Potatoes," *Bulletin of the Entomological Society of America* 31 (1985): 27–29, and entry on the beetle in *Encyclopedia of Entomology,* ed. John L. Capinera (Boston: Kluwer, 2004), 1:583–585; Benjamin D. Walsh, "The New Potato Bug, and Its Natural History," *Practical Entomologist,* October 30, 1865, 2.

18. Riley, *Potato Pests,* 12–16, 22–27; Tower, "Colorado Potato Beetle," 438–440; Arthur W. Gilbert, *The Potato* (New York: Macmillan, 1921), 170–171; Treat, *Injurious Insects of the Farm and Garden,* 95.

19. "Potato Bugs in New Jersey," *New York Times,* May 27, 1884, 5.

20. Riley, *Potato Pests,* 23; Felt, "Why Do Insects Become Pests?" 438.

21. Comstock, quoted in Leland O. Howard, "Striking Entomological Events of the Last Decade of the Nineteenth Century," *Scientific Monthly,* July 1930, 9; Daniel William Coquillet, *Report on the Scale Insects of California,* Department of Agriculture, Division of Entomology, Bulletin no. 26 (Washington, D.C.: Government Printing Office, 1892), 21–23, who wrote: "It is to be regretted that any locality should be thus stigmatized by having its name applied to a pestiferous scale insect"; Felt, "Why Do Insects Become Pests?" 439; T. D. A. Cockerell, "The Kieffer Pear and the San Jose Scale," *Science,* September 28, 1900, 488; John B. Smith, *The San Jose Scale and How It May Be Controlled,* New Jersey Agricultural Experiment Station, Bulletin no. 125 (New Brunswick: New Jersey Agricultural Experiment Station, 1897), 4.

22. Smith, "San Jose Scale," 14; "A Scale Destroyer," *San Jose Mercury News,* May 11, 1888, 3; "The San Jose Scale: Pest Brought to Tacoma in Oranges," *Tacoma Daily News,* April 17, 1891, 1.

23. "The Scale and the Beetle," *New York Times,* January 13, 1895, 4; "A New Fruit Tree Pest," *Wheeling Register,* March 10, 1896, 2; "Insect Pests," *Dallas Morning News,* June 23, 1900, 4; Howard, "Striking Entomological Events," 9; Harold Morrison, "Scale Insects," *Scientific Monthly,* March 1926, 245.

24. "American Fruit in Germany," *New York Times,* February 13, 1898, 19; Smith, "San Jose Scale," 4.

25. T. D. A. Cockerell, "Some Western Weeds and Alien Weeds in the West," *Botanical Gazette* 20, no. 11 (1895): 503.

26. The article was reprinted as "The Crops: A General Review of the Condition and Prospects," *New York Times,* July 30, 1874, 6.

27. "The Chinch-Bug in Illinois," *Science,* April 12, 1889, 275–276; "Agriculture at Yale," *New York Times,* February 3, 1860, 5; C.L. Marlatt, *The Principle Insect Enemies of Growing Wheat,* Department of Agriculture, Bulletin no. 132 (Washington, D.C.: Government Printing Office, 1901), 6; Charles V. Riley, "The Chinch Bug," *American Agriculturalist,* November 1881, 476.

28. *Independent Chronicle and Boston Patriot,* June 30, 1821, 1; "Wheat Harvest," *Alexandria Herald,* May 23, 1823, 3; "Chinch Bug," *Boston Daily Advertiser,* May 21, 1823, 2; on the outbreak in North Carolina, see Victor E. Shelford and W. P. Flint, "Populations of the Chinch Bug in the Upper Mississippi Valley from 1823 to 1940," *Ecology* 24, no. 4 (1943): 435; "The Crops," *New Bedford Mercury,* July 19, 1839, 2; Stephen A. Forbes, *On the Chinch Bug in Illinois: Present Condition and Prospects for 1887 and 1888,* Office of the State Entomologist of Illinois, Bulletin no. 2 (Champaign: Gazette Steam Print, 1887), 28.

29. Shelford and Flint, "Populations of the Chinch Bug in the Upper Mississippi Valley," 435–437; Forbes, *On the Chinch Bug in Illinois,* 33.

30. Forbes, *On the Chinch Bug in Illinois*, 28–31; Department of Agriculture, Division of Entomology, *Reports of Observations and Experiments in the Practical Work of the Division*, Bulletin no. 13 (Washington, D.C.: Government Printing Office, 1885), 35–36.

31. Ohio Agricultural Experiment Station, *The Chinch Bug*, Bulletin no. 69 (Columbus: State Printer, 1896), 18–20; William LeBaron, *Second Annual Report on the Noxious Insects of the State of Illinois* (Springfield: State Printing Office, 1872), 1–10.

32. Quoted in "The Hessian Fly's Ravages," *New York Times*, November 20, 1879, 1; F. M. Webster, *The Hessian Fly*, Ohio Agricultural Experiment Station, Bulletin no. 7 (Columbus: State Printer, 1891), 136–137.

33. "Prices," *New Orleans Times*, June 18, 1878, 2; *Dallas Weekly Herald*, May 26, 1877, 31; "Hessian Fly Ravages," 1; "Illinois Crops," *Philadelphia Inquirer*, August 15, 1881, 1; "The Crops," *Philadelphia Inquirer*, July 26, 1881, 8; "The Hessian Fly Pest Spreading," *New York Times*, June 9, 1883, 1; "Bad for Illinois Wheat," *New York Times*, February 12, 1889, 1; "Missouri's Wheat Fields," *New York Times*, April 28, 1891, 1; "Work of the Hessian Fly," *New York Times*, May 17, 1891, 8; "Hessian Fly Near Geneva, N.Y.," *New York Times*, June 2, 1899, 5.

34. F. M. Webster, "The Early History of the Hessian Fly in America" (paper presented at the annual meeting of the Society for the Promotion of Agricultural Science, Philadelphia, 1905), 1–7.

35. "Missouri," *Daily Constitution*, July 29, 1874, 2.

36. Lockwood, *Locust*, 14–16; Charles V. Riley, "The Locust Plague: How to Avert It," *Proceedings of the American Association for the Advancement of Science* 24 (1875): 215–221, and *The Locust Plague in the United States: Being More Particularly a Treatise on the Rocky Mountain Locust or So-called Grasshopper, as It Occurs East of the Rocky Mountains, with Practical Recommendations for Its Destruction* (Chicago: Rand McNally, 1877); Charles Bomar, "The Rocky Mountain Locust: Extinction and the American Experience" (available at www.sciences.org/locusts/locusts.asp); Entomological Commission, *First Annual Report for the Year 1877 Relating to the Rocky Mountain Locust and the Best Means of Preventing Its Injuries and of Guarding Against Its Invasions* (Washington, D.C.: Government Printing Office, 1878), 86–88.

37. "Minnesota," *Wheeling Register*, May 7, 1877, 1; John T. Schlebecker, "Grasshoppers in American Agricultural History," *Agricultural History* 27, no. 3 (1953): 87; Entomological Commission, *First Annual Report for the Year 1877 Relating to the Rocky Mountain Locust*; Lockwood, *Locust*, 211.

38. Laura Ingalls Wilder, quoted in Bomar, "Rocky Mountain Locust"; E. O. C. Ord to William W. Belknap, "Letter from the Secretary of War."

39. Lisa Levitt Ryckman, "The Great Locust Mystery: Grasshoppers that Ate the West Became Extinct," *Rocky Mountain News*, July 29, 1999.

40. Schlebecker, "Grasshoppers in American Agricultural History," 90–92.

4. Charles V. Riley and the Broken Promises of Early Insecticides

1. Leland O. Howard, "The Entomological Society of Washington," *Proceedings of the Entomological Society of Washington* 9 (1909): 78–79.

2. Quoted in Arnold Mallis, *American Entomologists* (New Brunswick, N.J.: Rutgers University Press, 1971), 71.

3. Mallis, *American Entomologists*, 69–71; G. [sic] V. Riley, "Friendly Notes," *Canadian Entomologist* 3 (1871): 117; W. Conner Sorenson, *Brethren of the Net: American Entomology, 1840–1880* (Tuscaloosa: University of Alabama Press, 1995), 77; Hae-Gyung Geong, "Exerting Control: Biology and Bureaucracy in the Development of American Entomology, 1870–1930" (Ph.D. diss., University of Wisconsin, 1999), 24.

4. Mallis, *American Entomologists*, 72–73; Geong, "Exerting Control," 25; Margaret W. Rossiter, "The Organization of the Agricultural Sciences," in *The Organization of Knowledge in Modern America, 1860–1920*, ed. Alexandra Oleson and John Voss (Baltimore: Johns Hopkins University Press, 1979), 220–222. In somewhat backhanded fashion, Leland O. Howard called Riley's reports "monuments to the State of Missouri" before agreeing that they "formed the basis for the new economic entomology of the world" ("Progress in Economic Entomology in the United States," in *Yearbook of the United States Department of Agriculture* [Washington, D.C.: Government Printing Office, 1899], 140).

5. Charles V. Riley, *Sixth Annual Report on the Noxious, Beneficial, and Other Insects of the State of Missouri* (Jefferson City: Regan and Carter, 1874), 30–40, 52, and *Eighth Annual Report on the Noxious, Beneficial, and Other Insects of the State of Missouri* (Jefferson City: Regan and Carter, 1876), 160–165; Sorenson, *Brethren of the Net*, 99, 114–116; Geong, "Exerting Control," 25; Charles V. Riley, *Annual Address as President of the Entomological Society of Washington for the Year 1884* (Washington, D.C.: Gibson, 1886), 6, 8. This report is useful in that it reveals Riley's openness to all forms of control and preference for cultural control: "With the improved methods of applying bisulphide of carbon . . . the French grape grower can measurably protect his vineyards . . . but the chief reliance is on the resistant American vines" (6). Riley's primary concern with entomologists who spent too much time on description was that they became obsessed with declaring new species based on "every little individual variation," an impulse that "will never give us a natural system" ("Friendly Notes," 119). On his work with carnivorous plants, see Charles V. Riley, "Descriptions and Natural History of Two Insects Which Brave the Dangers of *Saarracena variolaris*," *Transactions of the Academy of Science of St. Louis* 4 (1874): 235.

6. It is worth noting that Riley himself lobbied hard for the formation of an entomological commission, in one speech declaring, after a detailed overview of the locust ravages, that "the time is most opportune, for a commission created this winter would have an opportunity, that may not again occur for years, on studying and experimenting with the young insects" (*Address Before the Academy of Science of St. Louis at Its Annual Meeting for 1877* [St. Louis: Studley, 1877], 8); Sorenson, *Brethren of the Net*, 141–143; Edmund P. Russell, "War on Insects: Warfare, Insecticides, and Environmental Change in the United States, 1870–1945" (Ph.D. diss., University of Michigan, 1993), 20–25; Entomological Commission, *First Annual Report for the Year 1877 Relating to the Rocky Mountain Locust and the Best Means of Preventing Its Injuries and of Guarding Against Its Invasions* (Washington, D.C.: Government Printing Office, 1878), and *Second Report, for the Years 1878 and 1879; Relating to the Rocky Mountain Locust, and the Western Cricket, and Treating of the Best Means of Subduing the Locust in Its Permanent Breeding Grounds, with a View of Preventing Its Migrations into the More Fertile Portions of the Trans-Mississippi Country* (Washington, D.C.: Government Printing Office, 1880).

7. Entomological Commission, *First Annual Report for the Year 1877 Relating to the Rocky Mountain Locust,* 79, 124–128, 284, 399; Charles V. Riley, *Destructive Locusts: A Popular Consideration of a Few of the More Injurious Locusts (or "Grasshoppers") in the United States, Together with the Best Means of Destroying Them,* Department of Agriculture, Division of Entomology, Bulletin no. 25 (Washington, D.C.: Government Printing Office, 1891), 9–26; Geong, "Exerting Control," 26. On his close attention to local conditions as he conducted his experiments, see Charles V. Riley, "Entomological Papers," *Proceedings of the American Association for the Advancement of Science* 27 (1878): 5–7.

8. One can see Riley struggling to balance the local with the national implications of the findings of the Division of Entomology in the reports he published. For example, in a study of insects that attack the hackberry tree, he felt compelled to add, "What is said of the forms growing in Missouri in my sixth report on the insects of Missouri (1874, p. 137) will apply to other parts of the country in the same latitudes—in fact, throughout its range" (Charles V. Riley, "Insects Injurious to the Hackberry," in *Fifth Report of the United States Entomological Commission, on Insects Injurious to Forest and Shade Trees* [Washington, D.C.: Government Printing Office, 1890], 601).

9. Geong, "Exerting Control," 26–28. For evidence that he did not reject chemical options in choosing to favor biological ones, see Charles V. Riley, "Improved Methods for Spraying Trees Against Insects," *Proceedings of the American Association for the Advancement of Science* 32 (1884): 466–467.

10. As a firm advocate of biological control, Benjamin Walsh dedicated the bulk of his research to distinguishing native and introduced insect species, as evidenced in "On Certain Entomological Speculations of the New England School of Naturalists," *Proceedings of the Entomological Society of Philadelphia* 3 (1864): 207–210. The professorial target of Walsh's ire was James Dwight Dana: "Professor Dana and His Entomological Speculations," *Proceedings of the Entomological Society of Philadelphia* 6 (1866): 119. For background on Walsh, see Carol A. Sheppard, "Benjamin Dann Walsh: Pioneer Entomologist and Promoter of Darwinian Theory," *Annual Review of Entomology* 49 (2004): 1–25.

11. Benjamin D. Walsh, "Imported Insects," *Practical Entomologist,* September 29, 1866, 119. Riley may have also learned about the importance of local investigation from Walsh, who continually remarked that the findings in one state mattered little for the findings in another: "Insects Injurious to Vegetation in Illinois," *Transactions of the Illinois State Agricultural Society* 4 (1861): 336–339.

12. Charles V. Riley, "Darwin's Work in Entomology," *Proceedings of the Biological Society of Washington, D.C.* 1 (1882): 70–80, and *Address Before the St. Louis Academy of Science,* 9; Leland O. Howard, *A Study in Insect Parasitism* (Washington, D.C.: Government Printing Office, 1897), 6; Walsh, "Insects Injurious to Vegetation in Illinois," 339. According to Paul D. Harwood, "The generalizations Darwin formed from observations in that land [Galápagos] and elsewhere might have enabled a keen analyst to predict the results of our modern uses of insecticides" ("Is Natural Selection an Outworn Term?" *Ohio Journal of Science* 50, no. 6 [1950]: 281).

13. Charles V. Riley, "Recent Advances in Economic Entomology," *Bulletin of the Philosophical Society of Washington* 5 (1886): 11–12, and *Remarks on the Insect Defoliators of Our Shade Trees* (New York: Globe Stationary and Printing, 1887), 7–9. Riley was quick to point out the flaws of biological control as well, as when he

declared of the English sparrow that "there is no evidence that the bird has been instrumental in checking any of our insect pests" ("The Insectivorous Habits of the English Sparrow," in *The English Sparrow in America,* Department of Agriculture, Division of Economic Ornithology and Mammalogy, Bulletin no. 1 [Washington, D.C.: Government Printing Office, 1889], 132).

14. Charles V. Riley, "Parasites of the Cotton Worm," *Canadian Entomologist* 11, no. 9 (1879): 161–162; "On the Parasites of the Hessian Fly," *Proceedings of the American Association for the Advancements of Science* 34 (1885): 332; and "Insecticides and Means of Applying Them to Shade and Forest Trees," in *Fifth Report of the United States Entomological Commission,* 34.

15. V. G. Dethier, *Man's Plague? Insects and Agriculture* (Princeton, N.J.: Darwin Press, 1976), 144–145; *Insect Life,* quoted in Albert Koebele, *Report of a Trip to Australia . . . to Investigate the Natural Enemies of the Fluted Scale,* Department of Agriculture, Division of Entomology, Bulletin no. 21 (Washington, D.C.: Government Printing Office, 1890), 6; E. O. Essig, *A History of Entomology* (New York: Macmillan, 1931), 125; "How Hawaii Got Rid of Insects," *New York Times,* August 8, 1897, 18.

16. Charles V. Riley, "Parasitism in Insects" [annual address of the president], *Proceedings of the Entomological Society of Washington* 2, no. 4 (1893): 2; F. B. Webster, *The Hessian Fly,* Ohio Agricultural Experiment Station, Bulletin no. 7 (Columbus: State Printer, 1891): 133; Frederick W. Malley, *The Boll Worm of Cotton,* Department of Agriculture, Division of Entomology, Bulletin no. 24 (Washington, D.C.: Government Printing Office, 1891), 27; Charles V. Riley, *The Icerya or Fluted Scale, Otherwise Known as the Cottony Cushion Scale,* Department of Agriculture, Division of Entomology, Bulletin no. 15 (Washington, D.C.: Government Printing Office, 1887), 13; Stephen A. Forbes, "On a Bacterial Disease of the Larger Corn Root Worm," in *Seventeenth Report of the State Entomologist of Illinois on the Noxious and Beneficial Insects* (Springfield: Rokker, 1891), 72–73; Charles V. Riley, "Parasitic and Predacious Insects in Applied Entomology," *Insect Life* 6, no. 2 (1893): 131–141.

17. Riley, *Icerya or Fluted Scale,* 25; Webster, *Hessian Fly,* 136; C. L. Marlatt, *The Principal Insect Enemies of Growing Wheat,* Department of Agriculture, Farmer's Bulletin no. 132 (Washington, D.C.: Government Printing Office, 1901), 19, 24; Charles V. Riley, *Reports of Observations and Experiments in the Practical Work of the Division,* Department of Agriculture, Division of Entomology, Bulletin no. 26 (Washington, D.C.: Government Printing Office, 1891), 48; "Insect Pests, a Noted Entomologist Expresses Some Opinions," *San Jose Mercury News,* January 14, 1888, 3. A newspaper feature on failed attempts to control the gypsy moth recounted that "the thick underbrush of an infested forest was sprayed with Paris green [a popular synthetic insecticide]," and "cocoons were collected and placed in barrels covered over with gauze." The point of this procedure "was to prevent the escape of the gypsy moths." But, in a telling indication of where biological control stood, the reporter could not conceal his disappointment that "the parasites that prey upon them were permitted to get away" (*Dallas Morning News,* January 6, 1895, 19).

18. Act of 1887 Establishing Agricultural Experiment Stations" (available at www.oardc.ohio-state.edu/www/hatch.html). For an institutional look at the rise of agricultural experiment stations and the Hatch Act, see Alan I. Marcus, *Agricultural Science and the Quest for Legitimacy: Farmers, Agricultural Colleges, and Experiment Stations, 1870–1890* (Ames: Iowa State University Press, 1985), 59–61, 171–185; Geong,

"Exerting Control," 27–28; and A.B. Ward, "An Agriculture Experiment Station," *New England Magazine,* 1893, 65.

19. Descriptions of how to mix several insecticides are in Hatch Experiment Station of the Massachusetts Agricultural College, *Fungicides, Insecticides, Spraying Calendar,* Bulletin no. 80 (Amherst: Carpenter and Morehouse, 1902), 9–12. See also Howard Evarts Weed, *Insecticides and Their Application,* Mississippi Agricultural and Mechanical College, Experiment Station, Bulletin no. 27 (Starkville: Mississippi Agricultural and Mechanical College, 1893), 2–4; F.H. Fowler, *Insecticides and Their Application* (Boston: Wright and Potter, 1891), 6–9; John H. Perkins, *Insects, Experts, and the Insecticide Crisis: The Quest for New Pest Management Strategies* (New York: Plenum, 1982), 3–4; Adelynne Hiller Whitaker, "A History of Federal Pesticide Regulation in the United States to 1947" (Ph.D. diss., Emory University, 1974), 5; Essig, *History of Entomology,* 423–431; Howard, "Progress in Economic Entomology in the United States," 145–150.

20. Essig, *History of Entomology,* 403–405, 438–450; E.G. Lodeman, *The Spraying of Plants: A Succinct Account of the History, Principles, and Practice of the Application of Liquids and Powders to Plants, for the Purpose of Destroying Insects and Fungi* (New York: Macmillan, 1897), 5, 148; J.E. Dudley, *Nicotine Dust Kills Cucumber Beetles,* Wisconsin Agricultural Experiment Station, Bulletin no. 355 (Madison: Wisconsin Agricultural Experiment Station, 1923), 1–10; Joel Gillingham Account Book, 1890–1895, Manuscript 0.313; Philip Martino Account Book, 1909–1911, Manuscript 0.238, both in the Phillips Library, Peabody Essex Museum, Salem, Mass.

21. "Giving Insects Fits," *New Hampshire Patriot,* February 25, 1874, 4; "Chloride of Lime as an Insecticide," *Farmer's Cabinet,* July 10, 1862, 4; "Abolition of Insect Pests," *Macon Daily Telegraph,* December 7, 1861, 2; Essig, *History of Entomology,* 406; Russell, "War on Insects," 55.

22. Department of Agriculture, *Report of the Commissioner of Agriculture for 1881–1882* (Washington, D.C.: Government Printing Office, 1882), 158–159; Matthew Cooke, *Injurious Insects of the Orchard, Vineyard, Field, Garden, Conservatory, Household, Storehouse, Domestic Animals, with Remedies for Their Extermination* (Sacramento, Calif.: Crocker, 1883), 402, 420.

23. Benjamin D. Walsh, editorial, *Practical Entomologist,* May 28, 1866, 74; Whitaker, "History of Federal Pesticide Regulation," 2–9; Perkins, *Insects, Experts, and the Insecticide Crisis,* 4–5; John B. Smith, *Raupenlein and Dendrolene,* New Jersey Agricultural Experiment Station, Bulletin no. 111 (New Brunswick: New Jersey Agricultural Experiment Station, 1895), 34. For a list and evaluation of bogus elixirs, see J.K. Haywood, *Insecticides and Fungicides,* Department of Agriculture, Farmers' Bulletin no. 146 (Washington, D.C.: Government Printing Office, 1902), 10–13.

24. In retrospect, it should have been. See http://www.epa.gov; http://environmentalchemistry.com/yogi/hazmat/placards/class6.html; www.scorecard.org/chemical-profiles/summary; "Arsenic for Cotton Worms," *Fort Worth Gazette,* June 28, 1891, 3; and "Insects and Insecticides," *Knoxville Journal,* August 22, 1894, 6. Contemporary studies have revealed chilling information that nineteenth-century farmers and insecticide pushers did not know. "Acute exposure to Paris green," according to the Environmental Protection Agency, "may require decontamination and life support for the victims" (http://www.epa.gov). Less acute exposure, however, is no picnic. Signs of the slightest ingestion include "a sweetish metallic taste and garlic odor to breath and feces, difficulty in swallowing, vomiting, diarrhea,

and dehydration" (http://www.epa.gov). The Environmental Protection Agency rates London purple a "division 6.1 poison," meaning that it is "presumed to be toxic to humans" and is "a material . . . which causes extreme irritation" (http://www.epa.gov). Kerosene is a suspected endocrine toxicant, gastrointestinal intoxicant, neurotoxicant, and respiratory intoxicant. Lead arsenite is a proven carcinogen and reproductive intoxicant. Pyrethrum, although obtained from the innocent chrysanthemum, is a suspected carcinogen, immunotoxicant, and liver toxicant.

25. C. H. Tyler Townsend, *Insecticides and Their Applications,* New Mexico College of Agriculture and Mechanic Arts, Agricultural Experiment Station, Bulletin no. 9 (Las Cruces: Rio Grande Republican, 1892), 8–10; Fowler, *Insecticides and Their Application,* 13, 15; *Insect Life* 4, nos. 5–6 (1891): 204–205; E. V. Wilcox, "Some Results of Experiment Station Work with Insecticides," in *Annual Report of the Office of Experimental Stations for the Year Ended June 30, 1905* (Washington, D.C.: Government Printing Office, 1906), 259; Leland O. Howard, *The San Jose Scale in 1896–1897,* Department of Agriculture, Division of Entomology, Bulletin no. 12 [new series] (Washington, D.C.: Government Printing Office, 1898), 16–17; John B. Smith, *Cut-Worms: The Sinuate Pear-Borer; the Potato Stalk-Borer; Bisulphide of Carbon as an Insecticide,* New Jersey Agricultural Experiment Station, Bulletin no. 109 (New Brunswick: New Jersey Agricultural Experiment Station, 1895), 36. When a farmer in San Marcos, Texas, discovered that "his plantation hands" would have nothing to do with the arsenic he asked them to apply to his cotton crop ("not a mother's son would have anything to do with it"), "he hauled off his boots, placed half a pint of the poisonous solution into each, and then drew them on again." His insistence on the safety of arsenic rested on the evidence that "no injurious effect whatsoever resulted." Too often, this kind of proof, buttressed by the claim of infallibility, sufficed.

26. Stephen A. Forbes, "Synopsis of Recent Work with Arsenical Insecticides," *Transactions of the Illinois Horticultural Society* 23 (1890): 318–319; Virginia Agricultural Experiment Station, *Lime Sulphur Wash,* Bulletin no. 141 (Blacksburg: Virginia Agricultural Experiment Station, 1902), 235.

27. Wilcox, "Some Results of Experiment Station Work with Insecticides," 250, who concluded that "it is not possible to kill all of the scales on infested trees for the reason that the spray does not come into contact with all of the insects"; Townsend, *Insecticides and Their Applications,* 5.

28. Quoted in Howard, *San Jose Scale in 1896–1897,* 20–21.

29. According to Townsend, "It is better to mix for yourself an efficient insecticide, whose ingredients you know, than to use a manufactured one which is likely to prove useless" (*Insecticides and Their Applications,* 4).

30. Townsend, *Insecticides and Their Applications,* 7–8; Herbert Osborn, "An Experiment with Kerosene Emulsions," *Insect Life* 4 (1891–1892): 63; Daniel William Coquillet, "Reports on the Various Methods of Destroying the Red Scale of California," in Department of Agriculture, Division of Entomology, *Reports of Observations and Experiments in the Practical Work of the Division,* Bulletin no. 22 (Washington, D.C.: Government Printing Office, 1890), 17; C. H. Fernald, "The Association of Economic Entomologists—Address by the President—The Evolution of Economic Entomology," *Science,* October 16, 1896, 546.

31. Townsend, *Insecticides and Their Applications,* 6–7; Fowler, *Insecticides and Their Application,* 11; Idaho entomologist, quoted in Wilcox, "Some Results of Experiment Station Work with Insecticides," 262.

32. Forbes, "Synopsis of Recent Work with Arsenical Insecticides," 319; Wilcox, "Some Results of Experiment Station Work with Insecticides," 250; *Insect Life* 4, nos. 3–4 (1891): 154; John B. Smith, *Insecticide Experiments for 1904*, New Jersey Agricultural Experiment Station, Bulletin no. 178 (New Brunswick: New Jersey Agricultural Experiment Station, 1904), 3–5; Malley, *Boll Worm of Cotton*, 41–43; Department of Agriculture, Division of Entomology, *Reports of Observations and Experiments in the Practical Work of the Division*, Bulletin no. 3 (Washington, D.C.: Government Printing Office, 1883), 20–23; "Dosing Trees with Sulphur and Other Substances," *Insect Life* 1, no. 7 (1889): 223; "Alum as a Current Worm Remedy," *Insect Life* 1, no. 7 1889): 229–230; Haywood, *Insecticides and Fungicides*, 12.

33. Smith, *Insecticide Experiments for 1904*, 3–5; Forbes, "Synopsis of Recent Work with Arsenical Insecticides," 315–317; John B. Smith, *Crude Petroleum as an Insecticide*, New Jersey Agricultural Experiment Station, Bulletin no. 138 (New Brunswick: New Jersey Agricultural Experiment Station, 1899), 5; Wilcox, "Some Results of Experiment Station Work with Insecticides," 250; *Insect Life* 4, nos. 3–4 (1891): 154.

34. F. M. Webster, *Reports of Experiments with Various Insecticide Substances*, Department of Agriculture, Division of Entomology, Bulletin no. 11 (Washington, D.C.: Government Printing Office, 1886), 9–22; Fernald, "Evolution of Economic Entomology," 544.

35. Howard, *Fighting the Insects*, 29; Mallis, *American Entomologists*, 878–883; Leland O. Howard, *A History of Applied Entomology (Somewhat Anecdotal)*, Smithsonian Miscellaneous Collections, no. 84 (Washington, D.C.: Smithsonian Institution, 1930), 90; Russell, "War on Insects," 28.

36. Howard, *Fighting the Insects*, 233.

37. Mallis, *American Entomologists*, 83; Geong, "Exerting Control," 30–35; Russell, "War on Insects," 40–45.

5: Mosquitoes, War, and Chemicals

1. Arnold Mallis, *American Entomologists* (New Brunswick, N.J.: Rutgers University Press, 1971), 79–92; Leland O. Howard, *Fighting the Insects: The Story of an Entomologist* (New York: Macmillan, 1933), 3, 214; L. M. Russell, "Leland Ossian Howard: A Historical Review," *Annual Review of Entomology* 23 (1978): 3; Charles V. Riley, "Recent Advances in Economic Entomology," *Bulletin of the Philosophical Society of Washington* 5 (1886): 10.

2. Charles E. Rosenberg, "Rationalization and Reality in the Shaping of American Agricultural Research, 1875–1914," *Social Studies of Science* 7, no. 4 (1977): 401. An example of how pervasive this insecurity was comes from J. C. Neal, a Florida entomologist, who wrote, "It is very common to ridicule the efforts of practical entomologists, and belittle the results obtained by their methods of preventing or mitigating the ravages of insects" (quoted in F. H. Fowler, *Insecticides and Their Application* [Boston: Wright and Potter, 1891], 3).

3. One example of the imperative to bring the government to bear on national problems came from Seaman A. Knapp, a professor of agriculture at Iowa State University, who wrote that the "magnitude" of agriculture's problems "is too great for individual enterprise, or even associated enterprise . . . national government is the only party which has the means to carry out successful termination investigations upon a scale commensurate with the interests involved" ("Agricultural Experi-

ment Stations," *Western Stock Journal and Farmer* 6 [1877]: 246). On progressivism and populism in general, a bibliography would consume an entire volume. My own thoughts on the populism–progressive elision come from a critical reading of Richard Hofstadter, *The Age of Reform: Bryan to F.D.R.* (New York: Vintage, 1955), 148–164. See also Elizabeth Sanders, *Roots of Reform: Farmers, Workers, and the American State, 1877–1917* (Chicago: University of Chicago Press, 1999), and Robert H. Wiebe, *Search for Order, 1877–1920* (New York: Hill and Wang, 1967).

4. Leland O. Howard, *The Principal Insects Affecting the Tobacco Plant*, Department of Agriculture, Farmer's Bulletin no. 120 (Washington, D.C.: Government Printing Office, 1900), 6–9; C. L. Marlatt, *The Principal Insect Enemies of Growing Wheat*, Department of Agriculture, Farmer's Bulletin no. 132 (Washington, D.C.: Government Printing Office, 1901), 10–12, 19, and *Scale Insects and Mites on Citrus Trees*, Department of Agriculture, Farmer's Bulletin no. 172 (Washington, D.C.: Government Printing Office, 1903), 6–9; Leland O. Howard, *Three Insect Enemies of Shade Trees*, Department of Agriculture, Farmer's Bulletin no. 99 (Washington, D.C.: Government Printing Office, 1899), 11–12, and *Fighting the Insects*, 53.

5. H. Kirkland, Superintendent for the Suppressing of the Gypsy and Brown Tail Moth, to General Appleton, January 15, 1906, manuscript N 517, box 10, folder 118; H. M. Aiken to "the editor," February 9, 1906, box 10, folder 112; and C. S. Saergent to General Appleton, April 12, 1905, box 15, all in papers of the Massachusetts Society for the Propagation of Agriculture, Massachusetts Historical Society, Boston.

6. Thomas R. Dunlap, *DDT: Scientists, Citizens, and Public Policy* (Princeton, N.J.: Princeton University Press, 1981), 33–35; Edmund P. Russell, "War on Insects: Warfare, Insecticides, and Environmental Change in the United States, 1870–1945" (Ph.D. diss., University of Michigan, 1993), 42. In 1911, Leland O. Howard and William F. Fiske wrote that "the published information concerning these enemies [imported to control the gypsy moth] was deficient and unreliable" (*The Importation into the United States of the Parasites of the Gypsy Moth and the Brown-Tail Moth*, Department of Agriculture, Bureau of Entomology, Bulletin no. 91 [Washington, D.C.: Government Printing Office, 1911], 305). See also Richard C. Sawyer, *To Make a Spotless Orange: Biological Control in California* (Ames: Iowa State University Press, 1996), 62–63.

7. Russell, "War on Insects," 42–47; Mallis, *American Entomologists*, 391; W. D. Hunter, *The Boll Weevil Problem*, Department of Agriculture, Farmer's Bulletin no. 344 (Washington, D.C.: Government Printing Office, 1909), 46; Richard C. Sawyer, "Monopolizing the Insect Trade: Biological Control in the USDA, 1888–1951," *Agricultural History* 64, no. 2 (1990): 276; Frederick W. Malley, *The Mexican Cotton Boll Weevil*, Department of Agriculture, Farmer's Bulletin no. 130 (Washington, D.C.: Government Printing Office, 1901), 23; Howard, *Fighting the Insects*, 156.

8. Howard, *Three Insect Enemies of Shade Trees*, 10–25, and "Danger of Importing Insect Pests," in *Yearbook of the United States Department of Agriculture for 1897* (Washington, D.C.: Government Printing Office, 1897), 529–552. Support for biological control of the tussock moth comes from Leland O. Howard, *A Study in Insect Parasitism*, Department of Agriculture, Division of Entomology, Technical Series no. 5 (Washington, D.C.: Government Printing Office, 1897), 6.

9. Mallis, *American Entomologists*, 79–86; Leland O. Howard, "Progress in Economic Entomology in the United States," in *Yearbook of the United States Department of Agriculture* (Washington, D.C.: Government Printing Office, 1899), 135–156.

10. Howard, *Fighting the Insects*, 116–138; Robert A. Matheson, *A Handbook of Mosquitoes of North America* (Springfield, Ill.: Thomas, 1929), 66–71; Leland O. Howard, *Mosquitoes: How They Live; How They Carry Disease; How They Are Classified; How They May Be Destroyed* (New York: McClure, Phillips, 1901), 167–173; E. O. Essig, *A History of Entomology* (New York: Macmillan, 1931), 909; Gordon Harrison, *Mosquitoes, Malaria, and Man: A History of the Hostilities Since 1880* (New York: Dutton, 1978), 167–168; Howard, *Fighting the Insects*, 53.

11. The story of Ross's discovery has been often told and, naturally, often embellished, most often by Ross himself, who, predictably, was running neck and neck with other scientists who were aiming to pinpoint the same vector. Sober accounts of this famous, somewhat cut-throat event in the annals of science can be found in Andrew Spielman and Michael D'Antonio, *Mosquito: A Natural History of Our Most Persistent and Deadly Foe* (New York: Hyperion, 2001), 86–92, and Harrison, *Mosquitoes, Malaria, and Man*, 72–80. Also useful is Centers for Disease Control and Prevention, "Ross and the Discovery that Mosquitoes Transmit Malaria Parasites" (available at www.cdc.gov/malaria/history/ross.htm).

12. Howard, *Fighting the Insects*, 53; Mallis, *American Entomologists*, 81. Howard explained, "We are fond of saying that circumstances have conspired to bring about such-and-such a result, and surely a series of different things led me naturally to take a great interest in the subject of the insect carriers of disease" (*Fighting the Insects*, 116).

13. Howard, *Fighting the Insects*, 123, 116–117, 53; Harrison, *Mosquitoes, Malaria, and Man*, 161–162; Mallis, *American Entomologists*, 79–85.

14. Leland O. Howard, "An Experiment Against Mosquitoes," *Insect Life* 5 (1893): 12–14; *Notes on the Mosquitoes of the United States: Giving Some Account of Their Structure and Biology, with Remarks on Remedies*, Department of Agriculture, Bureau of Entomology, Bulletin no. 25 (Washington, D.C.: Government Printing Office, 1900); and *How to Distinguish the Different Mosquitoes of North America*, Department of Agriculture, Bureau of Entomology, Circular no. 40 (Washington, D.C.: Department of Agriculture, 1900).

15. Smith had, however, expressed his skepticism about the use of kerosene as an oil-spray insecticide by writing extensively about the damage and death it caused to so many trees. See "Oil Spray Insecticides," *San Jose Mercury News*, February 2, 1913, 11.

16. Mallis, *American Entomologists*, 317–319; John B. Smith, *Economic Entomology for the Farmer and Fruit-Grower* (Philadelphia: Lippincott, 1896), 423; H. S. Barber, "Dr. John Bernhard Smith," *Proceedings of the Entomological Society of Washington* 14 (1912): 111–117; R. Hunt, "Highlights," *Proceedings of the Fifteenth Annual Meeting of the New Jersey Mosquito Extermination Association* (1916): 21–30; Smith, quoted in Thomas J. Headlee, *The Mosquitoes of New Jersey and Their Control* (New Brunswick, N.J.: Rutgers University Press, 1945), 264. The friend also recalled Smith "walking through meadows with a net, always smiling" (quoted in Mallis, *American Entomologists*, 318).

17. John B. Smith, "Concerning Certain Mosquitoes," *Science*, January 3, 1902, 13–15.

18. "Object Lessons in Controlling Mosquitoes," *Dallas Morning News*, April 6, 1902, 11; "Mosquitoes and How to Prevent Them," *Grand Forks Herald*, May 25, 1904, 2; "Male Skeeter Harmless: A Scientific Way of Ridding a Community of the

Pest," *Baltimore Sun*, reprinted in *The State*, September 20, 1900, 4; John H. McCollom, "The Role of Insects in the Propagation of Disease," *American Journal of Nursing* 2, no. 3 (1901): 192; John B. Smith, *The Common Mosquitoes of New Jersey*, New Jersey Agricultural Experiment Station, Bulletin no. 171 (New Brunswick: New Jersey Agricultural Experiment Station, 1904).

19. Headlee, *Mosquitoes of New Jersey and Their Control*, 264–270.

20. "Mosquito War Profitable," *New York Times*, May 11, 1919, 31; quoted in "Emerson Acquits New Jersey Mosquitoes," *New York Times*, September 16, 1916, 7.

21. Howard, *Mosquitoes*, 167–173. For a more thorough elaboration of the progressive idea of progress, see Robert M. Crunden, *Ministers of Reform: The Progressives' Achievement in American Civilization, 1889–1920* (New York: Basic Books, 1982). The popularity of kerosene's promise compared with the effectiveness of more natural measures is captured in A.S. Packard's glowing review of Howard's book: "It is refreshing to read of the immense inroads made by fishes upon the larva, by dragon flies and by birds. . . . [B]ut it will afford the reader still more satisfaction to know how easily these dangerous pests can be exterminated by the use of so simple a remedy as petroleum" ("Mosquitoes," *Science*, August 9, 1901, 218–219); Townsend, quoted in Dunlap, *DDT*, 113.

22. "Slay the Mosquitoes," *The State*, December 7, 1900, 2; "Mosquito Bulletin," *San Jose Mercury News*, August 29, 1900, 2; "State Health Board Discusses Control of Mosquitoes and Their Malarial Effects," *San Jose Mercury News*, August 16, 1910, 8; "Puts Blame on Mosquitoes—The United States Government Says They Transmit Yellow Fever," *Grand Forks Herald*, January 1, 1901, 3; Packard, "Mosquitoes"; "The *Anopheles* Must Go," *The State*, June 13, 1901, 4.

23. The most comprehensive work on insecticide legislation is Adelyne Hiller Whitaker, "A History of Federal Pesticide Regulation in the United States to 1947" (Ph.D. diss., Emory University, 1974), which discusses the Insecticide Act of 1910 on 81–85. Evidence of industry's input into the process of legislation is in A.B. Ansbacher & Co. to Alfred B. Kittredge, May 1, 1908, in "Correspondence," *Oil, Paint, and Drug Reporter*, May 4, 1908, 17. Farmers' support for the bill can be found in *Journal of the National Grange of the Patrons of Husbandry* (1908): 81, and (1909): 52. See also Frederick Knab, "A Proposal for the Control of Certain Mosquitoes," *Science*, January 24, 1913, 147–148; Allan K. Fitzsimmons, "Environmental Quality as a Theme in Federal Legislation," *Geographical Review* 70, no. 3 (1980): 318; and J.K. Haywood, "Insecticide and Fungicide Legislation in the United States with Especial Reference to the Federal Insecticide Act of 1910," *Journal of the Association of Official Agricultural Chemists* 4 (1920–1921): 19.

24. Walter C. O'Kane, *Injurious Insects: How to Recognize and Control Them* (New York: Macmillan, 1920), 351; B.W. Scheib, "Household Insects and Their Remedies," *American Midland Naturalist*, January 1920, 114; W. Alfred Bruce, "A Health Lesson for Upper Grades and High School," *Peabody Journal of Education* 7, no. 5 (1930): 288; Edgar E. Foster, "Mosquito Control on Hydroelectric Projects," *Scientific Monthly*, December 1933, 529. Some historians have claimed that Howard was not lukewarm toward biological controls. By 1930, however, he certainly was. He wrote in his memoir, "There has been a tendency for many years for persons with strange beliefs to migrate to California . . . and today southern California is known as the home of all the heterodoxies, and biological control in general: So great an enthusiasm for natural control was aroused in California by the success of the Austral-

ian ladybird that the state made apparently made no advances in her fight against insects for many years. Mechanical and chemical measures were abandoned. The subject of natural control held the floor. It is safe to say that a large share of the loss through insects suffered by California from 1888 until, let us say, 1898, was due to this prejudiced and badly based policy" (quoted in V. G. Dethier, *Man's Plague? Insects and Agriculture* [Princeton, N.J.: Darwin Press, 1976], 146).

25. Leland O. Howard, "Entomology and the War," *Scientific Monthly,* February 1919, 109–117, and *Fighting the Insects,* 138–139; J. S. Ames, "The Trained Man of Science in the War," *Science,* October 25, 1918, 403.

26. A. Hunter Dupree, *Science in the Federal Government: A History of Policies and Activities to 1940* (Cambridge, Mass.: Harvard University Press, 1957), 324.

27. Howard, "Entomology and the War," 110; A. E. Shipley, *The Minor Horrors of War* (London: Smith, Elder, 1915); Maurice C. Hall, "Parasites in War Time," *Scientific Monthly,* February 1918, 107. Shipley had an eye for the choice sensationalistic detail, peppering his account with references to "a clotted scab of lice" (4) and the "white, legless, repellent maggot" (14), and referring to the fly's reproduction capability as "appalling fecundity" (24), while including a choice drawing of "*M. domestica* in the act of regurgitating food" (30).

28. Howard, "Entomology and the War," 109–111.

29. Royal N. Chapman, "Measures for Protecting Wheat-Flour Substitutes from Insects," *Science,* January 14, 1918, 581; "Important in the Conduct of the War," *New York Times,* April 12, 1918, 12; "Castor Bean Being Raised to Help Aircraft Program," *Dallas Morning News,* January 6, 1918, 8; "Want Castor Beans Grown for Airplane Lubricant," *Dallas Morning News,* January 8, 1918, 7; "How Castor Beans Should Be Planted," *Dallas Morning News,* January 11, 1918, 6; "Willing to Plant Castor Beans on Twenty Acres," *Dallas Morning News,* January 10, 1918, 6; Howard, "Entomology and the War," 113; "That Oil for Airplanes," *New York Times,* September 1, 1918, 27.

30. Howard, "Entomology and the War," 114; Hall, "Parasites in War Time," 112.

31. William Moore, "Fumigation of Animals to Destroy Their External Parasites," *Journal of Economic Entomology* 9 (1916): 71–78.

32. Malcolm Evan MacGregor, "Insects as Carriers of Disease," *Transactions of the American Microscopical Society* 37, no. 1 (1918): 12; Hall, "Parasites in War Time," 108; Howard, "Entomology and the War," 114.

33. Howard, "Entomology and the War," 117.

34. Howard, "Entomology and the War," 111–112; Thomas Lee Guyton, "A Taxonomic, Ecologic, and Economic Study of Ohio Aphisidae," *Ohio Journal of Science* 24, no. 1 (1924): 28.

35. Howard, "Entomology and the War," 113; Hall, "Parasites in War Time," 112; "Microbe and Bug Killers," *New York Times,* December 13, 1917, 17; "Board Announces Enlarged Embargo," *New York Times,* October 22, 1917, 6; "Makers of Disinfectants Are Told to Excel If They Are Not to Fall Behind Foreigners," *New York Times,* December 12, 1916, 15; Ezra Dwight Sanderson, *Insect Pests of the Farm, Garden, and Orchard* (New York: Wiley, 1921), 670; information on Piver is from Lea S. Hitchner, "The Insecticide Industry," in *Insects: The Yearbook of Agriculture* (Washington, D.C.: Government Printing Office, 1952), 451; R. F. Bacon, "The Industrial Fellowships of the Mellon Institute," *Science,* April 27, 1917, 402. The domestic use

of insecticides increased despite the drop in the supply of arsenical insecticides because the United States stopped importing them and, at the same time, was using smelting technology (used to manufacture arsenical compounds) to make much-needed metal goods. Howard argued that this reduced supply demanded better organization and more efficient use of the limited supply.

36. Shipley, *Minor Horrors of War*, 20, 47; Howard, "Entomology and the War," 107–108; Bailey K. Ashford, "The Application of Sanitary Science to the Great War in the Zone of the Army," *Proceedings of the American Philosophical Society* 58, no. 5 (1919): 329–330; William Moore, "The Effect of Laundering upon Lice (*Pediculus corporis*) and Their Eggs," *Journal of Parasitology* 5, no. 2 (1918): 64–66; Scheib, "Household Insects and Their Remedies," 112–115; MacGregor, "Insects as Carriers of Disease," 10–14; Hall, "Parasites in Wartime," 107, 114.

37. Dunlap, *DDT*, 36–37; A.L. Melander, "Fighting Insects with Powder and Lead," *Scientific Monthly*, February 1933, 168; E.O. Essig, "Man's Influence on Insects," *Scientific Monthly*, June 1929, 506; Hall, "Parasites in War Time," 107.

38. Joseph M. Ginsburg, "Studies of Pyrethrum as a Mosquito Larvicide," *Proceedings of the Seventeenth Annual Meeting of the New Jersey Mosquito Extermination Association* (1930): 57, and "Airplane Oiling to Control Mosquitoes," *Science*, May 20, 1932, 542; L.L. Williams and S.S. Cook, "Paris Green Applied by Airplane in the Control of *Anopheles* Production," Public Health Report, reprint no. 1140 (Washington, D.C.: Government Printing Office, 1927), 459; Foster, "Mosquito Control on Hydroelectric Projects," 529. Further evidence of the chemical transition to larvicides can be found in Headlee, *Mosquitoes of New Jersey and Their Control*, 289–300.

39. Headlee, *Mosquitoes of New Jersey and Their Control*, 293–296.

40. Quoted in Hae-Gyung Geong, "Exerting Control: Biology and Bureaucracy in the Development of American Entomology, 1870–1930" (Ph.D. diss., University of Wisconsin, 1999), 83.

41. Department of Agriculture, *Chronological History of the Development of Insecticides and Control Equipment from 1854 Through 1954* (Washington, D.C.: Department of Agriculture, Agricultural Research Service, 1954). This document was published to honor the centennial of professional entomology in the United States. A summary of the report, "Fighting Our Insect Enemies: Achievements of Professional Entomo-logy (1854–1954)," is available at http://entweb.clemson.edu/pesticid/history/htm.

42. Howard, *Fighting the Insects*, 325.

6. Residues, Regulations, and the Politics of Protecting Insecticides

1. F.J. Francis, "Harvey W. Wiley: Pioneer in Food Science and Quality," in *A Century of Food Science* (Chicago: Institute of Food Technologists, 2000), 13–14; Thomas R. Dunlap, *DDT: Scientists, Citizens, and Public Policy* (Princeton, N.J.: Princeton University Press, 1981), 42; James Whorton, *Before "Silent Spring": Pesticides and Public Health in Pre-DDT America* (Princeton, N.J.: Princeton University Press, 1974), 99–121; Harvey W. Wiley, *The History of a Crime Against the Food Law: The Amazing Story of the National Food and Drug Law Intended to Protect the Health of the People, Perverted to Protect Adulteration of Foods and Drugs* (Washington, D.C.: Wiley, 1929).

2. Quoted in Dunlap, *DDT*, 40; Frank Yoder, "Anton J. Carlson: Physiology," in *The University of Chicago Faculty: A Centennial View* (Chicago: University of Chicago

Libraries, 1991); Ruth deForest Lamb, *American Chamber of Horrors* (New York: Grosset and Dunlap, 1936), 218–219.

3. Whorton, *Before "Silent Spring,"* 141–143, 115–117; Lamb, *American Chamber of Horrors*, 216–217, 219.

4. John H. Perkins, *Insects, Experts, and the Insecticide Crisis: The Quest for New Pest Management Strategies* (New York: Plenum, 1982), 15–20; Joseph Waldo Ellison, "Cooperative Movement in Oregon Apple Industry," *Agricultural History* 13 (1939): 77–96; Arthur D. Borden, "Control of Codling Moth on Pears with a DDT Spray," *Journal of Economic Entomology* 41 (1948): 118–120.

5. Dunlap, *DDT*, 40–47. The following analysis relies heavily on both Dunlap's *DDT* and Whorton's *Before "Silent Spring."*

6. P. J. Hanzlik, "Health Hazards of Chemo-Enemies in Contaminated Foods," *Scientific Monthly*, May 1937, 439.

7. *American Agriculturalist* 27, no. 231 (1868); F. M. Webster, *Insect Life* 7 (1894–1895): 132–134; W. H. White, *Country Gentleman* 48 (1883): 334–335; S. L. Allen, *Country Gentlemen* 42 (1877): 168; several examples are also mentioned in Whorton, *Before "Silent Spring,"* 24, 60.

8. Whorton, *Before "Silent Spring,"* chap. 2.

9. "Opens New Crusade Against Poor Food," *New York Times*, December 12, 1920, E20; Lamb, *American Chamber of Horrors*, 202–203.

10. "Find Arsenic in American Apples," *New York Times*, November 26, 1925, 2; "Warn British on Apples," *New York Times*, January 8, 1926, 23; "British Spurn Our Apples," *New York Times*, April 7, 1926, 4; Lamb, *American Chamber of Horrors*, 210–215.

11. Arthur Kallet, "Foods and Drugs for the Consumer," *Annals of the American Academy of Political and Social Science* 173 (1934): 26–34; Arthur Kallet and Frederick J. Schlink, *100,000,000 Guinea Pigs: Dangers in Everyday Foods, Drugs, and Cosmetics* (New York: Vanguard Press, 1933).

12. Hanzlik, "Health Hazards of Chemo-Enemies in Contaminated Foods," 435–439; see also E. M. Nelson, A. M. Hurd-Karrar, and W. O. Robinson, "Selenium as an Insecticide," *Science*, August 11, 1933, 124.

13. Lamb, *American Chamber of Horrors*, 5, 196–251.

14. *American Agriculturalist* 36 (1877): 274–275.

15. Editorial, *Medical Record* 59 (1891): 599–600; Whorton, *Before "Silent Spring,"* 58–59, 64.

16. Whorton, *Before "Silent Spring,"* 31–32.

17. Kallet, "Foods and Drugs for the Consumer," 27.

18. "State Protection of the Reputation of Its Products," *Yale Law Journal* 43, no. 8 (1934): 1274–1284.

19. David F. Cavers, "The Food, Drug, and Cosmetic Act of 1938: Its Legislative History and Its Substantive Provisions," *Law and Contemporary Problems* 6, no. 1 (1939): 2–13.

20. Kallet, "Foods and Drugs for the Consumer," 30; Dunlap, *DDT*, 52–53; Hanzlik, "Health Hazards of Chemo-Enemies in Contaminated Foods," 436; C. W. Crawford, "Technical Problems in Food and Drug Law Enforcement," *Law and Contemporary Problems* 1, no. 1 (1933): 36–43.

21. Dunlap, *DDT*, 52–53; Lamb, *American Chamber of Horrors*, 198.

22. Cavers, "Food, Drug, and Cosmetic Act of 1938," 3.

23. Raymond D. Tousley, "The Federal Food, Drug, and Cosmetic Act of 1938," *Journal of Marketing* 5, no. 3 (1941): 259.

7. A Year in the Life of an Insecticide Nation

1. The Bureau of Entomology and Plant Quarantine, a division of the Department of Agriculture, began to publish the *Insect Pest Survey Bulletin* with the cooperation of state entomological agencies in 1920. Unless otherwise indicated, the survey cited in this chapter, for the year 1938, is volume 18.

2. S.W. Frost, *Insect Life and Insect Natural History*, 2nd ed. rev. (New York: Dover, 1959), 52; the figures on insecticides are based on figures for 1934 cited in V. G. Dethier, *Man's Plague? Insects and Agriculture* (Princeton, N.J.: Darwin Press, 1976), 110.

3. Deborah Fitzgerald, *Every Farm a Factory: The Industrial Ideal in American Agriculture* (New Haven, Conn.: Yale University Press, 2003), 2–5; Alan Olmstead and Paul Rhode, "An Overview of California Agricultural Mechanization," *Agricultural History* 62 (1988): 86–112; Benton MacKaye, "The New Exploration: Charting the Industrial Wilderness," *Survey Graphic*, May 1, 1925, 155–192.

4. Herbert Osborn, *Fragments of Entomological History, Including Some Personal Recollections of Men and Events* (Columbus, Ohio: Osborn, 1937), 40–48.

5. Adelynne Hiller Whitaker, "A History of Federal Pesticide Regulation in the United States to 1947" (Ph.D. diss., Emory University, 1974); Lea S. Hitchner, "The Insecticide Industry," in *Insects: The Yearbook of Agriculture* (Washington D.C.: Government Printing Office, 1952), 451–453; John H. Perkins, *Insects, Experts, and the Insecticide Crisis: The Quest for New Pest Management Strategies* (New York: Plenum, 1982), 13. In 1949, the American chemical industry reported that "Atlas interests today are actually more closely related to insecticides and cosmetics than to gunpowder" (quoted in Williams Haynes, *American Chemical Industry*, vol. 6, *The Chemical Companies* [New York: Van Nostrand, 1949], 39).

6. E. O. Essig, "Man's Influence on Insects," *Scientific Monthly*, January 1929, 506; Department of Agriculture, Agricultural Research Service, "Fighting Our Insect Enemies: Achievements of Professional Entomology (1854–1954)" (available at: http://entweb.clemson.edu/pesticid/history/htm).

7. *Insect Pest Survey Bulletin*, no. 10 [summary for 1938], 657; no. 6 (August 1, 1938), 319.

8. *Insect Pest Survey Bulletin*, no. 3 (May 1, 1938), 115; no. 4 (June 1, 1938), 187; no. 5 (July 1, 1938), 277; no. 6 (August 1, 1938), 347; no. 7 (September 1, 1938), 475; no. 8 (October 1, 1938), 543.

9. *Insect Pest Survey Bulletin*, no. 1 (March 1, 1938), 10; no. 2 (April 1, 1938), 57; no. 3 (May 1, 1938), 97; no. 4 (June 1, 1938), 161; no. 5 (July 1, 1938), 252–253; no. 6 (August 1, 1938), 330–331; no. 7 (September 1, 1938), 459; no. 8 (October 1, 1938), 459, 531 (supplement).

10. *Insect Pest Survey Bulletin*, no. 1 (March 1, 1938), 13; no. 2 (April 1, 1938), 62–63; no. 3 (May 1, 1938), 103; no. 5 (July 1, 1938), 261; no. 6 (August 1, 1938), 333; no. 7 (September 1, 1938).

11. *Insect Pest Survey Bulletin*, no. 2 (April 1, 1938), 58; no. 3 (May 1, 1938), 97; no. 4 (June 1, 1938), 146, 162; no. 5 (July 1, 1938), 241, 253; no. 8 (November 1, 1938), 585–586.

12. *Insect Pest Survey Bulletin,* no. 6 (August 1, 1938), 321–322; Frost, *Insect Life and Insect Natural History,* 54–55; Daniel Ludwig, "Development of Cold Hardiness in the Larva of the Japanese Beetle (*Popilla japonica* Newm.)," *Ecology* 9, no. 3 (1928): 303–306; Stanley A. Kowalczyk, "A Report on the Intestinal Protozoa of the Larva of the Japanese Beetle," *Transactions of the American Microscopical Society* 17, no. 3 (1938): 229–244; Charles H. Hadley and Walter E. Fleming, "The Japanese Beetle," in *Insects,* 567.

13. Corn borer: *Insect Pest Survey Bulletin,* no. 3 (May 1, 1938), 99; no. 4 (June 1, 1938), 163; no. 5 (July 1, 1938), 256–257; no. 7 (October 1, 1938), 545. Cabbage worm: *Insect Pest Survey Bulletin,* no. 2 (April 1, 1938), 70; no. 3 (May 1, 1938), 117; no. 4 (June 1, 1938), 191. Codling moth: *Insect Pest Survey Bulletin,* no. 3 (May 1, 1938), 52; no. 4 (June 1, 1938), 88, 242; no. 7 (September 1, 1938), 446.

14. Harlequin bug: *Insect Pest Survey Bulletin,* no. 2 (April 1, 1938), 52; no. 3 (May 1, 1938), 117–118; no. 4 (June 1, 1938), 192; no. 7 (September 1, 1938), 481. Sugarcane borer: *Insect Pest Survey Bulletin,* no. 2 (April 1, 1938), 62; no. 4 (June 1, 1938), 167; no. 5 (July 1, 1938), 260. Potato tuber worm: *Insect Pest Survey Bulletin,* no. 6 (August 1, 1938), 347, 354. The potato tuber worm adapted so well to tobacco that it became commonly known as the tobacco splitworm: *Insect Pest Survey Bulletin,* no. 8 (October 1, 1938), 520. Boll weevil: *Insect Pest Survey Bulletin,* no. 10 [summary for 1938], 671.

15. *Insect Pest Survey Bulletin,* no. 10 [summary for 1938], 684–685.

16. *Insect Pest Survey Bulletin,* no. 1 (March 1, 1938), 27, 28; no. 2 (April 1, 1938), 52, 61–62, 73, 85; no. 4 (June 1, 1938), 209; no. 3 (May 1, 1938), 88, 91–92, 74.

17. *Insect Pest Survey Bulletin,* no. 1 (March 1, 1938), 29; no. 4 (June 1, 1938), 223; no. 2 (April 1, 1938), 86; no. 4 (June 1, 1938), 224; no. 2 (April 1, 1938), 80.

18. *Insect Pest Survey Bulletin,* no. 2 (April 1, 1938), 66; no. 3 (May 1, 1938), 108–109; no. 5 (June 1, 1938), 176; no. 2 (April 1, 1938), 68; no. 9 (October 1, 1938), 535; no. 6 (June 1, 1938), 221; no. 2 (April 1, 1938), 62.

19. *Insect Pest Survey Bulletin* 4, no. 3 (May 1, 1924), 31, 39; no. 4 (June 1, 1924), 82; no. 2 (April 1, 1924), 71; no. 5 (July 1, 1924), 133; no. 3 (May 1, 1924), 66.

20. *Insect Pest Survey Bulletin,* no. 1 (March 1, 1938), 36; no. 5 (July 1, 1938), 260; no. 2 (April 1, 1938), 62, 71; no. 5 (July 1, 1938), 256; no. 3 (May 1, 1938), 91; no. 10 [summary for 1938], 660; Dwight Isely, *Methods of Insect Control* (Minneapolis: Burgess, 1937), 1:1.

21. *Insect Pest Survey Bulletin,* no. 1 (March 1, 1938), 34; no. 4 (June 1, 1938), 160; no. 3 (May 1, 1938), 109, 99; no. 2 (April 1, 1938), 72.

22. Henry Joseph Quayle, "The Development of Resistance to Hydrocyanic Acid in Certain Scale Insects," *Hilgardia* 11, no. 5 (1938): 183–210; Harold R. Yust and R. L. Busby, "A Comparison of the Susceptibility of the So-Called Resistant and Non-Resistant Strains of California Red Scale to Methyl Bromide," *Journal of Economic Entomology* 35 (1942): 343–345; A. L. Melander, "Can Insects Become Resistant to Sprays?" *Journal of Economic Entomology* 7 (1914): 167–172; Harry S. Smith, "Racial Segregation in Insect Populations and Its Significance in Applied Entomology," *Journal of Economic Entomology* 34 (1941): 1–5; Perkins, *Insects, Experts, and the Insecticide Crisis,* 34–36.

23. Dethier, *Man's Plague?* 148; R. W. Glaser and Henry Fox, "A Nematode Parasite of the Japanese Beetle (*Popillia japonica* Newm.)," *Science,* January 3, 1930, 16;

R. W. Glaser, E. E. McCoy, and H. B. Girth, "The Biology and Economic Importance of a Nematode Parasite in Insects," *Journal of Parasitology* 26, no. 6 (1940): 479–495; American Chemical Society, *Chemistry in the Economy* (Washington, D.C.: National Science Foundation, 1973), 239.

24. Isely, *Methods of Insect Control*, 1:1.

25. American Chemical Society, *Chemistry in the Economy*, 231; Frank Graham Jr., *Since "Silent Spring"* (Boston: Houghton Mifflin, 1970); Sievert A. Rohwer, "Report of Special Committee on DDT, with S. A. Rohwer as Chairman," *Journal of Economic Entomology* 38 (1945): 144.

26. Perkins, *Insects, Experts, and the Insecticide Crisis*, 8–9; Edmund P. Russell, "'Speaking of Annihilation': Mobilizing for War Against Human and Insect Enemies, 1914–1945," *Journal of American History* 82, no. 4 (1996): 1506, 1524; William Herms, "Medical Entomology Meets the Impact of War," *Journal of Economic Entomology* 38 (1945): 8–11; E. O. Essig, "An All Out Entomological Program" [annual address of the president], *Journal of Economic Entomology* 38 (1945): 1.

27. Essig, "All Out Entomological Program," 6; R. C. Roark and N. E. McAdoo, *A Digest of the Literature on DDT Through April 30, 1944* (Washington, D.C.: Department of Agriculture, Bureau of Entomology and Plant Quarantine, 1944), 4; Perkins, *Insects, Experts, and the Insecticide Crisis*, 10.

28. Perkins, *Insects, Experts, and the Insecticide Crisis*, 10–13; Russell, "'Speaking of Annihilation,'" 1526; Haynes, *American Chemical Industry*, 6:274–275.

29. Rohwer, "Report of Special Committee on DDT," 144; Edmund P. Russell, "War on Insects: Warfare, Insecticides, and Environmental Change in the United States, 1870–1945" (Ph.D. diss., University of Michigan, 1993), 49.

30. Rachel Carson to Miss Roxanne Delano, May 26, 1961, series II, box 102, folder 1940, Rachel Carson Papers, Beinecke Rare Book and Manuscript Library, Yale University, New Haven, Conn.

8. Silent Springs and Loud Protests

1. Harold Fromm, "The Rhetoric and Politics of Environmentalism," *College English* 59, no. 8 (1997): 946–950. This article is a review of two books that also place Carson's work in a larger literary perspective: Carl G. Herndl and Stuart C. Brown, eds., *Green Culture: Environmental Rhetoric in Contemporary America* (Madison: University of Wisconsin Press, 1996), and Daniel G. Payne, *Voices in the Wilderness: American Nature Writing and Environmental Politics* (Hanover, N.H.: University Press of New England, 1996).

2. The chapter relies heavily on several superb studies of twentieth-century insecticides. The history of the insect wars before Carson's participation are best overviewed in James Whorton, *Before "Silent Spring": Pesticides and Public Health in Pre-DDT America* (Princeton, N.J.: Princeton University Press, 1974). See also John H. Perkins, *Insects, Experts, and the Insecticide Crisis: The Quest for New Pest Management Strategies* (New York: Plenum, 1982), 3–90, and Thomas R. Dunlap, *DDT: Scientists, Citizens, and Public Policy* (Princeton, N.J.: Princeton University Press, 1981).

3. Rachel Carson, *Silent Spring* (New York: Houghton Mifflin, 2002), 30, 38.

4. Roderick Nash, *Wilderness and the American Mind*, 3rd ed. (New Haven, Conn.: Yale University Press, 1982), 239–271; Robert Gottlieb, *Forcing the Spring: The Transformation of the American Environmental Movement* (Washington, D.C.: Island Press,

1993), 108–110; Ted Steinberg, *Down to Earth: Nature's Role in American History* (Oxford: Oxford University Press, 2002), 240–250.

5. Carson, *Silent Spring*, xv; Clarence Cottam and Thomas G. Scott, "A Commentary on *Silent Spring*," *Journal of Wildlife Management* 27, no. 1 (1963): 151–156.

6. Dunlap, *DDT*, 3, 59–62; Perkins, *Insects, Experts, and the Insecticide Crisis*, 5–10; Edmund P. Russell, *War and Nature: Fighting Humans and Insects with Chemicals from World War I to "Silent Spring"* (Cambridge: Cambridge University Press, 2001), 130–160.

7. Ernst T. Krebs Jr., "New Use for DDT," *Science*, April 12, 1946, 459–460; Paul D. Harwood, "Pseudoscience and the DDT Scandal," *Science*, November 30, 1951, 583–584; Edward F. Knipling, "The Greater Hazard—Insects or Insecticides," *Journal of Economic Entomology* 46 (1953): 1–7; Dunlap, *DDT*, 72; Edmund P. Russell, "'Speaking of Annihilation': Mobilizing for War Against Insect and Human Enemies, 1914–1945," *Journal of American History* 82, no. 4 (1996): 1505–1529.

8. W. C. Rhoades, "The History and Use of Agricultural Chemicals," *Florida Entomologist* 46, no. 4 (1963): 275–277; Dunlap, *DDT*, 69.

9. Dunlap, *DDT*, 68–71; W. W. Sutherland, "No More Legislation Needed," *Agricultural Chemicals* 7 (1952): 43.

10. Leland O. Howard, *The Insect Menace* (New York: Century, 1931), x; Russell, "'Speaking of Annihilation,'" 1508–1515; Paolo Palladino, "Ecological Theory and Pest Control Practice: A Study of the Institutional and Conceptual Dimensions of a Scientific Debate," *Social Studies of Science* 20, no. 2 (1990): 255–281; Clarence Cottam to Rachel Carson, January 29, 1963, series II, box 102, folder 1937, Rachel Carson Papers, Beinecke Rare Book and Manuscript Library, Yale University, New Haven, Conn.

11. Perkins, *Insects, Experts, and the Insecticide Crisis*, 12–14; J. V. Sherman, "New Products Assure Growth in Chemical Industry," *Barron's*, February 19, 1945, 9–10.

12. Dunlap, *DDT*, 69–88; Wayland J. Hayes, testimony in House Select Committee to Investigate the Use of Chemicals in Food Products, *Chemicals in Food Products: Hearings*, 82nd Cong., 1st sess. (Washington, D.C.: Government Printing Office, 1953), 95.

13. Z. Vaz, Rubens S. Pereira, and Decio M. Malheiro, "Calcium in Prevention and Treatment of DDT Poisoning," *Science*, April 27, 1945, 434–436; C. F. Hoffmann and J. P. Linduska, "Some Considerations of the Biological Effects of DDT," *Scientific Monthly*, August 1949, 104–114; Charles F. Wurster, "DDT Goes on Trial in Madison," *BioScience* 19, no. 9 (1969): 809–813; Kenneth S. Davis, "The Deadly Dust: The Unhappy History of DDT," *American Heritage*, February 1971, 44–47.

14. Seymour H. Sohmer, Joel Warren, Daniel Smiley, and George J. Wallace, letters to the editor ["Still More on the Pesticide Controversy"], *BioScience* 18, no. 2 (1968): 80–82; Thomas R. Dunlap, "Science as a Guide in Regulating Technology: The Case of DDT in the United States," *Social Studies of Science* 8, no. 3 (1978): 265–285.

15. Joseph J. Hickey, J. A. Keith, and Francis B. Koon, "An Exploration of Insecticides in a Lake Michigan Ecosystem," *Journal of Applied Ecology* 3 (1966): 141–143; Ernst A. Boykins, "The Effects of DDT-Contaminated Earthworms in the Diet of Birds," *BioScience* 17, no. 1 (1967): 37–39; Wallace, letter to the editor, 82; Perkins, *Insects, Experts, and the Insecticide Crisis*, 50; Russell, *War and Nature*, 155–160; John

H. Draize, G. Woodard, O. G. Fitzhugh, A. A. Nelson, R. B. Smith, and H. O. Calvery, "Summary of Toxicological Studies of the Insecticide DDT," *Chemical and Engineering News* 22 (1944): 1503–1504; Jane Stafford, "Insect War May Backfire," *Science News Letter*, August 5, 1944, 91; Dunlap, *DDT*, 83.

16. Dunlap, *DDT*, 131–138; Hickey, Keith, and Koon, "Exploration of Insecticides in a Lake Michigan Ecosystem," 141–143.

17. Perkins, *Insects, Experts, and the Insecticide Crisis*, 34–36; A. L. Melander, "Can Insects Become Resistant to Sprays?" *Journal of Economic Entomology* 7 (1914): 167–172; Henry Joseph Quayle, "The Increase in Resistance in Insects to Insecticides," *Journal of Economic Entomology* 36 (1943): 493–500.

18. Quoted in Davis, "Deadly Dust," 4.

19. Robert Wernick, "Speaking Out: Let's Spoil the Wilderness," *Saturday Evening Post*, November 6, 1965, quoted in Nash, *Wilderness and the American Mind*, 238.

20. Barrington Moore, "The Beginnings of Ecology," *Ecology* 19, no. 4 (1938): 592, and "The Scope of Ecology," *Ecology* 1, no. 1 (1920): 3–5; Stephen A. Forbes, "The Humanizing of Ecology," *Ecology* 3, no. 2 (1922): 89–92; H. G. Duncan, "The Concept of Personal Ecology," *Social Forces* 6, no. 3 (1928): 426–429.

21. Francis Evans, "Ecosystem as a Basic Unit in Ecology," *Science*, June 22, 1956, 1127–1128; Eugene P. Odum, "Ecology Course at Woods Hole," *AIBS Bulletin* 8, no. 1 (1958): 36, and "The Ecosystem Approach in the Teaching of Ecology," *Ecology* 38, no. 3 (1957): 531–535.

22. Nash, *Wilderness and the American Mind*, 239–246, 157; "Naked He Plunges Alone to Live in the Woods for Two Months," *Boston Post*, August 10, 1913, 1; Roderick Nash, "The American Cult of the Primitive," *American Quarterly* 18, no. 3 (1966): 517–537.

23. Nash, *Wilderness and the American Mind*, 220–225.

24. George R. Healy, "French Jesuits and the Idea of a Noble Savage," *William and Mary Quarterly* 15, no. 2 (1958): 143–167; Wilcome Washburn, "A Moral History of Indian–White Relations: Needs and Opportunities for Study," *Ethnohistory* 4, no. 1 (1957): 47–61; Leslie Fiedler, "Mythicizing the Unspeakable," *Journal of American Folklore* 103, no. 410 (1990): 390–399; Francis Parkman, *The Oregon Trail* (New York: Penguin, 1982), 44.

25. Evon Z. Vogt, "The Acculturation of American Indians," *Annals of the American Academy of Political and Social Science* 311, no. 1 (1957): 139.

26. Harold E. Driver and William C. Massey, "Comparative Studies of North American Indians," *Transactions of the American Philosophical Society* 47, no. 2 (1957): 165–456.

27. Carson, *Silent Spring*, 6, 48, 69.

28. Carson, *Silent Spring*, 15, 189, 192–194, 198.

29. Carson, *Silent Spring*, 6, 3, 277.

30. Linda Lear, *Rachel Carson: Witness for Nature* (New York: Owl Books, 1998), 433.

31. Wurster, "DDT Goes on Trial in Madison," 809–813; Dunlap, *DDT*, 105–111.

32. Wernick, "Let's Spoil the Wilderness," quoted in Nash, *Wilderness and the American Mind*, 238; Carson, *Silent Spring*, 297; Thomas Jukes, "Silence, Miss Carson," *Chemical and Engineering News*, August 18, 1962, 5.

33. Cottam to Carson, January 29, 1963.

34. Steinberg, *Down to Earth,* 239; "The Cities: The Price of Optimism," *Time,* August 1, 1969, 41–43; William Ashworth, *The Late Great Lakes: An Environmental History* (Cleveland: Wayne State University Press, 1987), 143.

35. Steinberg, *Down to Earth,* 250–253; William Hazeltine, David B. Peakall, and Charles F. Wurster, letters to the editor ["Environmental Defense Fund, Anti-Pesticide"], *BioScience* 23, no. 1 (1973): 7; Smith, quoted in Steinberg, *Down to Earth,* 248.

36. Dunlap, *DDT,* 231.

37. Carson, *Silent Spring,* 18, 237.

38. Rachel Carson to Dr. George Crile, February 17, 1963, series II, box 2, folder 1938, Rachel Carson Papers.

BIBLIOGRAPHY

Books

Abercrombie, John. *Every Man His Own Gardener: The Complete Gardener.* London, 1832.

Abercrombie, John, and Thomas Mawe. *The Complete Gardener; Being a Gardener's Calendar.* London, 1832.

Acrelius, Reverend Israel. "Account of the Swedish Churches in New Sweden." In *Narratives of Early Pennsylvania, West New Jersey, and Delaware, 1630–1707*, edited by Albert Cook Myers, 51–82. New York: Scribner, 1912.

Allen, David Grayson. *In English Ways: The Movement of Societies and the Transferal of English Local Law and Custom to Massachusetts Bay in the Seventeenth Century.* Chapel Hill: University of North Carolina Press, 1981.

American Chemical Society. *Chemistry in the Economy.* Washington, D.C.: National Science Foundation, 1973.

Anderson, Virginia. *Creatures of Empire: How Domestic Animals Transformed Early America.* New York: Oxford University Press, 2004.

Andrews, Evangeline Walker. *Journal of a Lady of Quality: Being the Narrative of a Journey from Scotland to the West Indies, North Carolina, and Portugal, in the Years 1774 to 1776.* New Haven, Conn.: Yale University Press, 1922.

Archer, Sellers G. *Soil Conservation.* Norman: University of Oklahoma Press, 1956.

Ashworth, William. *The Late Great Lakes: An Environmental History.* Cleveland: Wayne State University Press, 1987.

Atkins, Michael D. *Insects in Perspective.* New York: Macmillan, 1978.

Austen, Ernest Edward. *Illustrations of British Blood-Sucking Flies.* London: British Museum, 1906.

Banister, John. *John Banister and His Natural History of Virginia, 1678–1692.* Edited by Joseph Ewan and Nesta Ewan. Urbana: University of Illinois Press, 1970.

Bartram, John, and William Bartram. *John and William Bartram's America: Selections from the Writings of the Philadelphia Naturalists.* Edited by Helen Gere Cruickshank. New York: Devin-Adair, 1957.

Bartram, William. *Travels of William Bartram.* Edited by Mark Van Doren. New York: Dover, 1955.

Beneden, Pierre Joseph van. *Animal Parasites and Messmates.* New York: Appleton, 1876.

Blanchard, Raphäel. *Les Moustiques, histoire naturelle et médicale.* Paris: Rudeval, 1905.

———. *Traité de zoologie médicale.* Vol. 2. Paris: Baillière, 1890.

Bogue, Allan G. *From Prairie to Corn Belt: Farming on the Illinois and Iowa Prairies in the Nineteenth Century.* Chicago: University of Chicago Press, 1963.

Boisduval, Jean-Baptiste Alphonse. *Essai sur l'entomologie horticole.* Paris: Donnaud, 1867.

Bordley, J. B. *Essays and Notes on Husbandry and Rural Affairs.* Philadelphia: Budd and Bartram, 1801.

Braun, Max. *The Animal Parasites of Man: A Handbook for Students and Medical Men.* Translated by Pauline Falcke. Edited by Louis W. Sambon and Fred V. Theobald. New York: Wood, 1907.

Bruce, Robert V. *The Launching of Modern American Science, 1846–1876.* New York: Knopf, 1987.

Brumwell, Stephen. *Redcoats: The British Soldier and War in the Americas, 1755–1763.* Cambridge: Cambridge University Press, 2002.

Buchan, William. *Domestic Medicine: Or, a Treatise on the Prevention and Cure of Diseases by Regimen and Simple Medicines.* Halifax: Hodge, 1801.

Byrd, William. *The Secret History of the Line.* Edited by Wendy Martin. New York: Penguin, 1994.

Calkins, Gary N. *Protozoölogy.* New York: Lea and Febiger, 1909.

Capinera, John L., ed. *Encyclopedia of Entomology.* Vol. 1. Boston: Kluwer, 2004.

Carroll, Charles F. *Timber Economy of Puritan New England.* Providence, R.I.: Brown University Press, 1974.

Carson, Rachel. *The Edge of the Sea.* Boston: Houghton Mifflin, 1955.

———. *The Sea Around Us.* New York: Oxford University Press, 1951.

———. *Silent Spring.* Boston: Houghton Mifflin, 1962.

———. *Under the Sea-Wind, a Naturalist's Picture of Ocean Life.* New York: Simon and Schuster, 1941.

Catesby, Mark. *The Natural History of Carolina, Florida, and the Bahama Islands.* Vol. 1. London: Royal Society, 1947.

Chittenden, F. H. *Insects Injurious to Vegetables.* New York: Judd, 1907.

Cobbett, William. *Cottage Economy.* New York: Gould, 1824.

The Compleat English and French Vermin-Killer: Being a Companion for All Families, with Some Direction for Gardiners, and the Prizes of Workmen's Labour. London: Conyers, 1710.

The Complete Pocket Ferrier. Fredericktown, Md.: Bartigas, 1801.

Conant, Helen S. *The Butterfly Hunters.* Boston: Ticknor and Fields, 1868.

Cooke, Matthew. *Injurious Insects of the Orchard, Vineyard, Field, Garden, Conservatory, Household, Storehouse, Domestic Animals, with Remedies for Their Extermination.* Sacramento, Calif.: Crocker, 1883.

Cresswell, Nicholas. *The Journal of Nicholas Cresswell, 1774–1777.* Edited by A. G. Bradley. New York: Dial Press, 1928.

Cronon, William. *Changes in the Land: Indians, Colonists, and the Ecology of New England.* New York: Hill and Wang, 1983.

———. *Nature's Metropolis: Chicago and the Great West.* New York: Norton, 1991.

Crosby, Alfred. *Ecological Imperialism: The Biological Expansion of Europe, 900–1900.* Cambridge: Cambridge University Press, 1986.

Crunden, Robert M. *Ministers of Reform: The Progressives' Achievement in American Civilization, 1889–1920.* New York: Basic Books, 1982.

Curtis, John. *Farm Insects: Being the Natural History and Economy of the Insects Injurious to the Field Crops of Great Britain and Ireland, and Also Those Which Infest Barns and Granaries.* Glasgow: Blackie, 1860.

Curtis, John T. *The Vegetation of Wisconsin: An Ordination of Plant Communities.* Madison: University of Wisconsin Press, 1959.

Danhof, Clarence H. *Change in Agriculture: The Northern United States.* Cambridge, Mass.: Harvard University Press, 1969.

Deane, Samuel. *The New-England Farmer: or, Georgical Dictionary. Containing a Compendious Account of the Ways and Methods in Which the Important Art of Husbandry, in All Its Various Branches, Is, or May Be, Practised, to the Greatest Advantage, in This Country.* 3rd ed. Boston: Wells and Lilly, 1822.

Dethier, V. G. *Man's Plague? Insects and Agriculture.* Princeton, N.J.: Darwin Press, 1976.

Doutt, Richard L. "The Historical Development of Biological Control." In *Biological Control of Insect Pests and Weeds,* edited by Paul DeBach, 21–42. New York: Reinhold, 1964.

Drayton, John. *A View of South Carolina, as Respects Her Natural and Civil Concerns.* Charleston: Young, 1802.

Dubreuilh, William Auguste, and Lucien Beille. *Les Parasites animaux de la peau humaine.* Paris: Masson, 1896.

Dunlap, Thomas R. *DDT: Scientists, Citizens, and Public Policy.* Princeton. N.J.: Princeton University Press, 1981.

———. *Saving America's Wildlife.* Princeton, N.J.: Princeton University Press, 1988.

Dupree, A. Hunter. "The Measuring Behavior of Americans." In *Nineteenth-Century American Science: A Reappraisal,* edited by George H. Daniels, 22–37. Evanston, Ill.: Northwestern University Press, 1972.

———. *Science in the Federal Government: A History of Policies and Activities to 1940.* Cambridge, Mass.: Harvard University Press, 1957.

Dwight, Timothy. *Travels in New-England and New-York.* 4 vols. New Haven, Conn.: Dwight, 1821–1822.

Eddowes, Ralph. *Account of the Wheat Moth or Virginia Fly as It Appears in France in the Year 1755.* Philadelphia: Bartram, 1805.

Elton, Charles S. *The Ecology of Invasions by Animals and Plants.* London: Methuen, 1958.

Essig, E. O. *A History of Entomology.* New York: Macmillan, 1931.

Fisher, Jabez. "The War with Insects." In *Twenty-sixth Annual Report of the Secretary of the Massachusetts Board of Agriculture for 1878.* Boston: Rand, Avery, 1879.

Fitch, Asa. *The Wheat Fly.* Albany, N.Y.: Niles, 1843.

Fitzgerald, Deborah. *Every Farm a Factory: The Industrial Ideal in American Agriculture.* New Haven, Conn.: Yale University Press, 2003.

Fitzgerald, Terrence D. *The Tent Caterpillars.* Ithaca, N.Y.: Cornell University Press, 1995.

Flagg, Wilson. "Plan of a Series of Experiments for Investigating the Potato Disease." In *Transactions of the Essex Agricultural Society for 1859.* Newbury, Mass.: Huse, 1859.

———. "Utility of Birds." In *Abstract of Returns of the Agricultural Societies of Massachusetts for 1861.* Boston: White, 1862.

Flannery, Tim. *The Weather Makers: How Man Is Changing the Climate and What It Means for Life in Earth.* New York: Atlantic Monthly Press, 2005.

Flint, Charles L., ed. *Abstract of Returns of the Agricultural Societies of Massachusetts.* Boston: Wright and Potter, 1872.

———. *Abstract of Returns of the Agricultural Societies of Massachusetts, 1856.* Boston: White, 1857.

Flores, Dan. *The Natural West: Environmental History in the Great Plains and Rocky Mountains.* Norman: University of Oklahoma Press, 2001.

Forbes, Stephen A. "Notes from Livingston and Adjacent Counties." In *Miscellaneous Essays on Economic Entomology,* edited by Stephen A. Forbes, 2–13. Springfield, Ill.: Rokker, 1886.

Fowler, F. H. *Insecticides and Their Application.* Boston: Wright and Potter, 1891.

Francis, F. J. "Harvey W. Wiley: Pioneer in Food Science and Quality." In *A Century of Food Science,* 13–14. Chicago: Institute of Food Technologists, 2000.

Frost, S. W. *Insect Life and Insect Natural History.* 2nd ed. rev. New York: Dover, 1959.

Gilbert, Arthur W. *The Potato.* New York: Macmillan, 1921.

Gillanders, A. T. *Forest Entomology.* Edinburgh: Blackwood, 1908.

Goldsmith, Oliver. *An History of the Earth, and Animated Nature.* Philadelphia: Carey, 1795.

Gottlieb, Robert. *Forcing the Spring: The Transformation of the American Environmental Movement.* Washington, D.C.: Island Press, 1993.

Graham, Frank, Jr. *Since "Silent Spring."* Boston: Houghton Mifflin, 1970.

Greene, John C. *American Science in the Age of Jefferson.* Ames: Iowa State University Press, 1984.

Griffith, Mary. *Our Neighborhood, or Letters on Horticulture and Natural Phenomena: Interspersed with Opinions on Domestic and Moral Economy.* New York: Bliss, 1831.

Grünberg, Karl. *Die blutsaugenden Diptern.* Jena: Fischer, 1907.

Hagan, K. S., and J. M. Franz, "A History of Biological Control." In *History of Entomology,* edited by Ray F. Smith, Thomas E. Mittler, and Carroll N. Smith, 433–476. Palo Alto, Calif.: Annual Reviews, 1973.

Hamilton, Alexander. *Gentleman's Progress: The Itinerarium of Dr. Alexander Hamilton, 1744.* Edited by Wendy Martin. New York: Penguin, 1994.

Hammond, Samuel H., and L. W. Mansfield. *Country Margins and Rambles of a Journalist.* New York: Derby, 1855.

Hardeman, Nicholas P. *Shucks, Shocks, and Hominy Blocks: Corn as a Way of Life in Colonial America.* Baton Rouge: Louisiana State University Press, 1981.

Harlan, Jack R. *Theory and Dynamics of Grassland Agriculture.* Princeton, N.J.: Van Nostrand, 1956.

Harris, Thaddeus William. *Entomological Correspondence of Thaddeus William Harris, M.D.* Edited by Samuel H. Scudder. Occasional Papers of the Boston Society of Natural History, no. 1. Boston: Boston Society of Natural History, 1869.

———. *A Report of the Insects of Massachusetts Injurious to Vegetation.* Cambridge, Mass.: Folsom, Wells, and Thurston, 1841.

Harrison, Gordon. *Mosquitoes, Malaria, and Man: A History of the Hostilities Since 1880.* New York: Dutton, 1978.

Haynes, Williams. *American Chemical Industry.* 6 vols. New York: Van Nostrand, 1945–1954.

Headlee, Thomas J. *The Mosquitoes of New Jersey and Their Control.* New Brunswick, N.J.: Rutgers University Press, 1945.
Heidenreich, Conrad. *Huronia: A History and Geography of the Huron Indians, 1600–1650.* Toronto: McClelland and Stewart, 1971.
Herndl, Carl G., and Stuart C. Brown, eds. *Green Culture: Environmental Rhetoric in Contemporary America.* Madison: University of Wisconsin Press, 1996.
Herndon, G. Melvin. *William Tatham and the Culture of Tobacco, Including a Facsimile Reprint of "An Historical and Practical Essay on the Culture and Commerce of Tobacco" by William Tatham.* Coral Gables, Fla.: University of Miami Press, 1969.
Hofstadter, Richard. *The Age of Reform: From Bryan to F.D.R.* New York: Vintage, 1955.
Howard, Leland O. *Fighting the Insects: The Story of an Entomologist.* New York: Macmillan, 1933.
———. *The Insect Menace.* New York: Century, 1931.
———. *Mosquitoes: How They Live; How They Carry Disease; How They Are Classified; How They May Be Destroyed.* New York: McClure, Phillips, 1901.
Howe, Joseph S. "Progressive Farming." In *Abstract of Returns of the Agricultural Societies of Massachusetts,* edited by Charles L. Flint. Boston: Wright and Potter, 1872.
Huber, J. C. *Bibliographie der klinischen entomologie.* 4 parts. Jena: Pohle, 1899–1900.
Hudson, John C. *Making the Corn Belt: A Geographical History of Middle Western Agriculture.* Bloomington: Indiana University Press, 1994.
Hudson, Norman. *Soil Conservation.* 2nd ed. Ithaca, N.Y.: Cornell University Press, 1981.
Isely, Dwight. *Methods of Insect Control.* 2 vols. Minneapolis: Burgess, 1937.
Jefferson, Thomas. *Thomas Jefferson's Farm Book.* Edited by Edwin Morris Betts. Charlottesville: University of Virginia Press, 1976.
Johnson, Willis Grant. *Fumigation Methods: A Practical Treatise for Farmers, Fruit Growers, Nurserymen, Gardeners, Florists, Millers, Grain Dealers, Transportation Companies, College and Experiment Station Workers, etc.* New York: Judd, 1902.
Kallet, Arthur, and Frederick J. Schlink. *100,000,000 Guinea Pigs: Dangers in Everyday Foods, Drugs, and Cosmetics.* New York: Vanguard Press, 1933.
Kalm, Peter. *The America of 1750: Peter Kalm's Travels in North America: The English Version of 1770.* Edited by Adolph B. Benson. Vol. 2. New York: Wilson-Erickson, 1937.
Kaltenbach, J. H. *Die Pflanzenfeinde aus der Classe der Insekten.* Stuttgart: Hoffmann, 1874.
Kastner, Joseph. *A Species of Eternity.* New York: Knopf, 1977.
Kirby, William, and William Spence. *An Introduction to Entomology; or, Elements of the Natural History of Insects.* 4 vols. London: Longman, Hurst, Rees, Orme, and Brown, 1816–1826.
Kollar, Vincent. *A Treatise on Insects Injurious to Gardeners, Foresters, and Farmers.* Translated by Jane Loudon and Mary Loudon. London: Smith, 1840.
Kretch, Shepherd. *The Ecological Indian: Myth and History.* New York: Norton, 1999.
Kroeber, A. L. *Cultural and Natural Areas of Native North America.* Berkeley: University of California Press, 1939.
Lamb, Ruth deForest. *American Chamber of Horrors.* New York: Grosset and Dunlap, 1936.

Latrobe, Benjamin Henry. *The Virginia Journals of Benjamin Henry Latrobe, 1795–1798.* Vol. 1. Edited by Edward C. Carter II. New Haven, Conn.: Yale University Press, 1977.

Lawson, John. *A New Voyage to Carolina: Containing the Exact Description and Natural History of That Country.* London, 1709.

Lear, Linda. *Rachel Carson: Witness for Nature.* New York: Owl Books, 1998.

Leighton, Ann. *Early American Gardens: For Meate or Medicine.* Amherst: University of Massachusetts Press, 1986.

Lintner, Joseph A. "Entomological Contributions." In *New York State Museum of Natural History, Twenty-fourth Annual Report,* 154–167. Albany: Angus, 1872.

Lockwood, Jeffrey A. *Locust: The Devastating Rise and Mysterious Disappearance of the Insect that Shaped the American Frontier.* New York: Basic Books, 2004.

Lodeman, E. G. *The Spraying of Plants: A Succinct Account of the History, Principles, and Practice of the Application of Liquids and Powders to Plants, for the Purpose of Destroying Insects and Fungi.* New York: Macmillan, 1897.

Lucet, E. *Les Insectes nuisibles aux rosiers sauvages et cultivés en France.* 2nd ed. Paris: Klinksieck, 1900.

MacKenzie, Alexander. *Voyages from Montreal on the River St. Lawrence, Through the Continent of North America, to the Frozen and Pacific Oceans, in the Years 1789 and 1793.* New York: Hopkins, 1802.

Mallis, Arnold. *American Entomologists.* New Brunswick, N.J.: Rutgers University Press, 1971.

Manson, Patrick. *Tropical Diseases: A Manual of the Diseases of Warm Climates.* New York, 1904.

Marcus, Alan L. *Agricultural Science and the Quest for Legitimacy: Farmers, Agricultural Colleges, and Experiment Stations, 1870–1890.* Ames: Iowa State University Press, 1985.

Martin, John Frederick. *Profits in the Wilderness.* Cambridge, Mass.: Harvard University Press, 1991.

Massey, J. Earl. "Early Money Substitutes." In *Studies on Money in Early America,* edited by Eric Newman and Richard Doty, 15–24. New York: American Numismatic Society, 1976.

Mather, Cotton. *The Christian Philosopher: A Collection of the Best Discoveries in Nature, with Religious Improvements.* London: Matthews, 1721.

Matheson, Robert. *A Handbook of Mosquitoes of North America.* Springfield, Ill.: Thomas, 1929.

McCusker, John J., and Russell R. Menard. *The Economy of British America, 1607–1789.* Chapel Hill: University of North Carolina Press, 1985.

M'Clure, David. *Memoirs of the Rev. Eleazer Wheelock: Founder and President of Dartmouth College and Moor's Charity School.* Newburyport, Mass.: Little, 1811.

Mégnin, Jean-Pierre. *La Faune des cadavres: Application de l'entomologie à la médicine légale.* Paris: Masson, 1894.

Merrell, James. *Into the American Woods: Negotiators on the Pennsylvania Frontier.* New York: Norton, 1999.

Mooney, James. *Myths of the Cherokee and Sacred Formulas of the Cherokee.* Nashville, Tenn.: Elder, 1972.

Muhlenberg, Henry Melchoir. *The Notebook of a Colonial Clergyman, Condensed from the Journals of Henry Melchior Muhlenberg.* Edited and translated by Theodore G. Tappert and John W. Doberstein. Minneapolis: Fortress Press, 1998.

Nash, Roderick. *Wilderness and the American Mind.* 3rd ed. New Haven, Conn.: Yale University Press, 1982.

Oddy, S. A. *The Housekeeper's Receipt Book.* London: Oddy, 1813.

O'Kane, Walter C. *Injurious Insects: How to Recognize and Control Them.* New York: Macmillan, 1920.

Ormerod, Eleanor. A. *Handbook of Insects Injurious to Orchard and Bush Fruits.* London: Simkin, Marshall, 1898.

———. *A Manual of Injurious Insects with Methods of Prevention and Remedy for Their Attacks on Food Crops, Forest Trees, and Fruit, and with Short Introduction to Entomology.* London: Sonnenschein & Allen, 1881.

Packard, A. S., Jr., ed. *Record of "American Entomology" for the Year 1868 [–1873].* Salem, Mass.: Naturalist's Book Agency, Essex Institute Press, 1869–1877.

Packard, Clarissa. *Recollections of a Housekeeper.* New York: Harper, 1934.

Payne, Daniel G. *Voices in the Wilderness: American Nature Writing and Environmental Politics.* Hanover, N.H.: University Press of New England, 1996.

Perkins, John H. *Insects, Experts, and the Insecticide Crisis: The Quest for New Pest Management Strategies.* New York: Plenum, 1982.

Price, Peter W. *Insect Ecology.* New York: Wiley, 1975.

Quick, Herbert. *Vandemark's Folly.* Indianapolis: Bobbs-Merrill, 1922.

Riley, Charles V. *Address Before the Academy of Science of St. Louis at Its Annual Meeting for 1877.* St. Louis: Studley, 1877.

———. *Annual Address as President of the Entomological Society of Washington for the Year 1884.* Washington, D.C.: Gibson, 1886.

———. *The Locust Plague in the United States: Being More Particularly a Treatise on the Rocky Mountain Locust or So-called Grasshopper, as It Occurs East of the Rocky Mountains, with Practical Recommendations for Its Destruction.* Chicago: Rand McNally, 1877.

———. *Potato Pests: Being an Illustrated Account of the Colorado Potato-Beetle and the Other Insect Foes of the Potato in North America. With Suggestions for Their Repression and Methods for Their Destruction.* New York: Orange Judd, 1876.

———. *Remarks on the Insect Defoliators of Our Shade Trees.* New York: Globe Stationary and Printing, 1887.

Robert, Joseph Clarke. *The Story of Tobacco in America.* Chapel Hill: University of North Carolina Press, 1949.

———. *The Tobacco Kingdom: Plantation, Market, and Factory in Virginia and North Carolina, 1800–1860.* Gloucester, Mass.: Smith, 1965.

Robinson, Solon. "Something About the Western Prairies." In *Solon Robinson: Pioneer and Agriculturalist,* edited by Herbert Anthony Kellar, 1:41–43. New York: Da Capo Press, 1968.

———. "Wheat Versus Cattle: Which Is the Most Profitable for the Western Farmer?" In *Solon Robinson: Pioneer and Agriculturalist,* edited by Herbert Anthony Keller. New York: Da Capo Press, 1968.

Rountree, Helen C. The *Powhatan Indians of Virginia: Their Traditional Culture.* Norman: University of Oklahoma Press, 1989.

Rowlandson, Mary. *A True History of the Captivity and Restoration of Mrs. Mary Rowlandson.* Edited by Wendy Martin. New York: Penguin, 1994.
Russell, Edmund P. *War and Nature: Fighting Humans and Insects with Chemicals from World War I to "Silent Spring."* Cambridge: Cambridge University Press, 2001.
Sanders, Elizabeth. *Roots of Reform: Farmers, Workers, and the American State, 1877–1917.* Chicago: University of Chicago Press, 1999.
Sanderson, Ezra Dwight. *Insect Pests of the Farm, Garden, and Orchard.* New York: Wiley, 1921.
———. *Insects Injurious to Staple Crops.* New York: Wiley, 1902.
Saunders, William. *Insects Injurious to Fruits.* 2nd ed. Philadelphia: Lippincott, 1909.
Sawyer, Richard C. *To Make a Spotless Orange: Biological Control in California.* Ames: Iowa State University Press, 1996.
Schoolcraft, Henry. *Travels in the Central Portion of the Mississippi Valley: Comprising Observations of Its Mineral Geography, Internal Resources, and Aboriginal Population.* New York: Collins and Hannay, 1825.
Scott, Roy V. *The Reluctant Farmer: The Rise of Agricultural Extension to 1914.* Urbana: University of Illinois Press, 1970.
Seed, Patricia. *Ceremonies of Possession in Europe's Conquest of the New World, 1492–1640.* Cambridge: Cambridge University Press, 1995.
Sergent, Edmond. *Détermination des insectes piqueurs et suceurs de sang.* Paris: Doin, 1909.
Shipley, A. E. *The Minor Horrors of War.* London: Smith, Elder, 1915.
Smellie, William. *The Philosophy of Natural History.* Dover, N.H.: Thomas and Tappan, 1808.
Smith, D. M. "Changes in Eastern Forests Since 1600 and Possible Effects." In *Perspectives in Forest Entomology,* edited by John F. Anderson and Harry K. Kaya, 3–20. New York: Academic Press, 1976.
Smith, John. *The Generall Historie of Virginia, New England, and the Summer Isles.* Cleveland: World, 1966.
Smith, John B. *Economic Entomology for the Farmer and Fruit-Grower.* Philadelphia: Lippincott, 1896.
———. *Our Insect Friends and Enemies: The Relation of Insects to Man, to Other Animals, to One Another, and to Plants, with a Chapter on the War Against Insects.* Philadelphia: Lippincott, 1909.
Smith, Ray F., Thomas E. Mittler, and Carroll N. Smith, eds. *History of Entomology.* Palo Alto, Calif.: Annual Reviews, 1973.
Sorenson, W. Conner. *Brethren of the Net: American Entomology, 1840–1880.* Tuscaloosa: University of Alabama Press, 1995.
Spielman, Andrew, and Michael D'Antonio. *Mosquito: A Natural History of Our Most Persistent and Deadly Foe.* New York: Hyperion, 2001.
Steinberg, Ted. *Down to Earth: Nature's Role in American History.* Oxford: Oxford University Press, 2002.
Stephens, John William Watson, and S. R. Christophers. *The Practical Study of Malaria and Other Blood Parasites.* London: Longmans, Green, 1903.
Taylor, Ronald L. *Butterflies in My Stomach: or, Insects in Human Nutrition.* Santa Barbara, Calif.: Woodbridge, 1975.

Thwaites, Reuben Gold, ed. *The Jesuit Relations and Allied Documents*. Vol. 6, *Travels and Explorations of the Jesuit Missionaries in New France, 1610–1791*. New York: Pageant, 1959.

Tober, James A. *Who Owns the Wildlife? The Political Economy of Conservation in Nineteenth-Century America*. Westport, Conn.: Greenwood Press, 1981.

Todd, Kim. *Tinkering with Eden: A Natural History of Exotics in America*. New York: Norton, 2001.

Treat, Mary. *Chapters on Ants*. New York: Harper, 1879.

———. *Injurious Insects of the Farm and Garden*. New York: Orange Judd, 1887.

Underwood, Michael. *A Treatise on the Dieases [sic] of Children, with General Directions for the Management of Infants from the Birth*. Philadelphia: Dobson, 1793.

Weed, Clarence M. *Insects and Insecticides: A Practical Manual Concerning Noxious Insects and the Methods of Preventing Their Injuries*. New York: Orange Judd, 1908.

———. "On the Injurious Locusts of Central Illinois." In *Miscellaneous Essays on Economic Entomology*, edited by Stephen A. Forbes, 6–9. Springfield, Ill.: Rokker, 1886.

Wehner, Rüdiger, ed. *Information Processing in the Visual Systems of Arthropods*. New York: Springer, 1972.

Whitaker, James W. *Feedlot Empire: Beef Cattle Feeding in Illinois and Iowa, 1840–1900*. Ames: Iowa State University Press, 1975.

White, William N. *Gardening for the South*. New York: Moore, 1859.

Whitney, Gordon G. *From Coastal Wilderness to Fruited Plain: A History of Environmental Change in Temperate North America, 1500 to the Present*. Cambridge: Cambridge University Press, 1994.

Whorton, James. *Before "Silent Spring": Pesticides and Public Health in Pre-DDT America*. Princeton, N.J.: Princeton University Press, 1974.

Wiebe, Robert H. *Search for Order, 1877–1920*. New York: Hill and Wang, 1967.

Wiley, Harvey W. *The History of a Crime Against the Food Law: The Amazing Story of the National Food and Drug Law Intended to Protect the Health of the People, Perverted to Protect Adulteration of Foods and Drugs*. Washington, D.C.: Wiley, 1929.

William, Samuel. *History of Vermont*. Burlington: Mills, 1809.

Williamson, Charles. *A Description of the Genesee Country, in the State of New York*. [Canandaigua,] N.Y.: Printed for the author, 1804.

Woodmason, Charles. *The Carolina Backcountry on the Eve of the Revolution: The Journal and Other Writings of Charles Woodmason, Anglican Itinerant*. Edited by Richard J. Hooker. Chapel Hill: University of North Carolina Press, 1953.

Yoder, Frank. "Anton J. Carlson: Physiology." In *The University of Chicago Faculty: A Centennial View*. Chicago: University of Chicago Libraries, 1991.

Articles and Dissertations

Abril, A., and E. H. Bucher. "The Effects of Overgrazing on Soil Microbial Community and Fertility in the Chaco Dry Savannahs of Argentina." *Applied Soil Ecology* 12, no. 2 (1999): 159–167.

"Agricultural Quackery." *Southern Planter*, August 1857, 477.

The Agriculturalist. *Yankee Farmer*, January 6, 1838, 1.

An Agriculturist of Maryland. *Farmer's Cabinet,* September 15, 1836, 73.
Allen, S. L. *Country Gentleman* 42 (1877): 168.
"Alum as a Current Worm Remedy." *Insect Life* 1, no. 7 (1889): 229–230.
Ames, J. S. "The Trained Man of Science in the War." *Science,* October 25, 1918, 401–410.
"Annual Address of the President." *Proceedings of the Entomological Society of Washington* 1, no. 1 (1885): 3.
Ashford, Bailey K. "The Application of Sanitary Science to the Great War in the Zone of the Army." *Proceedings of the American Philosophical Society* 58, no. 5 (1919): 329–330.
Bacon, R. F. "The Industrial Fellowships of the Mellon Institute." *Science,* April 27, 1917, 399–403.
Baker, H. T. "Lice on Fruit Trees." *Michigan Farmer,* October 1851, 182.
Barber, H. S. "Dr. John Bernhard Smith." *Proceedings of the Entomological Society of Washington* 14 (1912): 111–117.
"The Bark Louse." *Prairie Farmer,* March 1851, 107.
Bassett, Yves, E. Charles, D. S. Hammond, and V. K. Brown. "Short-Term Effects of Canopy Openness on Insect Herbivores in a Rain Forest in Guyana." *Journal of Applied Ecology* 38, no. 5 (2001): 1045–1058.
Baumberger, J. P. "The Food of *Drosophila melanogaster* Meigen." *Proceedings of the National Academy of Sciences of the United States of America* 3, no. 2 (1917): 122–126.
Baxter, R. M. "Environmental Effects of Dams and Impoundments." *Annual Review of Ecology and Systematics* 8 (1977): 255–283.
Benke, Arthur C., and David I. Jacobi. "Production Dynamics and Resource Utilization of Snag-Dwelling Mayflies in a Blackwater River." *Ecology* 75, no. 5 (1994): 1219–1232.
Bennett, Hugh H. "Soil Erosion: A National Menace." *Scientific Monthly,* November 1934, 385–404.
"Boil Your Garden Before You Plant It." *Maine Farmer,* February 3, 1838, 34–35.
Bomar, Charles. "The Rocky Mountain Locust: Extinction and the American Experience." Available at: http://www.sciencecases.org/locusts/locusts.asp.
Borchert, John R. "The Climate of North American Grasslands." *Annals of the Association of American Geographers* 40 (1950): 1–30.
Borden, Arthur D. "Control of Codling Moth on Pears with a DDT Spray." *Journal of Economic Entomology* 41 (1948): 118–120.
"The Borer." *Prairie Farmer,* August 1851, 336.
Boykins, Ernst A. "The Effects of DDT-Contaminated Earthworms in the Diet of Birds." *BioScience* 17, no. 1 (1967): 37–39.
Bruce, W. Alfred. "A Health Lesson for Upper Grades and High School." *Peabody Journal of Education* 7, no. 5 (1930): 285–289.
"Bucks County Intelligencer." *Farmer's Cabinet,* January 1, 1837, 185.
"Bugs on Apple Buds." *New England Farmer,* July 1867, 332, 346–347.
Carter, Landon. "Observations Concerning the Fly-Weevil, that Destroys the Wheat." *Transactions of the American Philosophical Society* 1 (1771): 205–217.
Casagrande, R. A. "The Iowa Potato Beetle, Its Discovery and Spread to Potatoes." *Bulletin of the Entomological Society of America* 31 (1985): 27–29.

Cavers, David F. "The Food, Drug, and Cosmetic Act of 1938: Its Legislative History and Its Substantive Provisions." *Law and Contemporary Problems* 6, no. 1 (1939): 2–42.
Chapman, Royal. "Measures for Protecting Wheat-Flour Substitutes from Insects." *Science*, January 14, 1918, 579–581.
"Chemistry as Applied to Agriculture." *Southern Planter*, October 1856, 298.
"The Chinch-Bug in Illinois." *Science*, April 12, 1889, 275–276.
"Chloride of Lime as an Insecticide." *Farmer's Cabinet*, July 10, 1862, 4.
Cockerell, T. D. A. "The Kieffer Pear and the San Jose Scale." *Science*, September 28, 1900, 488–489.
———. "Some Western Weeds and Alien Weeds in the West." *Botanical Gazette* 20, no. 11 (1895): 503.
Connecticut River Vermont Farmer. "The Grain Worm." *Yankee Farmer*, March 17, 1838, 81–82.
"Correspondence." *Practical Entomologist*, October 30, 1865, 6.
Cottam, Clarence, and Thomas G. Scott. "A Commentary on *Silent Spring*." *Journal of Wildlife Management* 27, no. 1 (1963): 151–156.
Crawford, C. W. "Technical Problems in Food and Drug Law Enforcement." *Law and Contemporary Problems* 1, no. 1 (1933): 36–43.
"The Curculio." *Michigan Farmer*, June 1858, 175.
"The Curculio." *New England Farmer*, July 1867, 346–347.
"The Curculio." *Prairie Farmer*, February 1851, 60.
"The Cut Worm." *New England Farmer*, August 1867, 377.
Davis, Kenneth S. "The Deadly Dust: The Unhappy History of DDT." *American Heritage*, February 1971, 44–47.
del Rosario, Rosalie B., Emily A. Betts, and Vincent H. Resh. "Cow Manure in Headwater Streams: Aquatic Insect Responses to Organic Enrichment." *Journal of the North American Benthological Society* 21, no. 2 (2002): 278–289.
Denevan, William M. "The Pristine Myth: The Landscape of the Americans in 1492." *Annals of the Association of American Geographers* 82, no. 3 (1992): 369–385.
"Destructive Insects." *Farmer's Cabinet*, July 1, 1836, 5–6.
Didham, Raphael K., Peter M. Hammond, John H. Lawton, Paul Eggleton, and Nigel E. Stork. "Beetle Species Responses to Tropical Forest Fragmentation." *Ecological Monographs* 68, no. 3 (1998): 295–323.
Dodge, Harold R., and John M. Seago. "Sarcophagidae and Other Diptera Taken by Trap and Net on Georgia Mountain Summits in 1952." *Ecology* 35, no. 1 (1954): 50–59.
"Dosing Trees with Sulphur and Other Substances." *Insect Life* 1, no. 7 (1889): 223.
"Dr. Fitch's Report on the Insects of New York." *Southern Planter*, July 1857, 437.
Draize, John H., G. Woodard, O. G. Fitzhugh, A. A. Nelson, R. B. Smith, and H. O. Calvery. "Summary of Toxicological Studies of the Insecticide DDT." *Chemical and Engineering News* 22 (1944): 1503–1504.
Dreistadt, Steve H., Donald L. Dahlsten, and Gordon W. Frankie. "Urban Forests and Insect Ecology: Complex Interactions Among Trees, Insects, and People." *BioScience* 40, no. 3 (1990): 192–198.
Duncan, H. G. "The Concept of Personal Ecology." *Social Forces* 6, no. 3 (1928): 426–429.

Dunlap, Thomas R. "Science as a Guide in Regulating Technology: The Case of DDT in the United States." *Social Studies of Science* 8, no. 3 (1978): 265–285.

———. "The Triumph of Chemical Pesticides in Insect Control, 1890–1920." *Environmental Review* 1, no. 5 (1978): 38–47.

Dyer, M. I., C. L. Turner, and T. R. Seastedt. "Herbivory and Its Consequences." *Ecological Applications* 3, no. 1 (1993): 10–16.

Egerton, Frank N. "Changing Concepts of the Balance of Nature." *Quarterly Review of Biology* 48, no. 2 (1973): 322–350.

Ellison, Joseph Waldo. "Cooperative Movement in Oregon Apple Industry." *Agricultural History* 13 (1939): 77–96.

"Enemies of the Plant Louse." *Science*, August 9, 1889, 100–101.

Essig, E. O. "An All Out Entomological Program" [annual address of the president]. *Journal of Economic Entomology* 38 (1945): 1–8.

———. "Man's Influence on Insects." *Scientific Monthly*, June 1929, 499–506.

Euler, Robert C., and Volney H. Jones. "Hermetic Sealing as a Technique of Food Preservation Among the Indians of the American Southwest." *Proceedings of the American Philosophical Society* 100, no. 1 (1956): 87–99.

Evans, Francis. "Ecosystem as a Basic Unit in Ecology," *Science*, June 22, 1956, 1127–1128.

"Farm Economy." *New England Farmer*, June 1864, 183.

Farmer's Almanac, August 1, 1836, 24.

Felt, E. Porter. "Why Do Insects Become Pests?" *Scientific Monthly*, May 1938, 437–440.

Fernald, C. H. "The Association of Economic Entomologists—Address by the President—The Evolution of Economic Entomology." *Science*, October 16, 1896, 541–547.

Fitch, Asa. "The Diaries of Asa Fitch, M.D." Edited by Arnold Mallis. *Bulletin of the Entomological Society of America* 9 (1963): 20–29.

———. "The Hessian Fly: Its History, Character, Transformation, and Habits." *Transactions of the New York State Agricultural Society* 6 (1846): 316–372.

Fitzsimmons, Allan K. "Environmental Quality as a Theme in Federal Legislation." *Geographical Review* 70, no. 3 (1980): 314–327.

Forbes, Stephen A. "The Food of Birds." *Transactions of the Illinois Horticultural Society* 12 (1879): 140–145.

———. "The Humanizing of Ecology." *Ecology* 3, no. 2 (1922): 89–92.

———. "The Regulative Action of Birds upon Insect Oscillations." *Bulletin of the Illinois State Laboratory of Natural History* 1, no. 6 (1883): 3–32.

———. "Synopsis of Recent Work with Arsenical Insecticides." *Transactions of the Illinois Horticultural Society* 23 (1890): 310–324.

Foster, Edgar E. "Mosquito Control on Hydroelectric Projects." *Scientific Monthly*, December 1933, 522–530.

Fromm, Harold. "The Rhetoric and Politics of Environmentalism." *College English* 59, no. 8 (1997): 946–950.

Genesee Farmer. *Farmer's Cabinet,* August 1, 1836, 25–26.

———. "Things a Farmer Should Not Do." *Farmer's Cabinet,* July 1830, 13.

Geong, Hae-Gyung. "Exerting Control: Biology and Bureaucracy in the Development of American Entomology, 1870–1930." Ph.D. diss., University of Wisconsin, 1999.

Gerhard, W. P. "Bibliography of Flies and Mosquitoes in Relation to Disease." *Entomological News*, 1909, 84–89, 207–211.

Giddens, H. C. "To Kill Lice on Swine—Destroy Poisonous Mushrooms." *Southern Cultivator*, July 1852, 294.

Ginsburg, Joseph M. "Airplane Oiling to Control Mosquitoes." *Science*, May 20, 1932, 542.

———. "Studies of Pyrethrum as a Mosquito Larvicide." *Proceedings of the Seventeenth Annual Meeting of the New Jersey Mosquito Extermination Association* (1930): 57–72.

Glaser, R. W., and Henry Fox. "A Nematode Parasite of the Japanese Beetle (*Popillia japonica* Newm.)." *Science*, January 3, 1930, 16–17.

Glaser, R. W., E. E. McCoy, and H. B. Girth. "The Biology and Economic Importance of a Nematode Parasite in Insects." *Journal of Parasitology* 26, no. 6 (1940): 479–495.

Graham, Samuel A. "Forest Insect Populations." *Ecological Monographs* 9, no. 3 (1939): 301–310.

"The Grass Caterpillar." *Southern Planter*, July 1857, 439.

"Grasshoppers." *Yankee Farmer*, October 20, 1838, 331.

Greenham, P. M. "The Effects of the Variability of Cattle Dung on the Multiplication of the Bushfly (*Musca vetustissma* Walk)." *Journal of Animal Ecology* 41, no. 1 (1972): 153–166.

Guyton, Thomas Lee. "A Taxonomic, Ecologic, and Economic Study of Ohio Aphisidae." *Ohio Journal of Science* 24, no. 1 (1924): 1–30.

Hagen, H. A. "The Oldest Fossil Insects." *Nature*, March 24, 1881, 483–484.

Hall, Maurice C. "Parasites in War Time." *Scientific Monthly*, February 1918, 106–115.

Hanzlik, P. J. "Health Hazards of Chemo-Enemies in Contaminated Foods." *Scientific Monthly*, May 1937, 435–439.

Harwood, Paul D. "Is Natural Selection an Outworn Term?" *Ohio Journal of Science* 50, no. 6 (1950): 278–280.

———. "Pseudoscience and the DDT Scandal." *Science*, November 30, 1951, 583–584.

Haywood, J. K. "Insecticide and Fungicide Legislation in the United States with Especial Reference to the Federal Insecticide Act of 1910." *Journal of the Association of Official Agricultural Chemists* 4 (1920–1921): 19.

Helms, Douglas. "Technological Methods for Boll Weevil Control." *Agricultural History* 53 (1979): 286–299.

Hentz, N. M. "Description of Eleven New Species of North American Insects." *Transactions of the American Philosophical Society* 3 (1830): 253–258.

Herms, William. "Medical Entomology Meets the Impact of War." *Journal of Economic Entomology* 38 (1945): 8–11.

Hickey, Joseph J., J. A. Keith, and Francis B. Koon. "An Exploration of Insecticides in a Lake Michigan Ecosystem." *Journal of Applied Ecology* 3 (1966): 141–143.

Hoffmann, C. F., and J. P. Linduska. "Some Considerations of the Biological Effects of DDT." *Scientific Monthly*, August 1949, 104–114.

Howard, Leland O. "The Entomological Society of Washington." *Proceedings of the Entomological Society of Washington* 9 (1909): 78–79.

———. "Entomology and the War." *Scientific Monthly*, February 1919, 109–117.

———. "An Experiment Against Mosquitoes." *Insect Life* 5 (1893): 12–14.

———. "The Parasite Element of Natural Control of Injurious Insects and Its Control by Man." *Journal of Economic Entomology* 19 (1926): 271–282.

———. "Striking Entomological Events of the Last Decade of the Nineteenth Century." *Scientific Monthly*, July 1930, 5–18.

Hunt, R. "Highlights." *Proceedings of the Fifteenth Annual Meeting of the New Jersey Mosquito Extermination Association* (1916): 21–30.

Insect Life 4, nos. 3–4 (1891): 154.

"Insects." *Southern Cultivator*, July 1852, 109.

"Introduction." *Practical Entomologist*, October 30, 1865, 4.

"Introductory." *Practical Entomologist*, October 30, 1865, 6.

Jakle, John A. "The American Bison and the Human Occupance of the Ohio Valley." *Proceedings of the American Philosophical Society* 112 (1968): 299–305.

Janzen, David H. "Sweep Samples of Tropical Foliage Insects: Effects of Seasons, Vegetation Types, Elevation, Time of Day, and Insularity." *Ecology* 54, no. 3 (1973): 687–708.

"Japanese Parasite of the Gypsy Moth." *Insect Life* 4, nos. 5–6 (1891): 227.

Johnson, E. A. "Effects of Farm Woodland Grazing on Watershed Values in the Southern Appalachian Mountains." *Journal of Forestry* 50, no. 2 (1952): 109–113.

Johnson, James W. "The Neo-Classical Bee." *Journal of the History of Ideas* 22, no. 2 (1961): 262–266.

Johnston, J. Spencer, and W. B. Heed. "Dispersal of Desert Adapted *Drosophila*: The Saguro-Breeding *D. nigrospiracule*." *American Naturalist*, July–August 1976, 629–651.

Jones, E. L. "Creative Disruptions in American Agriculture, 1620–1820." *Agricultural History* 48, no. 4 (1974): 510–528.

Jukes, Thomas. "Silence, Miss Carson." *Chemical and Engineering News*, August 18, 1962, 5.

Kallet, Arthur. "Foods and Drugs for the Consumer." *Annals of the American Academy of Political and Social Science* 173 (1934): 26–34.

King, Walter. "How High Is Too High? Disposing of Dung in Seventeenth-Century Prescot." *Sixteenth Century Journal* 23, no. 3 (1992): 443–457.

Knab, Frederick. "A Proposal for the Control of Certain Mosquitoes." *Science*, January 24, 1913, 147–148.

Knapp, Seaman A. "Agricultural Experiment Stations." *Western Stock Journal and Farmer* 6 (1877): 243–258.

Knipling, Edward F. "The Greater Hazard—Insects or Insecticides." *Journal of Economic Entomology* 46 (1953): 1–7.

Knox, J. C. "Human Impacts on Wisconsin Stream Channels." *Annals of the Association of American Geographers* 67 (1977): 323–342.

Kowalczyk, Stanley A. "A Report on the Intestinal Protozoa of the Larva of the Japanese Beetle." *Transactions of the American Microscopical Society* 17, no. 3 (1938): 229–244.

Krebs, Ernst T., Jr. "New Use for DDT." *Science*, April 12, 1946, 459–460.

Lemmer, George F. "Early Agricultural Editors and Their Farm Philosophies." *Agricultural History* 31 (1957): 3–17.

Louda, Svata M., and James E. Rodman. "Insect Herbivory as a Major Factor in the Shade Distribution of a Native Crucifer (*Cardamine cordifolia* A. Gray, Bittercress)." *Journal of Ecology* 84, no. 2 (1996): 229–237.

Lovely, Robert Allyn. "Mastering Nature's Harmony: Stephen Forbes and the Roots of American Ecology." Ph.D. diss., University of Wisconsin, 1955.

Ludwig, Daniel. "Development of Cold Hardiness in the Larva of the Japanese Beetle (*Popilla japonica* Newm.)." *Ecology* 9, no. 3 (1928): 303–306.

Lundberg, Jakob, and Fredrick Moberg. "Mobile Link Organisms and Ecosystem Functioning: Implications for Ecosystem Resilience and Management." *Ecosystems* 6, no. 1 (2003): 87–98.

MacGregor, Malcolm Evan. "Insects as Carriers of Disease." *Transactions of the American Microscopical Society* 37, no. 1 (1918): 7–17.

MacKaye, Benton. "The New Exploration: Charting the Industrial Wilderness." *Survey Graphic*, May 1, 1925, 155–192.

Maine Farmer. "Bugs—Bugs—'O! the Bugs.'" *Farmer's Cabinet,* June 15, 1837, 366.

Malo, J. E., and F. Suarez. "Establishment of Pasture Species on Cattle Dung: The Role of Endozoochorus Seeds." *Journal of Vegetation Science* 6, no. 169 (1995): 169–174.

Manners, Ian R. "The Persistent Problem of the Boll Weevil: Pest Control in Principle and in Practice." *Geographical Review* 68 (1979): 25–42.

Marks, P. L. "The Role of Pin Cherry (*Prunus penslyvanica* L.) in the Maintenance of Stability in Northern Hardwood Ecosystems." *Ecological Monographs* 44 (1974): 73–88.

Marmor, Theodore R. "Anti-Industrialism and the Old South: The Agrarian Perspective of John C. Calhoun." *Comparative Studies in Society and History* 9, no. 4 (1967): 377–406.

Marotel, G. "Le Rôle actuel des Arthropodes en pathologie." *Annales de la Société d'agriculture, sciences et industrie de Lyon* (1907): 279–302.

McCay, Timothy S., and Gerald L. Storm. "Masked Shrew (*Sorex cinerus*) Abundance, Diet, and Prey Selection in an Irrigated Forest." *American Midland Naturalist*, October 1997, 268–275.

McCollom, John H. "The Role of Insects in the Propagation of Disease." *American Journal of Nursing* 2, no. 3 (1901): 181–193.

Melander, A. L. "Can Insects Become Resistant to Sprays?" *Journal of Economic Entomology* 7 (1914): 167–172.

———. "Fighting Insects with Powder and Lead." *Scientific Monthly*, February 1933, 168–172.

Mellers, P. "Fire Ecology, Animal Populations, and Man: A Study of Some Ecological Relationships in Prehistory." *Proceedings of the Prehistoric Society* 42 (1976): 15–45.

"Miscellaneous." *Yankee Farmer*, September 14, 1838, 99.

Moore, Barrington. "The Beginnings of Ecology." *Ecology* 19, no. 4 (1938): 592.

———. "The Scope of Ecology." *Ecology* 1, no. 1 (1920): 3–5.

Moore, William. "The Effect of Laundering upon Lice (*Pediculus corporis*) and Their Eggs." *Journal of Parasitology* 5, no. 2 (1918): 61–68.

———. "Fumigation of Animals to Destroy Their External Parasites." *Journal of Economic Entomology* 9 (1916): 71–78.

Morrison, Harold. "Scale Insects." *Scientific Monthly*, March 1926, 245.
Mundahl, Neal D., and Kenneth J. Kraft. "Abundance and Growth of Three Species of Aquatic Insects Exposed to Surface-Release Hydropower Flows." *Journal of the North American Benthological Society* 7, no. 2 (1988): 100–108.
Myers, Judith H. "Synchrony in Outbreaks of Forest Lepidoptera: A Possible Example of the Moran Effect." *Ecology* 79, no. 3 (1998): 1111–1117.
N. E. Farmer. "Cut Worm." *Farmer's Cabinet*, August 1, 1836, 28.
Nash, Roderick. "The American Cult of the Primitive." *American Quarterly* 18, no. 3 (1966): 517–537.
Natural History 1, no. 3 (1880): 80–148.
"Natural Sciences." *Southern Cultivator*, July 1852, 89.
Nelson, E. M., A. M. Hurd-Karrar, and W. O. Robinson. "Selenium as an Insecticide." *Science*, August 11, 1933, 124.
Olmstead, Alan, and Paul Rhode. "An Overview of California Agricultural Mechanization." *Agricultural History* 62 (1988): 86–112.
Osborn, Herbert. "An Experiment with Kerosene Emulsions." *Insect Life* 4 (1891–1892): 63–64.
Osborne, P. J. "An Insect Fauna from the Roman Site at Alcester, Warwickshire." *Britannia* 2 (1971): 156–165.
Overfield, Richard A. "Charles E. Bessey: The Impact of the 'New' Botany on American Agriculture, 1880–1910." *Technology and Culture* 16, no. 2 (1975): 162–181.
Packard, A. S., Jr. "Injurious and Beneficial Insects." *American Naturalist*, September 1873, 524–548.
———. "Mosquitoes." *Science*, August 9, 1901, 218–219.
———. "Moths Entrapped by an Asclepaid Plant (*Physianthus*) and Killed by Honey Bees." *Botanical Gazette* 5, no. 2 (1880): 17–20.
———. "Nature's Means of Limiting the Numbers of Insects." *American Naturalist*, May 1874, 270–282.
Palladino, Paolo. "Ecological Theory and Pest Control Practice: A Study of the Institutional and Conceptual Dimensions of a Scientific Debate." *Social Studies of Science* 20, no. 2 (1990): 255–281.
Pauley, Philip J. "Fighting the Hessian Fly: American and British Responses to Insect Invasion, 1776–1789." *Environmental History* 7, no. 3 (2002): 485–507.
Payne, Jerry A. "A Summer Carrion Study of the Baby Pig *Sus scrofa* Linneaus." *Ecology* 46, no. 5 (1965): 592–602.
Payne, Nellie. "The Effect of Environmental Temperatures upon Insect Freezing Points." *Ecology* 7, no. 1 (1926): 99–106.
"Pissant vs. Chinch Bug." *Southern Planter*, July 1857, 437.
Popenoe, Paul. "Biological Control of Destructive Insects." *Science*, August 5, 1821, 113–114.
Posey, Darrell A. "Entomological Considerations in Southern Aboriginal Demography." *Ethnohistory* 23, no. 2 (1976): 147–160.
Quayle, Joseph Henry. "The Development of Resistance to Hydrocyanic Acid in Certain Scale Insects." *Hilgardia* 11, no. 5 (1938): 183–210.
———. "The Increase in Resistance in Insects to Insecticides." *Journal of Economic Entomology* 36 (1943): 493–500.
"The Red Humped Caterpillar Killed by Parasites." *Insect Life* 4, nos. 5–6 (1891): 207.

"Remedy for the Culicue, or Plum Weevil." *Southern Cultivator,* July 1852, 301.

Rhoades, W. C. "The History and Use of Agricultural Chemicals." *Florida Entomologist* 46, no. 4 (1963): 275–277.

Riley, Charles V. "The Chinch Bug." *American Agriculturalist,* November 1881, 476.

———. "Controlling Sex in Butterflies." *American Naturalist,* September 1873, 513–521.

———. "Darwin's Work in Entomology." *Proceedings of the Biological Society of Washington, D.C.* 1 (1882): 70–80.

———. "Descriptions and Natural History of Two Insects Which Brave the Dangers of *Saarracena variolaris.*" *Transactions of the Academy of Science of St. Louis* 4 (1874): 235–240.

———. "Entomological Papers." *Proceedings of the American Association for the Advancement of Science* 27 (1878): 5–7.

———. "Further Notes on the Pollination of Yucca and on *Pronuba* and *Prodoxus.*" *Proceedings of the American Association for the Advancement of Science* 29 (1880): 617–639.

———. "Further Notes on Yucca Insects and Yucca Pollination." *Proceedings of the Biological Society of Washington, D.C.* 12 (1893): 41–54.

———. "Further Remarks on *Pronuba yuccasella* and on the Pollination of Yucca." *Transactions of the Academy of Science of St. Louis* 8 (1878): 568–573.

———. "Improved Methods for Spraying Trees Against Insects." *Proceedings of the American Association for the Advancement of Science* 32 (1884): 466–467.

———. "In Memoriam [Benjamin D. Walsh]." *American Entomologist* 2 (1869–1870): 65–68.

———. "The Locust Plague: How to Avert It." *Proceedings of the American Association for the Advancement of Science* 24 (1875): 215–221.

———. "On the Parasites of the Hessian Fly." *Proceedings of the American Association for the Advancements of Science* 34 (1885): 332–334.

———. "Parasites of the Cotton Worm." *Canadian Entomologist* 11, no. 9 (1879): 161–162.

———. "Parasitic and Predacious Insects in Applied Entomology." *Insect Life* 6, no. 2 (1893): 131–141.

———. "Parasitism in Insects" [annual address of the president]. *Proceedings of the Entomological Society of Washington* 2, no. 4 (1893): 1–35.

———. "Recent Advances in Economic Entomology." *Bulletin of the Philosophical Society of Washington* 5 (1886): 10–12.

Roberts, M. E. B. "The Hessian Fly." *Farmer's Review* 44, no. 442 (1881): 14.

Roeding, Cheryl E., and Leonard A. Smock. "Ecology of Macroinvertebrate Shredders in a Low-Gradient Sandy Bottomed Stream." *Journal of North American Bentological Society* 8, no. 2 (1989): 149–160.

Rohwer, Sievert A. "An Appraisal of Entomology and Entomologists." *Journal of Economic Entomology* 42 (1949): 5–9.

———. "Report of Special Committee on DDT, with S. A. Rohwer as Chairman." *Journal of Economic Entomology* 38 (1945): 144

Rosenberg, Charles E. "Rationalization and Reality in the Shaping of American Agricultural Research, 1875–1914." *Social Studies of Science* 7, no. 4 (1977): 401–422.

Russell, Edmund P. "'Speaking of Annihilation': Mobilizing for War Against Human and Insect Enemies, 1914–1945." *Journal of American History* 82, no. 4 (1996): 1505–1529.

———. "War on Insects: Warfare, Insecticides, and Environmental Change in the United States, 1870–1945." Ph.D. diss., University of Michigan, 1993.

Russell, Emily W. B., Ronald B. Davis, R. Scott Anderson, Thomas E. Rhodes, and Dennis S. Anderson. "Recent Centuries of Vegetational Change in the Glaciated Northeastern United States." *Journal of Ecology* 81, no. 4 (1993): 647–664.

Russell, L. M. "Leland Ossian Howard: A Historical Review." *Annual Review of Entomology* 23 (1978): 1–17.

Rusticus. "Diversify Your Products." *Southern Cultivator,* July 1852, 113.

Sambon, Louis A. "Ticks and Tick Fevers." *Journal of Tropical Medicine* 2 (1900): 217–223.

Sanders, D. A. "*Musca domestica* and Hippelates Flies—Vectors of Bovine Mastits." *Science,* September 27, 1940, 286.

Sawyer, Richard C. "Monopolizing the Insect Trade: Biological Control in the USDA, 1888–1951." *Agricultural History* 64, no. 2 (1990): 271–290.

Say, Thomas. "Descriptions of Insects of the Families of Carabici and Hydrocanthari of Latreille, Inhabiting North America." *Transactions of the American Philosophical Society* 2 (1825): 1–109.

Scheib, B. W. "Household Insects and Their Remedies." *American Midland Naturalist,* January 1920, 111–127.

Schlebecker, John T. "Grasshoppers in American Agricultural History." *Agricultural History* 27, no. 3 (1953): 85–93.

Schuyler, William R. "Wheat—The Hessian Fly." *Michigan Farmer,* July 1863, 9.

Scudder, Samuel H. "Address of Mr. Samuel H. Scudder." *Proceedings of the American Association for the Advancement of Science* 29 (1880): 3.

———. "Canons of Systematic Nomenclature for the Higher Groups." *Canadian Entomologist,* March 1873, 55–59.

———. "Recent Progress of Entomology in North America." *Pysch: Organ of the Cambridge Entomological Club* 2, nos. 45–46 (1978): 97–116.

Seftel, Howard. "Government Regulation and the Rise of the California Fruit Industry: The Entrepreneurial Attack on Fruit Pests, 1880–1920." *Business History Review* 59, no. 3 (1985): 369–402.

Senex. "Destructive Insects." *Farmer's Cabinet,* July 1, 1836, 36.

———. "Turnep Fly." *Farmer's Cabinet,* July 1, 1836, 6.

Shelford, Victor E., and W. P. Flint. "Populations of the Chinch Bug in the Upper Mississippi Valley from 1823 to 1940." *Ecology* 24, no. 4 (1943): 435–455.

Sheppard, Carol A. "Benjamin Dann Walsh: Pioneer Entomologist and Promoter of Darwinian Theory." *Annual Review of Entomology* 49 (2004): 1–25.

Sherman, J. V. "New Products Assure Growth in Chemical Industry." *Barron's,* February 19, 1945, 9–10.

Shimer, Henry. "Insects Injurious to the Potato." *American Naturalist,* April 1869, 91–99.

Smith, Charles Clinton. "The Effect of Overgrazing and Erosion upon the Biota of the Mixed-Grass Prairie of Oklahoma." *Ecology* 21, no. 3 (1940): 381–397.

Smith, Harry S. "Racial Segregation in Insect Populations and Its Significance in Applied Entomology." *Journal of Economic Entomology* 34 (1941): 1–5.

Smith, John B. "Concerning Certain Mosquitoes." *Science,* January 3, 1902, 13–15.
Solberg, Winton V. "Science and Religion in Early America: Cotton Mather's 'Christian Philosopher.'" *Church History* 56, no. 1 (1987): 73–92.
Souza, Wayne P. "The Role of Disturbance in Natural Communities." *Annual Review of Ecology and Systematics* 15 (1984): 353–391.
Starna, William A., George R. Hammel, and William L. Butts. "Northern Iroquoian Horticulture and Insect Infestation: A Cause for Village Removal." *Ethnohistory* 3, no. 31 (1984): 197–207.
"State Protection of the Reputation of Its Products." *Yale Law Journal* 43, no. 8 (1934): 1274–1284.
Stirrat, J. H., J. McLintock, G. W. Schwindt, and K. R. Depner. "Bacteria Associated with Wild and Laboratory-Reared Horn Flies, *Siphona irritans* (L.) (Diptera: Muscidae)." *Journal of Parisitology* 41, no. 4 (1955): 398–406.
"A String of Bugs." *Prairie Farmer,* August 1851, 1.
Swetham, Thomas W., and Ann M. Lynch. "Multicentury, Regional-Scale Patterns of Western Spruce Budworm Outbreaks." *Ecological Monographs* 63, no. 4 (1993): 399–424.
Taylor, John. "Ducks and the Colorado Potato Beetle." *Insect Life* 4, nos. 1–2 (1891): 76.
"To Drive Bugs from Vines." *Farmer's Cabinet,* July 1, 1836, 13.
Tousley, Rayburn D. "The Federal Food, Drug, and Cosmetic Act of 1938." *Journal of Marketing* 5, no. 3 (1941): 259–269.
Tower, W. L. "The Colorado Potato Beetle." *Science,* September 21, 1900, 438–440.
Treat, Mary. "Controlling Sex in Butterflies." *American Naturalist,* March 1873, 129–132.
Trelease, Sam F. "Bad Earth." *Scientific Monthly,* June 1942, 12–24.
"Turnep Fly." *Farmer's Cabinet,* October 1, 1836, 88.
Uvarov, B. P. "Problems of Insect Ecology in Developing Countries." *Journal of Applied Ecology* 1, no. 1 (1964): 159–168.
Van Dine, D. L. "The Relation of Malaria to Crop Production." *Scientific Monthly,* November 1916, 431–435.
Vaz, Z., Rubens S. Pereira, and Decio M. Malheiro. "Calcium in Prevention and Treatment of DDT Poisoning." *Science,* April 27, 1945, 434–436.
Wallace, J. Bruce, and Martin E. Gurtz. "Response of Baetis Mayflies (Ephemeroptera) to Catchment Logging." *American Midland Naturalist,* January 1986, 25–41.
Walsh, Benjamin D. Editorial. *Practical Entomologist,* May 28, 1866, 74.
———. "Imported Insects." *Practical Entomologist,* September 29, 1866, 119.
———. "Insects Injurious to Vegetation in Illinois." *Transactions of the Illinois State Agricultural Society* 4 (1861): 335–378.
———. "The New Potato Bug, and Its Natural History." *Practical Entomologist,* October 30, 1865, 1–4.
———. "On Certain Entomological Speculations of the New England School of Naturalists." *Proceedings of the Entomological Society of Philadelphia* 3 (1864): 207–249.
———. "Self Taught Entomologists." *Practical Entomologist,* May 12, 1867, 91.
Walsh, Benjamin D., and Charles V. Riley. "Imitative Butterflies." *American Entomologist* 1 (1869): 189–193.

Webster, F. M. "The Early History of the Hessian Fly in America." Paper presented at the annual meeting of the Society for the Promotion of Agricultural Science, Philadelphia, 1905.

———. *Insect Life* 7 (1894–1895): 132–134.

"Weevils." *Yankee Farmer*, March 24, 1838, 90.

"The Wheat Aphis." *New England Farmer*, January 1864, 16.

"The Wheat Fly." *Farmer's Cabinet*, December 1, 1836, 145.

"A Wheat or Corn Crop Saved by Chickens." *Southern Planter*, July 1857, 400.

"Wheat 'Starved to Death.'" *American Farmer*, May 1870, 69.

"Wheat Worm." *Yankee Farmer*, March 10, 1838, 4.

Whitaker, Adelynne Hiller. "A History of Federal Pesticide Regulation in the United States to 1947." Ph.D. diss., Emory University, 1974.

White, W. H. *Country Gentleman* 48 (1883): 334–335.

Widga, Christopher. "Bison, Bogs, and Big Bluestem: The Subsistance Ecology of Middle Holocene Hunter-Gatherers in the Eastern Great Plains." Ph.D. diss., University of Kansas, 2007.

Wilentz, Sean. "What Was Liberal History?" *New Republic*, July 10 and 17, 2006, 27.

Wiley, H. W. "The Dignity of Chemistry." *Science*, May 10, 1901, 721–732.

Wilson, C. E. "The *Anopheles* Mosquito Question in Relation to Malaria and Agriculture." *Florida Buggist*, September 21, 1917, 22.

Wilson, M. L. "Thomas Jefferson—Farmer." *Proceedings of the American Philosophical Society* 87, no. 3 (1943): 216–222.

"Wire Worms." *Prairie Farmer*, May 1851, 209.

Wurster, Charles F. "DDT Goes on Trial in Madison." *BioScience* 19, no. 9 (1969): 809–813.

Yust, Harold R., and R. L. Busby. "A Comparison of the Susceptibility of the So-Called Resistant and Non-Resistant Strains of California Red Scale to Methyl Bromide." *Journal of Economic Entomology* 35 (1942): 343–345.

Ziegler, D. "Descriptions of New North American Coleoptera." *Proceedings of the Academy of Natural Sciences of Philadelphia* 2 (1844–1845): 43–47, 266–272.

Government and Academic Reports

Act of 1887 Establishing Agricultural Experiment Stations. Available at http://www.oardc.ohio-state.edu/www/hatch.html.

Agency for Toxic Substances and Disease Registry. *Toxicological Profile for Fuel Oils*. Atlanta: Department of Health and Human Services, 1995.

Campbell, John B., and Steve Ensley. "Economically Significant Parasites of Range Cattle." University of Nebraska West Central Research and Extension Center. Available at: http://www.admani.com/alliancebeef/TechnicalEdge/Parasites.htm.

Centers for Disease Control and Prevention. "Ross and the Discovery that Mosquitoes Transmit Malaria Parasites." Available at http://www.cdc.gov/malaria/history/ross.htm.

Congress. House. Select Committee to Investigate the Use of Chemicals in Food Products. *Chemicals in Food Products: Hearings*. 82nd Cong., 1st sess. Washington, D.C.: Government Printing Office, 1953.

Cooper, Ellwood. "Address of President of the State Board of Horticulture." *Bulletin of the California State Board of Horticulture* 69 (1895): 9–12.

Coquillet, Daniel William. *Report on the Scale Insects of California.* Department of Agriculture, Division of Entomology, Bulletin no. 26. Washington, D.C.: Government Printing Office, 1892.

Department of Agriculture. *Chronological History of the Development of Insecticides and Control Equipment from 1854 Through 1954.* Washington, D.C.: Department of Agriculture, Agricultural Research Service, 1954.

———. *Report of the Commissioner of Agriculture.* Washington, D.C.: Government Printing Office, 1862–1882.

Department of Agriculture, Division of Entomology. *Reports of Observations and Experiments in the Practical Work of the Division.* Bulletin no. 3. Washington, D.C.: Government Printing Office, 1883.

———. *Reports of Observations and Experiments in the Practical Work of the Division.* Bulletin no. 13. Washington, D.C.: Government Printing Office, 1885.

Department of the Interior. *Report on the Productions of Agriculture as Returned at the Tenth Census (June 1, 1880).* Washington, D.C.: Government Printing Office, 1883.

Dodge, Charles Richards. *The Life and Entomological Work of Townend Glover.* Department of Agriculture, Bulletin no. 18. Washington, D.C.: Government Printing Office, 1898.

Dreistadt, Steve H., Jack Kelly Clark, and Mary Louise Flint. *Pests of Landscape Trees and Shrubs: An Integrated Pest Management Guide.* 2nd ed. Agriculture and Natural Resources Publication, no. 3359. Oakland: University of California Statewide Integrated Pest Management Program, 2004.

Dudley, J. E. *Nicotine Dust Kills Cucumber Beetles.* Wisconsin Agricultural Experiment Station, Bulletin no. 355. Madison: Wisconsin Agricultural Experiment Station, 1923.

Engels, Chad L. "The Effect of Grazing Intensity on Soil Bulk Density." North Dakota State University, Department of Civil Engineering. Available at: http://www.ag.ndsu.nodak.edu/streeter/99report/soil_bulk.htm.

Entomological Commission. *Fifth Report of the United States Entomological Commission, on Insects Injurious to Forest and Shade Trees.* Washington, D.C.: Government Printing Office, 1890.

———. *First Annual Report for the Year 1877 Relating to the Rocky Mountain Locust and the Best Means of Preventing Its Injuries and of Guarding Against Its Invasions.* Washington, D.C.: Government Printing Office, 1878.

———. *Second Report, for the Years 1878 and 1879; Relating to the Rocky Mountain Locust, and the Western Cricket, and Treating of the Best Means of Subduing the Locust in Its Permanent Breeding Grounds, with a View of Preventing Its Migrations into the More Fertile Portions of the Trans-Mississippi Country.* Washington, D.C.: Government Printing Office, 1880.

Fitch, Asa. [*First–Fourteenth*] *Report on the Noxious, Beneficial, and Other Insects of the State of New York.* Albany: Van Benthuysen, 1855–1872. [The title varies. For exact titles and the publication history of Fitch's reports, see Jeffrey K. Barnes, *Asa Fitch and the Emergence of American Entomology: With an Entomological Bibliography and a Catalog of Taxonomic Names and Type Specimens* (Albany: University of the State of New York, State Education Department, Biological

Survey, 1988), Appendix A, "Entomological Publications by Dr. Asa Fitch," 76–82.]

Forbes, Stephen A. [*First–Twenty-fourth*] *Report of the State Entomologist of Illinois on the Noxious and Beneficial Insects*. Springfield: Rokker, 1833–1908.

———. On a Bacterial Disease of the Larger Corn Root Worm." In *Seventeenth Report of the State Entomologist of Illinois on the Noxious and Beneficial Insects*. Springfield: Rokker, 1891.

———. *On the Chinch Bug in Illinois: Present Condition and Prospects for 1887 and 1888*. Office of the State Entomologist of Illinois, Bulletin no. 2. Champaign: Gazette Steam Print, 1887.

Hatch Experiment Station of the Massachusetts Agricultural College. *Fungicides, Insecticides, Spraying Calendar*. Bulletin no. 80. Amherst: Carpenter and Morehouse, 1902.

Hawley, C. J. "Aphid (Homoptera, Aphididae) Behavior in Relation to Host-Plant Selection and Crop Protection." Available at http://www.colostate.edu/Depts/Entomology/courses/en507/papers_1995/hawley.html.

Haywood, J. K. *Insecticides and Fungicides*. Department of Agriculture, Farmer's Bulletin no. 146. Washington, D.C.: Government Printing Office, 1902.

Hitchner, Lea S. "The Insecticide Industry." In *Insects: The Yearbook of Agriculture*, 265–269. Washington, D.C.: Government Printing Office, 1952.

Howard, Leland O. *The Chinch Bug: General Summary of Its History, Habits, Enemies, and of Remedies and Preventives to Be Used Against It*. Department of Agriculture, Division of Entomology, Bulletin no. 17. Washington, D.C.: Government Printing Office, 1888.

———. *A Contribution to the Study of Insect Fauna of Human Excrement (with Especial Reference to the Spread of Typhoid Fever by Flies)*. Washington, D.C.: Washington Academy of Science, 1900.

———. "Danger of Importing Insect Pests." In *Yearbook of the United States Department of Agriculture for 1897*, 529–552. Washington, D.C.: Government Printing Office, 1897.

———. *A History of Applied Entomology (Somewhat Anecdotal)*. Smithsonian Miscellaneous Collections, no. 84. Washington, D.C.: Smithsonian Institution, 1930.

———. *How to Distinguish the Different Mosquitoes of North America*. Department of Agriculture, Bureau of Entomology, Circular no. 40. Washington, D.C.: Department of Agriculture, 1900.

———. *Notes on the Mosquitoes of the United States: Giving Some Account of Their Structure and Biology, with Remarks on Remedies*. Department of Agriculture, Bureau of Entomology, Bulletin no. 25. Washington, D.C.: Government Printing Office, 1900.

———. *The Principal Insects Affecting the Tobacco Plant*. Department of Agriculture, Farmer's Bulletin no. 120. Washington, D.C.: Government Printing Office, 1900.

———. "Progress in Economic Entomology in the United States." In *Yearbook of the United States Department of Agriculture*, 135–156. Washington, D.C.: Government Printing Office, 1899.

———. *The San Jose Scale in 1896–1897*. Department of Agriculture, Division of Entomology, Bulletin no. 12. Washington, D.C.: Government Printing Office, 1898.

———. *A Study in Insect Parasitism.* Department of Agriculture, Division of Entomology, Technical Series no. 5. Washington, D.C.: Government Printing Office, 1897.

———. *Three Insect Enemies of Shade Trees,* Department of Agriculture, Farmer's Bulletin no. 99. Washington, D.C.: Government Printing Office, 1899.

Howard, Leland O., and William F. Fiske. *The Importation into the United States of the Parasites of the Gypsy Moth and the Brown-Tail Moth.* Department of Agriculture, Bureau of Entomology, Bulletin no. 91. Washington, D.C.: Government Printing Office, 1911.

Hunter, W. D. *The Boll Weevil Problem.* Department of Agriculture, Farmer's Bulletin no. 344. Washington, D.C.: Government Printing Office, 1909.

Koebele, Albert. *Report of a Trip to Australia . . . to Investigate the Natural Enemies of the Fluted Scale.* Department of Agriculture, Division of Entomology, Bulletin no. 21. Washington, D.C.: Government Printing Office, 1890.

Koehler, P. G., and F. M. Oi. "Filth-Breeding Flies." University of Florida, Institute of Food and Agricultural Sciences. Available at: http://edis.ifas.ufl.edu/IG091.

LeBaron, William. *[First–Fourth] Annual Report on the Noxious Insects of the State of Illinois.* Springfield: Springfield Journal Printing, 1871–1874.

Lintner, Joseph A. *[First–Thirteenth] Report of the State Entomologist of New York.* Albany: Lyon, 1882–1897.

Malley, Frederick W. *The Boll Worm of Cotton.* Department of Agriculture, Division of Entomology, Bulletin no. 24. Washington, D.C.: Government Printing Office, 1891.

———. *The Mexican Cotton Boll Weevil,* Department of Agriculture, Farmer's Bulletin no. 130. Washington, D.C.: Government Printing Office, 1901.

———. *Report of Progress in the Investigation of the Cotton Boll Weevil.* Department of Agriculture, Division of Entomology, Bulletin no. 26. Washington, D.C.: Government Printing Office, 1892.

Marlatt, C. L. *The Principal Insect Enemies of Growing Wheat.* Department of Agriculture, Farmer's Bulletin no. 132. Washington, D.C.: Government Printing Office, 1901.

———. *Scale Insects and Mites on Citrus Trees.* Department of Agriculture, Farmer's Bulletin no. 172. Washington, D.C.: Government Printing Office, 1903.

Melsheimer, Frederick Ernst. *Catalogue of the Described Coleoptera of the United States.* Revised by S. S. Haldeman and J. L. LeConte. Smithsonian Institution Publication, no. 62. Washington, D.C.: Smithsonian Institution, 1853.

Minakawa, Noburu. "The Dynamics of Aquatic Insect Communities Associated with Salmon Spawning." Water Center, University of Washington. Available at: http://depts.washington.edu/cwws/Theses/minakawa.html.

Nuttal, George Henry Falkiner. *On the Role of Insects, Arachnids, and Myriapods, as Carriers in the Spread of Bacterial and Parasitic Diseases of Man and Animals: A Critical and Historical Study.* Johns Hopkins Hospital Reports 8. Baltimore: Johns Hopkins Press, 1899.

Ohio Agricultural Experiment Station. *The Chinch Bug.* Bulletin no. 69. Columbus: State Printer, 1896.

Packard, A. S., Jr. *First Annual Report of the Injurious and Beneficial Insects of Massachusetts.* Boston: Wright and Potter, 1871.

Patent Office. *Report of the Commissioner of Patents, for the Year 1849*. Part 2, *Agriculture*. Washington, D.C.: Office of Printers to the House of Representatives, 1850.

———. *Report of the Commissioner of Patents for the Year 1856*. Vol. 4, *Agriculture*. Washington, D.C.: Office of Printers to the House of Representatives, 1857.

Pierce, W. Dwight, Robert A. Cushman, and Clarence E. Hood. *The Insect Enemies of the Cotton Boll Weevil*. Bureau of Entomology Bulletin no. 100. Washington, D.C.: Government Printing Office, 1912.

Riley, Charles V. *Destructive Locusts: A Popular Consideration of a Few of the More Injurious Locusts (or "Grasshoppers") in the United States, Together with the Best Means of Destroying Them*. Department of Agriculture, Division of Entomology, Bulletin no. 25. Washington, D.C.: Government Printing Office, 1891.

———. *[First–Ninth] Annual Report on the Noxious, Beneficial, and Other Insects of the State of Missouri*. Jefferson City: Regan and Carter, 1869–1877.

———. "Great Truths in Applied Entomology; Address to the Georgia State Agricultural Society, February 12, 1884." In *Report of the Commissioner of Agriculture*, 323–229. Washington, D.C.: Government Printing Office, 1884.

———. *The Icerya or Fluted Scale, Otherwise Known as the Cottony Cushion Scale*. Department of Agriculture, Division of Entomology, Bulletin no. 15. Washington, D.C.: Government Printing Office, 1887.

———. *Parasitic and Predaceous Insects in Applied Entomology*. Washington, D.C.: Government Printing Office, 1893.

———. *Reports of the Observations and Experiments in the Practical Work of the Division*. Department of Agriculture, Division of Entomology, Bulletin no. 26. Washington, D.C. Government Printing Office, 1891.

Roark, R. C., and N. E. McAdoo. *A Digest of the Literature on DDT Through April 30, 1944*. Washington, D.C.: Department of Agriculture, Bureau of Entomology and Plant Quarantine, 1944.

Simpson, C. B. *Report on the Codling Moth Investigations in the Northwest During 1901*. Department of Agriculture, Division of Entomology, Bulletin no. 35. Washington, D.C.: Government Printing Office, 1902.

Smith, John B. *The Common Mosquitoes of New Jersey*. New Jersey Agricultural Experiment Station, Bulletin no. 171. New Brunswick: New Jersey Agricultural Experiment Station, 1904.

———. *Crude Petroleum as an Insecticide*. New Jersey Agricultural Experiment Station, Bulletin no. 138. New Brunswick: New Jersey Agricultural Experiment Station, 1899.

———. *Cut-Worms: The Sinuate Pear-Borer; the Potato Stalk-Borer; Bisulphide of Carbon as an Insecticide*. New Jersey Agricultural Experiment Station, Bulletin no. 109. New Brunswick: New Jersey Agricultural Experiment Station, 1895.

———. *Insecticide Experiments for 1904*. New Jersey Agricultural Experiment Station, Bulletin no. 178. New Brunswick: New Jersey Agricultural Experiment Station, 1904.

———. *Raupenlein and Dendrolene*. New Jersey Agricultural Experiment Station, Bulletin no. 111. New Brunswick: New Jersey Agricultural Experiment Station, 1895.

———. *The San Jose Scale and How It May Be Controlled*. New Jersey Agricultural Experiment Station, Bulletin no. 125. New Brunswick: New Jersey Agricultural Experiment Station, 1897.

Sternberg, George Miller. "Transmission of Yellow Fever by Mosquitoes." *Smithsonian Annual Report* (1900): 657–673.

Thomas, Cyrus. *[First–Sixth] Report of the State Entomologist of Illinois.* Springfield: Rokker, 1876–1881.

Townsend, C. H. Tyler. *Insecticides and Their Applications.* New Mexico College of Agriculture and Mechanic Arts, Agricultural Experiment Station, Bulletin no. 9. Las Cruces: Rio Grande Republican, 1892.

Trimble, Stanley W., and Steven W. Lund. *Soil Conservation and the Reduction of Erosion and Sedimentation in the Coon Creek Basin, Wisconsin.* U.S. Geological Survey Professional Paper, no. 1234. Washington, D.C.: Government Printing Office, 1976.

Virginia Agricultural Experiment Station. *Lime Sulphur Wash.* Bulletin no. 141. Blacksburg: Virginia Agricultural Experiment Station, 1902.

Walsh, B. D. *Annual Report on the Noxious Insects of the State of Illinois.* Chicago, 1868.

Webster, F. M. *The Cinch Bug: Its Probable Origin and Diffusion, Its Habits and Development, Natural Checks, and Remedial and Preventative Measures.* Washington, D.C.: Government Printing Office, 1898.

———. *The Hessian Fly.* Ohio Agricultural Experiment Station, Bulletin no. 7 Columbus: State Printer, 1891.

———. *Reports of Experiments with Various Insecticide Substances.* Department of Agriculture, Division of Entomology, Bulletin no. 11. Washington, D.C.: Government Printing Office, 1886.

Weed, Howard Evarts. *Insecticides and Their Application.* Mississippi Agricultural and Mechanical College, Experiment Station, Bulletin no. 27. Starkville: Mississippi Agricultural and Mechanical College, 1893.

Wilcox, E. V. "Some Results of Experiment Station Work with Insecticides." In *Annual Report of the Office of Experiment Stations for the Year Ended June 30, 1905.* Washington, D.C.: Government Printing Office, 1906.

Williams, L. L., and S. S. Cook. "Paris Green Applied by Airplane in the Control of *Anopheles* Production." *Public Health Report.* Reprint no. 1140. Washington, D.C.: Government Printing Office, 1927.

Periodicals

Newspapers
Alexandria Herald
American Watchman
Boston Daily Advertiser
Boston Evening Post
Boston Gazette, or Weekly Journal
Boston News-Letter
Boston Post-Boy
Charlotte Daily Observer
Cincinnati Price Currant
City Gazette and Daily Advertiser
The Clarion
Clarion Ledger
Columbia Herald

Commercial Advertiser
Connecticut Courant
Connecticut Magazine
Country Journal
Daily Constitution
Dallas Morning News
Dallas Weekly Herald
Denver Rocky Mountain News
Fairfield Gazette
Fort Worth Gazette
General Advertiser and Political, Commercial, Agricultural, and Literary Journal
Grand Forks Herald
Hagers-Town Gazette
Independent Chronicle and Boston Patriot
Independent Gazetteer
Knoxville Journal
Macon Daily Telegraph
Macon Weekly Telegraph
Maryland Chronicle
Milwaukee Sentinel
Morning Chronicle
New Bedford Mercury
New Hampshire Patriot
New Hampshire State Gazette
New Haven Gazette
New Jersey Journal
New Orleans Times
New York Gazette
New York Times
New York Weekly Journal
Pennsylvania Chronicle
Pennsylvania Evening Post
Pennsylvania Gazette
Philadelphia Gazette
Philadelphia Inquirer
Pittsfield Sun
Providence Gazette and County Journal
San Jose Mercury News
South Carolina Gazette
The State
Tacoma Daily News
Universal Advertiser
Weekly Mercury
Wheeling Register

Journals

American Entomologist, edited by Charles V. Riley. Vol. 3 [2nd ser., vol. 1]. New York, 1880.

American Entomologist and Botanist, edited by Charles V. Riley and George Vasey. Vol. 2. St. Louis, 1870.

American Naturalist [a monthly journal devoted to the natural sciences in their widest sense]. New York, 1867–.

Annals of the Entomological Society of America. Columbus, Ohio, 1908–.

Annals of the Lyceum of Natural History of New York. 8 vols. New York, 1824–1877.

Annals of the New York Academy of Science. New York, 1877–.

Bulletin of the American Museum of Natural History. New York, 1881–.

Bulletin of the Brooklyn Entomological Society. 7 vols. Brooklyn, N.Y., 1878–1885.

Bulletin of the Buffalo Society of Natural History.[Each volume covers several years.] Buffalo, N.Y., 1874–.

Bulletin of the Wisconsin Natural History Society. Milwaukee, 1900–.

Canadian Entomologist, edited by William Saunders and then C. J. S. Bethune. Guelph, Ont., 1868–.

Entomologica Americana. Brooklyn Entomological Society, Brooklyn, N.Y., 1885–1890.

Entomological News, and Proceedings of the Entomological Section of the Academy of Natural Sciences. Philadelphia, 1890–.

Entomological Student. 1 vol. Philadelphia, April 1900.

Journal of Economic Entomology. Concord, N.H., 1908–.

Journal of the Academy of Natural Sciences of Philadelphia. Philadelphia, 1817–.

Journal of the National Grange of the Patrons of Husbandry. Concord, N.H., 1908, 1909.

Journal of the New York Entomological Society. New York, 1893–.

Kansas University Quarterly. Lawrence, 1892–.

Memoirs of the Boston Society of Natural History. Boston, 1866–.

Occasional Memoirs of the Chicago Entomological Society. Chicago, 1900–.

Ohio Naturalist. Columbus, 1901–.

Papilio: Devoted Exclusively to Lepidoptera. 4 vols. New York Entomological Club, New York, 1881–1884.

Practical Entomologist. Vols. 1–2. Entomological Society of Philadelphia, Philadelphia, 1865–1867.

Prairie Farmer: Devoted to Western Agriculture, Mechanics, and Education. Vols. 1–15. Chicago, 1841–1855.

Proceedings of the Academy of Natural Sciences of Philadelphia. Philadelphia, 1860–.

Proceedings of the American Academy of the Arts and Sciences. Boston, 1846–.

Proceedings of the Boston Society of Natural History. Boston, 1841–.

Proceedings of the California Academy of Sciences. San Francisco, 1854–.

Proceedings of the Davenport Academy of Natural Sciences. Davenport, Iowa, 1876–.

Proceedings of the Entomological Society of Philadelphia. 6 vols. Philadelphia, 1861–1867.

Proceedings of the Entomological Society of Washington. Washington, D.C., 1884–.

Proceedings of the Iowa Academy of Sciences. Des Moines, 1887–.

Proceedings of the Washington Academy of Science. Washington, D.C., 1899–.

Psyche: Organ of the Cambridge Entomological Club. Cambridge, Mass., 1874–.

Publications of the Carnegie Institution of Washington. Washington, D.C., 1900–.

Transactions of the Academy of Science of St. Louis. St. Louis, 1856.

Transactions of the American Entomological Society, and Proceedings of the Entomological Section of the Academy of Natural Sciences. Philadelphia, 1868–.

Transactions of the American Philosophical Society of Philadelphia. 2nd ser. Philadelphia, Second 1818–.

Transactions of the Kansas Academy of Sciences. Topeka, 1874–.

Government Publications

Bulletin of the Illinois State Laboratory of Natural History. [Each volume covers several years.] Urbana, 1876–.

Bulletin of the United States National Museum. Washington, D.C., 1875–.

Bulletins of the United States Geological and Geographical Survey of the Territories. Department of the Interior [F. V. Hayden in charge], Washington, D.C., 1875–.

Insect Life: Devoted to the Economy and Life Habits of Insects, Especially in Their Relations to Agriculture. Vols. 1–7. Department of Agriculture, Washington, D.C., 1888–1895.

Proceedings of the Missouri State Horticultural Society. 1865–1870. In *Missouri State Board of Agriculture, Annual Reports.* Jefferson City, 1866–1871.

Reports of the United States entomologists of the Department of Agriculture: Townend Glover (1863–1878), Charles V. Riley (1878–1879, 1880–1894), John Henry Comstock (1879–1880), and Leland O. Howard (1894–1927).

Smithsonian Miscellaneous Collections. Smithsonian Institution, Washington, D.C., 1862–.

Transactions of the Illinois State Agricultural Society. Vols. 4–8. Springfield, 1861–1871.

Archives

Beinecke Rare Book and Manuscript Library, Yale University, New Haven, Connecticut

Rachel Carson Papers

Ernst Mayr Library, Special Collections, Museum of Comparative Zoology, Harvard University, Cambridge, Massachusetts

Archival materials relating to Asa Fitch and Thaddeus William Harris

Gray Herbarium Library, Harvard University, Cambridge, Massachusetts

Papers of Thaddeus William Harris

Massachusetts Historical Society, Boston

Papers of the Massachusetts Society for the Promotion of Agriculture

Papers relating to the gypsy moth

National Archives, College Park, Maryland

Papers of the Bureau of Entomology

Phillips Library, Peabody Essex Museum, Salem, Massachusetts

Miscellaneous collections

Sterling Memorial Library, Yale University, New Haven, Connecticut

Asa Fitch Papers

ILLUSTRATION CREDITS

Numbers following the sources are the pages on which the illustrations appear.

Arnold Mallis, *American Entomologists* (New Brunswick, N.J.: Rutgers University Press, 1971): 85, 112, 123.

Charles V. Riley, *Fourth Report of the U.S. Entomological Commission, Being a Revised Edition of Bulletin No. 3, and the Final Report on the Cotton Worm, Together with a Chapter on the Boll Worm* (Washington, D.C.: Government Printing Office, 1885): 48, 92, 99, 102, 103.

Charles V. Riley, *Third Report of the U.S. Entomological Commission, Relating to the Rocky Mountain Locust, the Western Cricket, the Army Worm, Canker Worms, and the Hessian Fly* (Washington, D.C.: Government Printing Office, 1883): 38, 39, 74, 77, 93.

D. L. Collins, "The Bug Catcher of Salem," *New York State Bulletin,* March 1954: 33.

(*Top*) Illinois Department of Agriculture (http://www.agr.ill.state.us/images/Gypsy MothFemale.gif); (*bottom*) United States Department of Agriculture, National Agricultural Library (http://www.nal.usda.gov/speccoll/collect/history/images/ext6.jpg): 117.

Illinois Natural History Survey: 88.

Library of Congress: 121.

New York State Museum: 32.

Rachel Carson Council: 196.

Richard C. Sawyer, *To Make a Spotless Orange: Biological Control in California* (Ames: Iowa State University Press, 1996): 115.

St. Louis Post-Dispatch: 131.

Thaddeus William Harris, *Entomological Correspondence of Thaddeus William Harris, M.D.,* ed. Samuel H. Scudder. Occasional Papers of the Boston Society of Natural History, no. 1 (Boston: Boston Society of Natural History, 1869), frontispiece: 30.

INDEX

Numbers in italics refer to pages on which figures appear.

Abercrombie, John, 21
accommodation, of planting/harvesting schedule, 22
agribusiness, 159, 186, 215, 223
agricultural imperialism, 64, 65
agriculture: agribusiness, 159, 169, 186, 215, 223; cash cropping, 59; "clean agriculture," 64, 184; clear-cutting for, 12; in colonial U.S., 5–7; diversification of products, 58; and early Americans, 5–25; experimental farms, 49; grassland to farmland transition, 57; and insect infestations during Depression, 172–176; monocultural, 57, 223; and Native Americans, 12, 15, 57. *See also* corn; cotton; rice; tobacco; wheat
aldrin, 191, 219
alfalfa weevil, 183
alkyl formates, 141
American history and insects. *See* insects and American history
Anopheles mosquito, 121, 189
ants, 44, 54, 97, 181
aphids, 14, 22, 45, 183; potato aphid, 135; woolly aphid, 107, 166–167
apple trees, 106, 107
apple worm, 107
apples, and arsenic poisoning, 147, 151–152
Armstrong, Neil, 218
armyworm, 21, *92–93*
arsemart, 22
arsenic, toxicity of, 101, 147, 150

arsenic poisoning, 146, 150–152; cumulative nature of, 161; prevention of, by washing produce, 153; public awareness of, 154–156; symptoms of, 154
arsenic-based insecticides, 95–96, 100, 106, 135, 140, 141, 147
arsenite of lead, 96
Asia, insects from, 176–177
Asiatic beetle, 176
Aspidiotus perniciosus, 67

bag worm, 43
balance of nature, 89
Banks, Llewellyn, 146–147, 149, 164
Barber, Herbert S., 122
barium fluosilicate, 141
bark louse, 50
bark-boring beetle, 132
bark-penetrating sprays, 172
barley, 174
Barton, Benjamin Smith, 15
Bartram, William, 65
Bassett, Benjamin, 20
bedbugs, 136
beetles: Asiatic beetle, 176; bark-boring beetle, 132; and bark-penetrating sprays, 172; blister beetle, 45, 86; Chinese ladybeetle, 183; Colorado potato beetle, 65–67, 68, 72, 173–174, 179, 183, 189, 199; elm leaf beetle, 114, 118; furniture beetle, 181; ground beetle, 45; Japanese beetle, 167, 177, 187; Mexican

beetles (*continued*)
  bean beetle, 140; nonnative, 13–14; saw-toothed grain beetle, 181; striped beetle, 43; striped flea beetle, 40; tobacco flea beetle, 6, 114; Vedalia beetle, 90
Benton, C., 174
benzene hexachloride, 191
biological control of insects, 114, 116, 119, 183; armyworm, *92–93;* by bacteria, 187; by birds, 21, 42–44; by parasitic insects, 45–46, 91, 183, 187; popularity of, 91; by predaceous insects, 44–45, 183; Riley's preference for, 87, 90, 94, 108
birds: effect of DDT on, 204, 205, 207–208; insect control by, 21, 42–44
biscuit moth, 130
Bissell, T. L., 175
bisulfide of carbon, 96, 100–101
black worm, 22
Blissus leucopterus, 73
blister beetle, 45, 86
blowfly, 130
boll weevil, 114, 116–118, *117,* 140, 178, 179, 207
boll worm, 91
Bomar, Charles, 78
Boos, William Frederick, 164
Bordley, J. B., 8
boxelder bug, 180, 181
Bradford, Perez, 20
broadcast sprayer, *103*
broom corn, 58
Bryson, H. R., 167
Bureau of Entomology and Plant Quarantine, 169–170, 183

cabbage, 40
cabbage looper, 207
cabbage worm, 107, 177, 178, 207
calcium arsenate, 140, 141
calcium cyanide, 140
Calhoun, John, 80
California red scale, 185
cancer, and chemicals, 220
cankerworm, 15, 20, 52, 180
Cannon, Clarence, 160–161, 164

carbolic acid, 97; emulsion of, as insecticide, 96
carbon bisulfide, 107
carbon tetrachloride, 141
Carlson, Anton J., 145, 146, 164
Carson, Rachel, 192–193, *196,* 203, 208, 211, 216; appeal of, 197, 213, 214, 215; death of, 220; impact of work of, 194, 220, 221, 223; inspiration for *Silent Spring,* 208
Carter, Landon, 6, 21
cash cropping, 59
castor bean, 132, 135
caterpillars, 15, 21, 22, 107
Catesby, Mark, 6
cattle: corn feed for, 63; grassland destruction by, 61; grazing by, 15–16, 57, 61, 62; and insect infestation, 15–16, 57; penned, 58, 60; World War I and diseases of, 133
cattle grub, 140
caustic soda, 105
Cavers, David F., 160, 163, 164
cereal crops, 58, 72
Chapman, Royal N., 131–132
charcoal dust, as insect repellant, 22
chemical insecticides, 95–110, 114. *See also* insecticides
chinch bug, 21, 45, 70–73, 97, 167, 173, 174–175, 184
Chinese ladybeetle, 183
chlordane, 191, 219
chloropicrin, 141
citrus industry, insecticide use by, 159
Clap, Ezra, 20
"clean agriculture," 64, 184
Clean Air Act (1970), 218
clear-cutting, by colonists, 12
climate, and insect growth, 14–15
clothes moth, 54
clover, 22
clover mite, 181
cloverleaf weevil, 141, 180
Coastal Zone Management Act (1972), 218
Cobbett, William, 53
Cockerell, T. D. A., 70
cockroach, 97

codling moth, 141, 167, 177, 178, 207
Colden, Cadwallader, 1
colonial America, insects and insect control in, 5–27
colonists, deforestation by, 11, 12–13
Colorado potato beetle, 65–67, 68, 72, 173–174, 178, 183
Committee on Medical Research, 204
Common Sense (Paine), 194, 195
Comstock, John Henry, 67, 112
conservation, 51
Cook, Albert, 34
Coon, B. F., 173
Copeland, Royal, 160, 164
Coquillet, Daniel William, 68
corn: for cattle, 62; and chinch bug, 21, 45, 70–73; and European corn borer, 177; insect control for, 21, 22, 172; insect infestations of, during Depression, 177–179; and Native Americans, 12
corn borer, 177, 184
corporate pragmatism, 158
Cottam, Clarence, 217
cotton, 58; insecticides for, 141
cotton fleahopper, 180
cotton leafworm, 182
cotton worm, 97
cottony-cushion scale, 90
Crosby, Alfred, 11
crucifer feeders, 14
cryolite, 141
Culex mosquito, 121, 122–125
Culpepper, Nicholas, 22
curculio, 41–42, 69, 181–182, 184
cutworm, 6, 23, 97, 180, 184

Darwin, Charles, 89
DDE, 219
DDT, 129, 167, 172, 189–192; ban on use of, 219; bio-concentration of, 206–207; and birds, 204, 205, 207–208; chronic toxicity of, 204; discovery of, 189; and dogs, 204; and fish, 206; in food chain, 206–207; and humans, 214; and malaria, 189; post–World War II use of, 199–204; and potato beetle, 189, 199; resistance to, 207; safety of, 192, 200, 203–204; unregulated use of, 203; and wildlife, 26, 192, 204, 205, 206
De Witt, Simeon, 29
Deane, Samuel, 6, 18, 21, 22, 24
deflector nozzles, *102*
deforestation: by colonists, 11, 12–13; by grazing cattle, 15–16; and insect infestation, 10–11, 13; by Native Americans, 10–11; sunlight exposure from, 14. *See also* forests
Deonier, C. C., 180
Depression: agricultural damage from insects during, 172–176; crop losses during, 176–179; federal bureaucracy during, 171; insecticide use during, 168–189
dieldrin, 191, 219
dogs, effect of DDT on, 204
Doryphora decemlineata, 66
Douglas, William O., 211
Drake, C. J., 166–167, 174
Draper, Frank Winthrop, 150, 156
Drayton, John, 8–9
Driver, Harold E., 212
drought, 217
Duncan, H. G., 209
Dunlap, Thomas, 137, 162, 204, 219
Dupree, A. Hunter, 52
dust bowl, 62, 64
Dutch elm disease, 205
Dwight, Timothy, 10–11, 14, 16
Dyar, Harrison G., 127

ecology, 198–199, 209, 210; and Carson, 193, 197, 219; and Native Americans, 10–12
economic entomologists, 29–34, 36, 37, 42, 58, 144, 216
Economic Entomology for the Farmer and Fruit-Grower (Smith), 122
economic ornithology, 44
ecosystem, 209, 211
Edge of the Sea, The (Carson), 208, 211
elm leaf beetle, 114, 118
Emerson, George, 31
Emmons, Ebenezer, 54

290 Index

Endangered Species Act (1972), 218
entomologists, 26–55, 113; economic, 29–34, 36, 37, 42, 58, 144, 216; second-generation, 81
environmental change: from deforestation, 10–16; from expansion to Great Plains, 57, 59
Environmental Defense Fund, 219
Environmental Protection Agency, 219
environmentalism, 210–211, 218
Essig, Edward O., 91, 137, 190
ethylene dibromide, 141
ethylene dichloride, 141
ethylene dichloride–carbon tetrachloride, 141
ethylene oxide, 141
Europe, insects from, 177–178
European corn borer, 177, 184
European elm scale, 180
Evans, Francis, 209

farmers: and entomologists, 26–55; insect control by, 17–22, 26. *See also* agriculture
feedlots, 60
Felt, E. Porter, 67, 68
Fernald, C. H., 51, 52, 108, 150
Fight for Conservation, The (Pinchot), 62
Fighting the Insects (Howard), 142
fire, use of, by Native Americans, 11, 12
fish, effect of DDT on, 206
fish oil, 141
Fisher, Jabez, 47
Fitch, Asa, 31, 32, *32,* 33, 34, 36, 41, 42, 58, 222
Flagg, Wilson, 44, 58
fleas, 12, 15
Fletcher, Colin, 211
flies, 6, 12, *131;* insecticides for, 136; mayfly, 13; sand fly, 180; stable fly, 180; tobacco fly, 23, 28; and typhoid, 136; wheat fly, 21. *See also* Hessian fly
Flint, Charles L., 43
fluted scale, 90
Food, Drug, and Cosmetic Act (1938), 149, 159, 160, 163, 166, 201

food, safety of, 145–146, 147, 148, 149, 161
Food and Drug Administration, insecticide regulation and testing by, 159–161, 164
food chain, 209; effect of DDT on, 206–207
Foods and Their Adulteration (Wiley), 145
Forbes, Stephen A., 41, 44, 101, 209
forests: canopy openness of, 13; and fire, 11; and grazing cattle, 15–16; replacement of, by agriculture, 15; replacement of, by grazing land, 15. *See also* deforestation
formaldehyde, 136
Foster, William H., 27
Fowler, F. H., 100, 105
Frost, S. W., 168–169
Frost, Samuel, 73
fruit industry, 147–149; and insecticide legislation, 159–161; insecticide use by, 140, 158–160
fruit trees, 15, 22, 23, 58; apple trees, 106, 107; effects of insecticides on, 107; and San Jose scale, 67–70, 173, 175, 182, 185, 207. *See also* orchards
fumigants, 141
furniture beetle, 181

Georgia State Agricultural Society, 49
Glover, Townend, 32–33, 34, 42, 86
gnat, 23
goat louse, 182
Goldsmith, Oliver, 5, 9–11
Goode, G. B., 82
Gould, G. E., 167
Graham, Frank, 189
grain, 57; and chinch bug, 72; and grain beetle, 181; and grain worm, 28; mills, 141; storage of, 132
grain beetle, 181
grain worm, 28
grape mealybug, 182
grape root louse, 84
grasshoppers, 12, 15, 22, 27, 28, 45, 173, 183

grasslands, destruction of, in Great Plains, 57, 61
grassworm, 44
grazing, 61; and insect infestation, 15–16, 57; and soil destruction, 62
Great Plains, 57; agricultural imperialism of, 64, 65; as dust bowl, 62, 64; expansion to, 56–80; grassland destruction in, 57, 61; insect pests in, 65–80; soil depletion in, 62–64
green leafhopper, 8
green tobacco worm, 6, 7
"greening of America," 219
Griffith, Mary, 22
Grote, A. R., 54
ground beetle, 45
grub worm, 106
gypsy moth, 114–116, 117, *117*, 199

Hall, Maurice C., 133, 137
Hanzlik, P. J., 150, 153–154
harlequin bug, 178
Harmen, S. W., 167
Harris, Thaddeus William, 1–3, 4, 29–31, *30*, 33–34, 36, 37, 40, 41, 54–56
Harwood, Paul D., 200
Hatch Act (1857), 94
Hawley, C. J., 14
Hawthorne, Nathaniel, 26–27
Hayes, Wayland J., 203
Haywood, J. K., 136
head louse, 207
Headlee, Thomas, 139
hellebore, 47, 96
hemlock looper, 183
heptachlor, 219
Hessian fly, 175–176; control of, 49, 91, 167, 173, 184; distribution of, in U.S., *74;* in early America, 7–8, 19, 23, 28; life cycle of, *38–39;* and wheat, 73–75, 76
Higginson, T. W., 2
home: furniture beetles in, 181; insects in, 52–54, 97; mosquito control in, 128
hornworm, 6, 7
Horton, J. R., 176

Housekeeper's Receipt Book, The (Oddy), 53
Howard, Leland O., 69, 81, 108–110, 111–121, *112*, *115*, 141–143, 144, 167, 195; on balance of nature, 89; career of, 141; and insect control during World War I, 129–132, 134, 136, 139, 202; on kerosene, 127–137; on mosquitoes, 125–127; retirement of, 143; and Riley, 83, 108–110, 111, 119
Howe, Joseph S., 49, 50
Huckins, Olga Owens, 207–209
Hueper, W., 220
Hull, John, 52
Hutchinson, Jonathan, 151, 156
hydrocyanic acid gas, 136, 207

IN-930 (DuPont), 172
Injurious Insects of the Farm and Garden (Treat), 40, 43
insect control: and agricultural schedule, 22; alternative methods of, 187–189; biological methods of, 42–46, 87, 90, 91–93, 94, 108, 114, 115, 119, 183, 187; by birds, 21, 42–44; with boiling water, 23–24; with brine, 23; with cow dung, 23; by crop dusting, 141; cultural methods of, 49, 59, 114, 119, 184, 188; in early America, 20–24; early insecticides for, 81–110; by farmers, 17–21; fumigants for, 141; and insect life cycle, 36–42; lures for, 23, 135; manual methods of, 20–21, 50; with microbes, 187; with mustard seed, 23; natural insecticides for, 22, 47, *48;* with parasitic insects, 45–46, 91, 183, 187; poisoned bait for, 135; with predaceous insects, 44–45, 183; repellants for, 21–22, 23, 47; with sheepskins, 23; soil manipulation for, 23–24, 47; with tar-soaked twine, 21; with tobacco, 22, 47; by wetlands drainage, 122–125; during World War I, 128–143. *See also* biological control of insects

Insect Pest Survey Bulletin, 168, 172, 186
Insecticide, Fungicide, and Rodenticide Act (1972), 219
Insecticide Act (1910), 127
insecticide poisoning, 101; and consumer-protection groups, 153; public awareness of, 154–156; recognition of, 151–153; symptoms of, 214
insecticides, 111–143; acceptance of, 114; accumulation of, 161; advocates of, 164–166; banning of, 219, 221; bio-concentration of, 206–207; and crop dusting, 141; in Depression, 168–189; efficacy of, 106–107, 183; Food and Drug Administration regulation of, 159–161; and food safety, 145–146, 147, 148, 149, 161; in fruit industry, 140, 158–160; fumigants as, 141; kerosene as, 96, 97, 101, 104, 105, 106, 107, 121, 124, 126–127, 128, 137, 139, 141; legislation for, 127–128, 148–149, 159–161, 218, 219; loss of effectiveness of, 182–183, 185; mixing and applying of, *102–103,* 104–106; natural, 22, 47–49, 96; in nineteenth century, 95–110; opposition to, 194–220; poisoning with, 150; and political trends, 113; post–World War I development of, 138–141; post–World War II development of, 191; proponents of, 158–159; Public Health Service regulation of, 161–162; and quack remedies, 97–98; residues of, 146, 147, 148, 149, 161; resistance to, 185, 192, 207; responsible use of, 158; safety of, 100–101, 127, 150; spraying of, *99,* 100, *102–103,* 104, 182, 184, 186–187; unregulated use of, 203; use of, in World War I, 135; use of, in World War II, 191. *See also* insecticide poisoning; natural insecticides
insects: attitude toward, 51–52; bacterial control of, 187; and birds, 21, 42–44; and climate, 14–15; in Great Plains, 65–80; in home, 52–54, 97, 128, 181; life cycle of, 36–42, 79; nonnative, 176–178; parasitic, 45–46, 91, 183, 187; predaceous, 44–45, 183; resistant strains of, 185, 192, 207; and wartime diseases, 133–134; weather and infestation of, 173, 184. *See also* entomologists; insect control; insects and American history
insects and American history: colonial America, 5–27; expansion to Great Plains, 56–80; rise of entomologists, 26–55
Insects and Insecticides (Weed), 54
International Apple Association, 164, 195
Iroquois, 12
Isely, Dwight, 184, 188

Japanese beetle, 167, 177, 187
Jefferson, Thomas, 8, 9, 22
"juise of arsemart," 22

Kallet, Arthur, 153, 158, 161, 195
Kedzie, Robert Clark, 150, 156
Kennicott, Robert, 51
kerosene, emulsion of, 97, 137, 141; application of, 104; efficacy of, 106, 107; for mosquito control, 96, 97, 101, 104, 105, 106, 107, 121, 124, 126–127, 128, 139; recipe for, 96; stability of, 101, 105
Knipling, Edward F., 200, 202
Knowles, Joe, 210
Koebele, Albert, 90, 115

ladybug, 90, 183
Lamb, Ruth deForest, 154–155, 195
lead arsenate, 140, 141, 147, 207
lead poisoning: cumulative nature of, 161; public awareness of, 154–156; symptoms of, 154
LeBaron, William, 37, 73
legislation, for insecticide use, 127–128, 148–149, 159–161, 218, 219
Leis dimidiatus, 183

lice: goat louse, 182; grape root louse, 84; head louse, 207; on humans, 130, 136, 207; on pigs, 49; plant louse, 44, 106; sheep-biting louse, 181; and typhus, 136, 189, 199
lime: mixture of, with salt and sulfur, 96, 106; slacked, 96; and sulfate of copper, 96
lime sulfur, 101
Lintner, Joseph Albert, 31–32
livestock. *See* cattle
Lockwood, Jeffrey, 76
locusts: infestations of, 58, 76, *77*, 78; red-legged locust, 12; Rocky Mountain locust, *77*, 78, 84
London purple, 95–96, 97, 100, 106, 107

maggots, 15, 130, 140
malaria: and DDT, 189; transmission of, by mosquitoes, 120–121
Malley, Frederick W., 118
Manifest Destiny, 58
Marchal, Paul, *115*
Marlatt, C. L., 71
Massachusetts Society for the Promotion of Agriculture, 20
Massey, William C., 212
mayfly, 13
McAdoo, N. E., 190–191
medical entomology, 133–134
Melander, A. L., 137, 185, 207
Merck, George W., 191
Methods of Insect Control (Isely), 184
Mexican bean beetle, 140
Mexico, insects from, 178–179
milky spore, 187
Miller Amendment (1948), 201
Minor Horrors of War, The (Shipley), 130
mirex, 219
mites, 84, 136; clover mite, 181; purple mite, 182; spider mite, 207
Moore, Barrington, 209
Morrill Land Grant College Act (1862), 49, 94
mosquito control: in home, 128; and insecticide resistance, 207; with kerosene emulsion, 96, 97, 101, 104, 105, 106, 107, 121, 124, 126–127, 128, 139; with Paris green, 138; by wetlands drainage, 122–125
mosquitoes, 12, 120; *Anopheles,* 121, 189; breeding of, 122–125; *Culex,* 121, 122–125; as malaria vector, 120–121; as yellow fever vector, 121
Mosquitoes: How They Live; How They Carry Disease; How They Are Classified; How They May Be Destroyed (Howard), 125
moths, 54; biscuit moth, 130; clothes moth, 54; codling moth, 141, 167, 177, 178, 207; gypsy moth, 114–116, 117, *117*, 199; satin moth, 176
Muhlenberg, Henry Melchior, 23
Muir, John, 197
Müller, Paul H., 188–189

naphthalene, 141
naphthol, 136
Nash, Roderick, 210, 218–219
National Wildlife Preservation System, 51
Native Americans: agricultural practices of, 15, 57; deforestation by, 10–11; and ecology, 10–12; and environment, 212–213; and insects, 12; land management by, 10–11; and myth of noble savage, 211–212
native grasses, 57, 61
natural insecticides, 22, 47–49, 96
natural selection, 89
Neoplactana, 187
nicotine dust, 47, 96
North American Entomologist (journal), 41

Oddy, S. A., 53
Odum, Eugene, 210
O'Kane, Walter C., 145, 195
orange tree scale, 97
orchards, 15; effects of insecticides on, 107; and San Jose scale, 67–70, 173,

orchards (*continued*)
  175, 182, 185, 207. *See also* fruit trees
Osborn, Herbert, 105

Packard, A. S., Jr., 35, 44, 45, 126
Packard, Clarissa, 53
Paine, Thomas, 194, 195, 219
Painter, R. H., 167
paradichlorobenzene, 141, 182
paraffin oil, 137
parasitic insects, for insect control, 45–46, 91, 183, 187
parathion, 191, 197
Paris, John Ayrton, 150–151
Paris green: efficacy of, 106; for mosquito control, 138, 141; safety of, 119, 150; stability of, 101; use of, 95, 96, 97, 138, 148
Parkman, Francis, 212
Parks, T. H., 182–183
pear slug, 43
peas, 58
Perkins, George, 191
pest management. *See* insect control
phylloxera, 84
pigs, 49, 133
Pinchot, Gifford, 62
Pinckney, Thomas, 22
pine weevil, 12
pinoleum, 107
pioneers, insect control by, 56–80
Piver, William C., 136
Plains. *See* Great Plains
plant louse, 44, 106
plow bellows sprayer, *99*
plum curculio, 69, 181–182, 184
populism, 113
potassium sulfide, 106
potato aphid, 135
potato beetle: Colorado potato beetle, 65–67, 68, 72, 173–174, 179; and DDT, 189, 199; insecticides for, 95, 96, 183
potato tuber worm, 178–179
Practical Entomologist (journal), 41, 43
pragmatism, 113, 158

predaceous insects, for insect control, 44–45, 183
progressivism, 113
public health, and food safety, 145–146, 147, 148, 149, 161
Public Health Service, 161–162
Pure Food and Drug Act (1906), 146, 147
purple mite, 182
Putnam, James J., 151, 156
pyrethrum, 47, 96, 106, 138, 141
Pyrethrum cinerariaefolium, 48

Quayle, Henry Joseph, 185
Quick, Herbert, 60

radish, 40
red-legged locust, 12
Reed, Walter, 121
Report of the Insects of Massachusetts Injurious to Vegetation, A (Harris), 2–3, 30–31, 37, 40, 54
Reppert, R. R., 183
Rhoades, W. C., 201
rice, 8–9, 15
Riley, Charles V., 34, *85*, 114; on biological control, 45, 87, 90, 94, 108; career of, 82–87, 89–91, 108; on chinch bug, 71; death of, 108; and Division of Entomology, 81, 82; and entomologists, 95; and Howard, 83, 108–110, 111, 119; on insecticide safety, 156, 195; on insecticides, 98, 108, 109, 119; on locusts, 76, 78
Roark, R. C., 190–191
Robert, Joseph Clarke, 6
Robinson, Solon, 3–4, 56, 57, 58, 65
Rocky Mountain locust, *77*, 78, 84
rose bug, 43
Ross, Ronald, 120, *121*
rotenone, 140, 141
Russell, Edmund P., 85, 202

salimene, 106
salt, as insect repellant, 22, 23, 47
San Jose scale, 67–70, 173, 175, 182, 185, 207

sand fly, 180
Sanderson, E. Dwight, 127, 128
satin moth, 176
Saunders, William, 34
saw-toothed grain beetle, 181
Say, Thomas, 65–66
Saylor, John P., 211
scabies, 133
scales: California red scale, 185; cottony-cushion scale, 90; European elm scale, 180; fluted scale, 90; orange tree scale, 97; San Jose scale, 67–70, 173, 175, 182, 185, 207
Schlebecker, John T., 80
Schuyler, William R., 49
Schwartz, E. A., 109
screwworm, 180
Sea Around Us, The (Carson), 208, 211
Sheals, R. A., 173
sheep, 133
sheep tick, 181
sheep-biting louse, 181
Shipley, A. E., 129–130
Silent Spring (Carson), 193, 194, 195, 197, 208–211, 213–215, 217, 220, 221
slugs, 43
Smith, Harry S., 185
Smith, John B., 68, *123*, 127; on bisulfide of carbon, 100–101; death of, 128, 143; on insecticide resistance, 185; on mosquito control, 121–125, 138; on San Jose scale, 68
Smith, Oliver, 20
Snider, E., 58
sodium arsenate, 141
soil, depletion of, in Great Plains, 62–64
soldiers, and insect-borne diseases, 136
Sorenson, W. Conner, 34
sorghum, 58
southern army worm, 132
Spencer, H., 183
spider mite, 207
spiders, 97
spindle worm, 22
sprayers, *99*, 100, *102–103*, 104, 182, 184, 186–187

stable fly, 180
Stearns, L. A., 167
Steinberg, Ted, 13
Stipe, Michael, 218
Stowe, Harriet Beecher, 194, 195, 219
strawberry weevil, 185
striped beetle, 43
striped flea beetle, 40
sugarcane borer, 178, 183
sulfate of copper and lime, 96
sulfate of iron, 136
sulfite of soda, 106
sulfur, 96, 136, 141
Sutherland, W. W., 201–202
sweet potatoes, 58

Tatham, William, 6, 7, 20–21, 28
"Texas fever," 133
thallium sulfate, 141
Thomas, W. A., 173
ticks, 12; dog tick, 181; scabies, 133; sheep tick, 181
tobacco: for insect control, 22, 47; and insects, 6–8; nicotine dust from, 47, 96; and potato tuber worm, 178–179; and tobacco flea beetle, 6, 114; and tobacco fly, 23, 28
tobacco flea beetle, 6, 114
tobacco fly, 23, 28
tomato hornworm, 207
tomatoes, 58, 179
Tousley, Raymond D., 163
Townsend, C. H. Tyler, 100, 104, 116, 126
toxaphene, 191
Toxic Control Substances Act (1976), 218
Treat, Mary, 40–41, 43, 66
trees: and bark-boring beetles, 132; Dutch elm disease, 205; and elm leaf beetle, 114, 118; and European elm scale, 180; and orange tree scale, 97; and San Jose scale, 67–70, 173, 175, 182, 185, 207; and tree borers, 133. *See also* fruit trees
trench fever, 133
Trouvelot, Étienne Léopold, 114
typhoid, 136

typhus, 136, 189, 199
Tyroglyphus phylloxerae, 84

Uncle Tom's Cabin (Stowe), 194, 195
Under the Sea-Wind (Carson), 208, 211

Van Deventer, W. C., 210
Van Dine, C. L., 128
Vance, A. M., 177
Vedalia beetle, 90
vegetable weevil, 180
Vogt, Evon Z., 212

Wallace, George J., 205
Walsh, Benjamin D., 34, 87–89, *88*, 90; on biological control, 87, 91; on Colorado potato beetle, 66; death of, 108; on insect life cycle, 41, 42; on insecticide safety, 156, 195; on quack remedies, 97–98; and Riley, 83
Walton, W. R., 140
wasps, 112
Water Pollution Control Act (1972), 218
weather, and insect infestation, 173, 184
Webster, F. M., 75, 91
Weed, Clarence M., 46, 54
weevils: alfalfa weevil, 183; boll weevil, 114, 116–118, *117*, 140, 178, 179, 207; cloverleaf weevil, 141, 180; control of, 179; pine weevil, 12; strawberry weevil, 185; vegetable weevil, 180; and weather, 184
Wernick, Robert, 209, 216
wetlands, drainage of, for insect control, 122–125
whale oil, 97
Wharton, W. R. M., 161

wheat: and chinch bug, 71, 72; in early America, 7–8, 17; and Hessian fly, 73–75, 76; insect control for, 22; and wheat fly, 21
"wheat aphis," 42
wheat fly, 21
white arsenic, 96, 106
Whorton, James, 151
Wilder, Laura Ingalls, 78
Wilderness Act (1964), 211
wildlife, effect of DDT on, 26, 192, 204, 205, 206
Wildlife Refuge System, 51
Wiley, Harvey W., 145, 156, 164
Williams, Samuel, 15
Wilson, James, 146, 147, 164
Winthrop, James, 20
woodpeckers, 44
woolly aphid, 107, 166–167
World War I, insect control during, 128–143
World War II, insect control during, 191
worms: apple worm, 107; armyworm, 21, *92–93;* bag worm, 43; black worm, 22; boll worm, 91; cabbage worm, 107, 178, 207; cankerworm, 15, 20, 52, 180; cotton leafworm, 182; cotton worm, 97; cutworm, 6, 23, 97, 180, 184; grain worm, 28; grassworm, 44; green tobacco worm, 6, 7; grub worm, 106; hornworm, 6, 7; potato tuber worm, 178–179; screwworm, 180; southern army worm, 132; spindle worm, 22; tomato hornworm, 207

xylol, 137

yellow fever, 121, 189

Zahniser, Howard, 211